AF574955

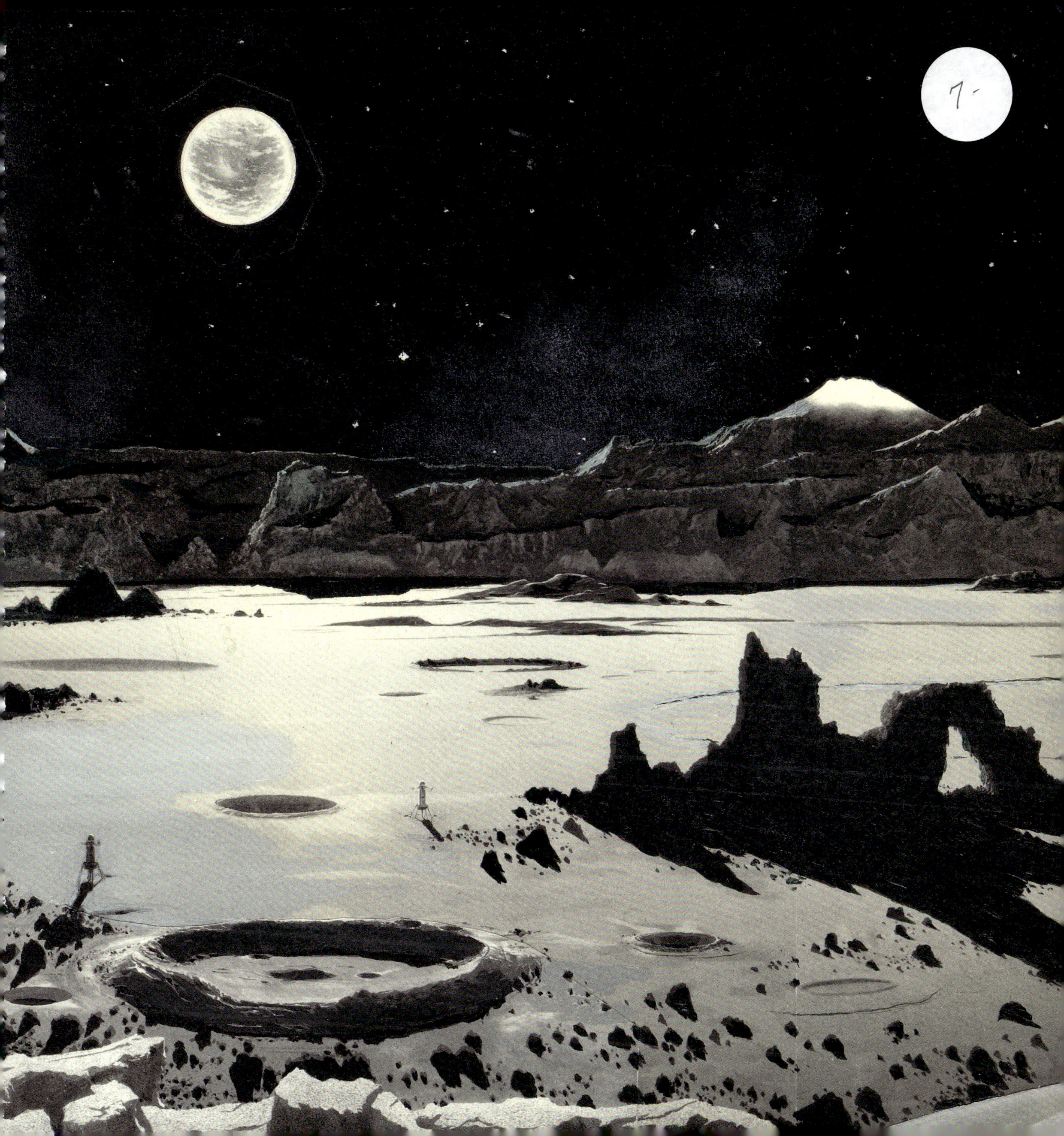

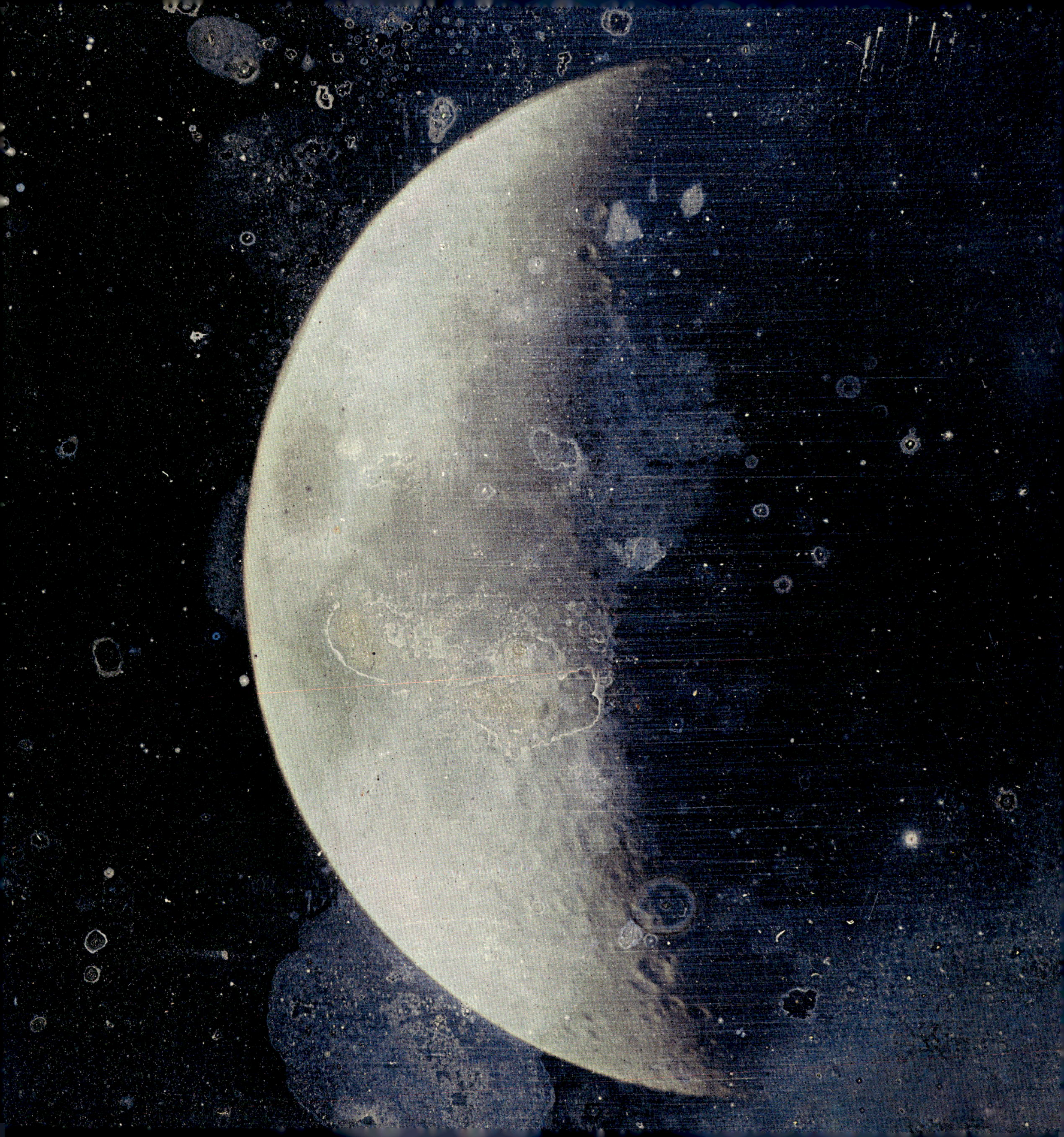

APOLLO'S MUSE

THE MOON IN THE AGE OF PHOTOGRAPHY

MIA FINEMAN AND BETH SAUNDERS

WITH AN INTRODUCTION BY TOM HANKS

THE MET

THE METROPOLITAN MUSEUM OF ART, NEW YORK

Distributed by Yale University Press, New Haven and London

This catalogue is published in conjunction with "Apollo's Muse: The Moon in the Age of Photography," on view at The Metropolitan Museum of Art, New York, from July 3 through September 22, 2019.

The exhibition is made possible by OMEGA.

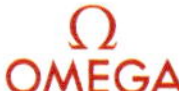

Additional support is provided by the Enterprise Holdings Endowment and The Horace W. Goldsmith Foundation.

The catalogue is made possible in part by The Andrew W. Mellon Foundation.

Published by The Metropolitan Museum of Art, New York
Mark Polizzotti, Publisher and Editor in Chief | Gwen Roginsky, Associate Publisher and General Manager
Peter Antony, Chief Production Manager | Michael Sittenfeld, Senior Managing Editor

Edited by Jennifer Bantz | Designed by Susan Marsh | Production by Sally VanDevanter with Peter Antony
Bibliography edited by Margaret Aspinwall | Image acquisitions and permissions by Jessica Palinski

Photographs of works in The Met collection are by Eugenia B. Tinsley, Imaging Department, The Metropolitan Museum of Art, unless otherwise noted. Additional photography credits appear on page 192.

Typeset in Trajan and Cardea by Matt Mayerchak | Printed on 135 gsm GardaPat Bianka | Separations by Altaimage, London
Printed and bound by VeronaLibri, Verona, Italy

Front cover: *Transparency of the Moon from Negatives Made at the Lick Observatory, Mount Hamilton, California*, ca. 1896 (detail, pl. 36)
Back cover: Neil Armstrong (NASA Apollo 11), *Buzz Aldrin Walking on the Surface of the Moon near a Leg of the Lunar Module*, 1969 (detail, pl. 85)

Frontispiece: Chesley Bonestell, *Study for A Lunar Landscape*, 1957 (detail, pl. 62); p. 2: John Adams Whipple, *View of the Moon*, 1852 (detail, pl. 14); p. 6: Caspar David Friedrich, *Two Men Contemplating the Moon*, ca. 1825–30 (detail, pl. 44); p. 10: *"Man in the Moon" postcard*, 1900s–1940s (detail, pl. 54); p. 12: Gustave Doré, *It Looked Round and Shining Like a Glittering Island*, 1862 (detail, pl. 42); p. 14: Edwin "Buzz" Aldrin, *Buzz Aldrin's Footprint on the Surface of the Moon*, 1969 (detail, fig. 16b); p. 16: John Russell, *Lunar Planisphere, Flat Light*, 1805 (detail, pl. 8); p. 70: *Drinking with the Moon*, 1910s (detail, pl. 55); p. 104: NASA Apollo 11, *Apollo 11 Command and Service Modules Photographed from the Lunar Module in Orbit*, 1969 (detail, pl. 81); p. 113: Neil Armstrong, *Buzz Aldrin Walking on the Surface of the Moon*, 1969 (detail, pl. 85); p. 136: Robert Rauschenberg, *Sky Garden (Stoned Moon)*, 1969 (detail, pl. 90); p. 172: Edward J. Steichen, *The Pond—Moonrise*, 1904 (detail, pl. 47)

FIRST PRINTING

The Metropolitan Museum of Art
1000 Fifth Avenue | New York, New York 10028 | metmuseum.org

Distributed by Yale University Press, New Haven and London | yalebooks.com/art | yalebooks.co.uk

Cataloguing-in-Publication Data is available from the Library of Congress.

ISBN 978-1-58839-684-6

CONTENTS

DIRECTOR'S FOREWORD | 7
ACKNOWLEDGMENTS | 8
LENDERS TO THE EXHIBITION | 11
INTRODUCTION | *Tom Hanks* | 13

MAPPING THE MOON | 17
Beth Saunders

DAYDREAMS BY MOONLIGHT | 71
Mia Fineman

MOONSHOT | 105
Mia Fineman

ART AFTER APOLLO | 137
Beth Saunders

WORKS IN THE EXHIBITION | 173
NOTES | 181
SELECTED BIBLIOGRAPHY | 186
INDEX | 188
PHOTOGRAPH CREDITS | 192

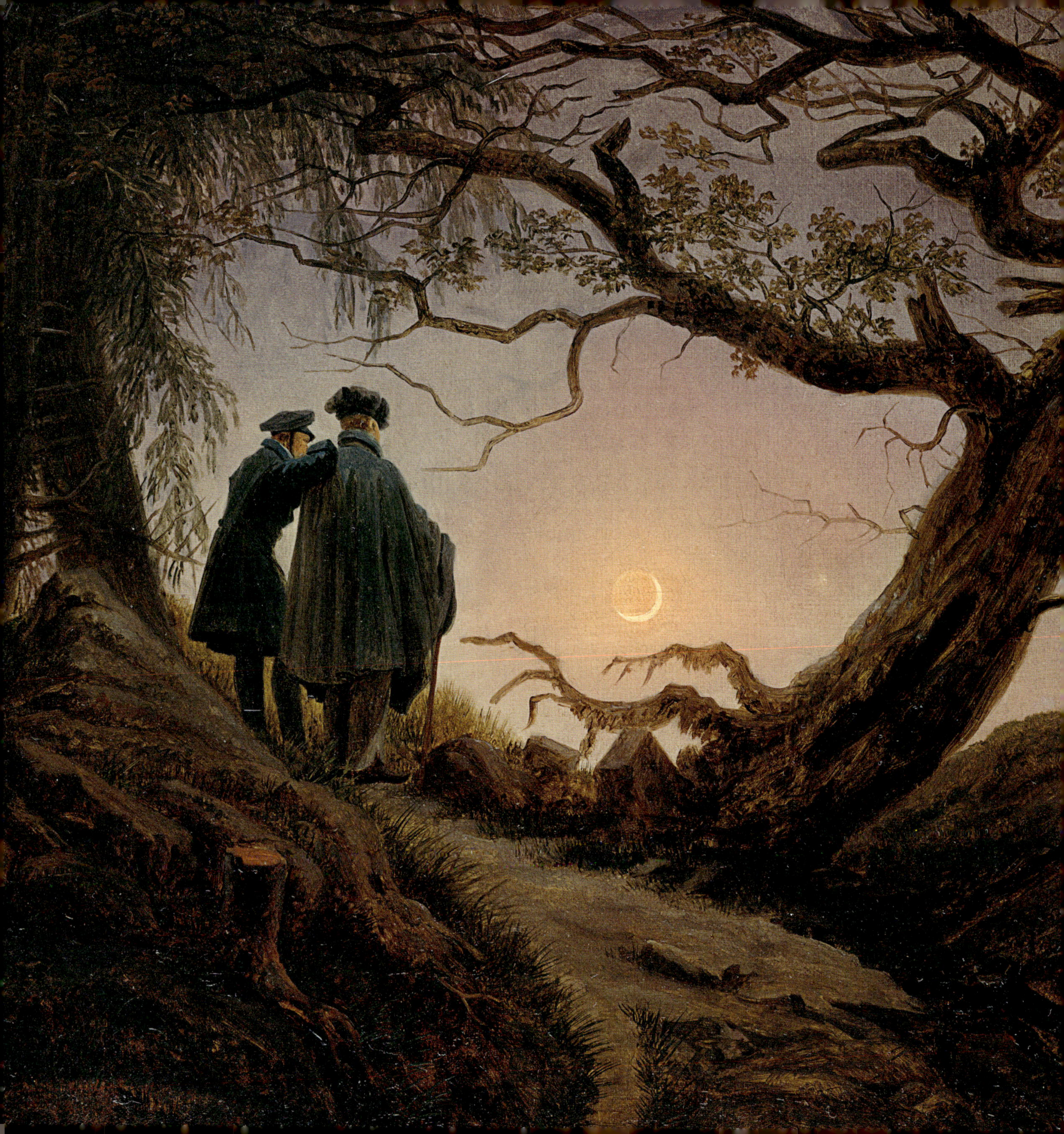

DIRECTOR'S FOREWORD

ON JULY 20, 1969, half a billion viewers around the world watched as the first television footage of American astronauts on the moon was beamed back to earth. This thrilling moment in the history of images radically expanded the limits of human vision, satisfying an age-old curiosity about our planet's only natural satellite. To celebrate the fiftieth anniversary of the Apollo 11 moon landing, this book and its accompanying exhibition survey the role photography has played in the scientific study and artistic interpretation of the moon from the dawn of the medium to the present.

When the astronomer François Arago announced the invention of the daguerreotype to the French parliament in 1839, he predicted the potential of photography to map the moon's surface quickly and accurately. The exhibition traces the progress of nineteenth-century astronomical photography, beginning with two newly discovered lunar daguerreotypes from the 1840s and culminating in a stunning presentation of a complete atlas of the moon produced at the Paris Observatory at the turn of the century. Alongside these scientific achievements, the show explores artists' fascination with the otherworldly effects of moonlight and the use of the camera to create convincing illusions of space travel and fantasies of life on the moon.

Advances in rocket science and the Cold War space race ushered in a new phase of exploration in which photography played a central role. Throughout the 1960s, crewless spacecraft captured ever clearer and more detailed photographs of the lunar surface in preparation for the astronauts of the Apollo program. The cultural impact of the 1969 moon landing is the focus of the exhibition's final section, which features works created during that era by such artists as Robert Rauschenberg and Nancy Graves, as well as the reflections of a new generation of artists.

The exhibition was conceived and organized by Mia Fineman, Curator, Department of Photographs, with contributions by former Met colleague Beth Saunders, now Curator and Head of Special Collections and Gallery at the Albin O. Kuhn Library and Gallery, University of Maryland, Baltimore County. Their essays in the pages that follow provide an eloquent overview of lunar imagery, highlighting photography and its interrelationships with works in a wide range of media, including paintings, drawings, prints, books, astronomical instruments, film, and video art. For his delightful introduction, we are grateful to actor and lifelong space enthusiast Tom Hanks.

I am deeply thankful to my colleagues at The Met who ensured this project's success. To the twenty-seven institutions and individuals who have generously agreed to share their holdings with the Museum's public, I extend our gratitude. Our heartfelt thanks go to the individual donors who have enriched The Met collection by helping to acquire numerous works featured in the exhibition. Finally, I gratefully acknowledge OMEGA for its lead sponsorship of the exhibition; the Enterprise Holdings Endowment, The Horace W. Goldsmith Foundation, and Mary C. Solomon for their additional support; and The Andrew W. Mellon Foundation for making possible this splendid catalogue.

MAX HOLLEIN
Director, The Metropolitan Museum of Art

ACKNOWLEDGMENTS

ORGANIZING AN EXHIBITION at The Met is a major undertaking—not quite as major as sending a human being to the moon, but challenging enough to require the dedication and expertise of many individuals. First, I am grateful to my coauthor and collaborator Beth Saunders for her crucial role in shaping the exhibition and her brilliant contributions to the catalogue. I extend my thanks to former Director Thomas P. Campbell for his initial encouragement, and for their continued support, Daniel H. Weiss, President and CEO; Max Hollein, Director; and Quincy Houghton, Deputy Director for Exhibitions. My learning curve on this project was unusually steep, but I was fortunate to find a few good teachers along the way. Utmost thanks go to Christopher Phillips for generously sharing his research on lunar photography and to Randy Liebermann for guiding me through his collection and beyond.

In the Department of Photographs, I am grateful to Jeff L. Rosenheim, Joyce Frank Menschel Curator in Charge, for his continuing support and encouragement, and to my research assistant Virginia McBride for her effective and enthusiastic aid with every aspect of the project, from loan requests to wall text. For coordinating the preparation of photographs from The Met collection, I thank Meredith Reiss, Karan Rinaldo, Robert Walsh, and Elena Tarchi. Our departmental technicians Predrag Dimitrijevic and Ryan Franklin attended to the framing and installation of the works with consummate skill. Rachel Mustalish in the Department of Paper Conservation and Katherine C. Sanderson, Georgia Southworth, and Alexandra Nichols in the Department of Photograph Conservation expertly prepared and monitored the works on display.

Although the exhibition traces the history of lunar photography, the narrative is enriched by works in many different media. For sharing important objects from their departmental collections, I thank my Met colleagues Andrew Bolton, Wendy Yu Curator in Charge, and Joyce Fung and Anna Yanofsky in the Costume Institute; Nadine M. Orenstein, Drue Heinz Curator in Charge, and Jennifer Farrell, Elizabeth Zanis, and David Del Gaizo in the Department of Drawings and Prints; and Keith Christiansen, John Pope-Hennessy Chairman, and Asher E. Miller in the Department of European Paintings. Martha Deese, Senior Administrator for Exhibitions and International Affairs, and Rachel Ferrante, Exhibitions Project Manager, deserve thanks for their help with many organizational and logistical matters. Senior Associate Registrar Nina S. Maruca coordinated the exhibition's loans with her usual efficiency and good humor. Jason Herrick, Marilyn B. Hernández, Kristin MacDonald, and John L. Wielk in Development found the resources to support the exhibition and its programs.

For their stunning design and warm collegiality, I thank Exhibition Designer Patrick K. Herron and Senior Graphic Designer Kamomi Solidum, along with Ria Roberts, Anna Rieger, and the entire production team. Thanks are also due to lighting designers Clint Ross Coller, Richard Lichte, Amy Nelson, and Andrew Zarou, who made the galleries glow. In the Digital Department,

Melissa Bell and Bryan Martin edited the film excerpts, and Paul Caro, Robin Schwalb, and Peter Berson beautifully installed and maintained all the time-based media works in the exhibition. I gratefully acknowledge my colleagues in the Education Department, especially Sandra Jackson-Dumont, Frederick P. and Sandra P. Rose Chairman, and Assistant Educator Marianna Siciliano, for conceiving and executing a wonderful program of talks, workshops, and activities for visitors of all ages. Erin Thompson and her colleagues in Merchandising and Retail designed and sourced a delightful array of moon-related products. Under the supervision of Kenneth Weine, Vice President of External Affairs and Chief Communications Officer, Alexandra Kozlakowski and Micol Spinazzi successfully promoted the exhibition worldwide.

For this exquisite book, I am sincerely grateful to my colleagues in the Publications and Editorial Department: Mark Polizzotti, Gwen Roginsky, Michael Sittenfeld, and Peter Antony. Sally VanDevanter, assisted by Sophia Bruneau, coordinated the production with meticulous attention to detail. Jessica Palinski adeptly handled image rights for the book. The Imaging Department, led by Barbara J. Bridgers, provided copy photography of the Museum's objects and several loans; special thanks go to Eugenia Burnett Tinsley, Juan Trujillo, and Jesse Ng for their excellent work. Senior Editor Jennifer Bantz shepherded this publication from beginning to end, providing expert guidance along the way. I extend my warmest thanks to Tom Hanks for his splendid introduction. Margaret Aspinwall edited the bibliography and notes, and Susan Marsh brought it all together with her elegant design.

The exhibition would not have been possible without the cooperation of the institutions and individuals who agreed to lend important works from their collections; they are listed on page 11. I am especially indebted to my colleagues at other institutions who have shared their expertise and facilitated the loans crucial to this endeavor: Giovanna Borasi at the Canadian Centre for Architecture; Françoise Lemerige and Isabelle Regelsperger at the Cinémathèque Française; Rebecca Cleman and Karl McCool at Electronic Arts Intermix; Christina Hunter at the Nancy Graves Foundation; Elizabeth Cronin at the New York Public Library; Janet Bunde at the New York University Archives; Jennifer Levasseur, Valerie Neal, and Margaret Weitekamp at the Smithsonian National Air and Space Museum; Morgan Aronson, Kirsten van der Veen, and Lilla Vekerdy at the Smithsonian Institution Libraries; Daina Bouquin and Maria McEachern at the John G. Wolbach Library, Harvard College Observatory; and Elisabeth Fairman and Scott Wilcox at the Yale Center for British Art.

Endeavors of this magnitude rely on the commitment of our generous donors. I thank OMEGA, as lead sponsor of the exhibition; the Enterprise Holdings Endowment, The Horace W. Goldsmith Foundation, and Mary C. Solomon, who provided important additional support; and The Andrew W. Mellon Foundation, whose generosity has helped bring this extraordinary catalogue to fruition.

For help of various kinds, I am grateful to Michael Benson, Peter C. Cohen, Richard Maurer, Abner Nolan, Carmen Pérez González, Christopher B. Steiner, Daniel Solomon, and Joni Moisant Weyl, with special thanks to Joel Smith and our daughter Sylvie for traveling with me to the moon and back.

MIA FINEMAN
Curator, Department of Photographs

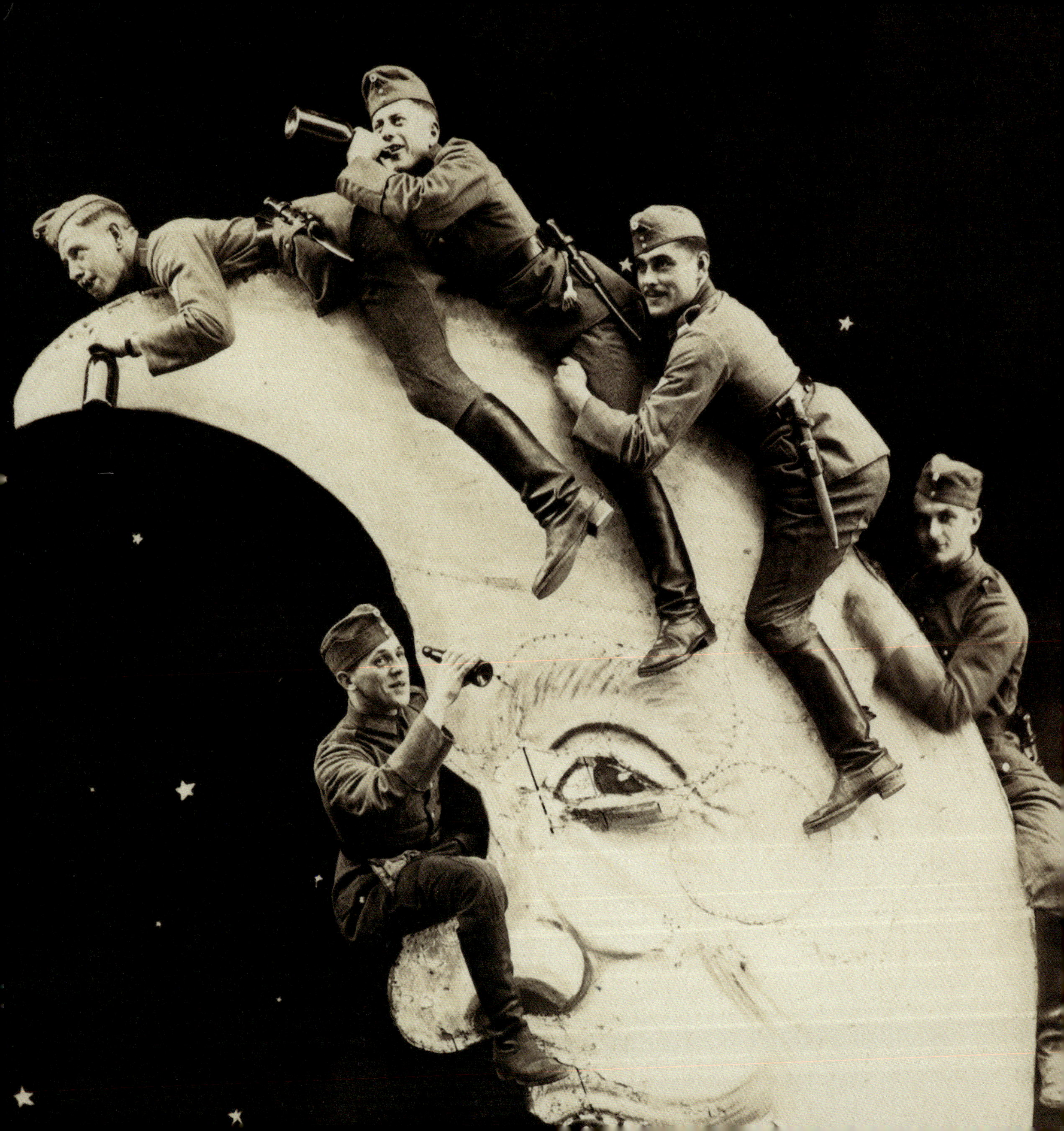

LENDERS TO THE EXHIBITION

Marc and Laura Andreessen
Canadian Centre for Architecture, Montreal
Cinémathèque Française, Paris
Paula Cooper Gallery, New York
Valerie Dillon
James and Abigail Draper
Electronic Arts Intermix, New York
Nancy Graves Foundation, New York
Harvard University Archives, Cambridge, Mass.
Houghton Library, Harvard University, Cambridge, Mass.
Hans P. Kraus Jr. Inc., New York
Randy and Yulia G. Liebermann Lunar and Planetary Exploration Collection
Bethany and Robert B. Millard
Aleksandra Mir
Forrest W. Myers
The New York Public Library
New York University Archives
Roberta J. M. Olson and Alexander B. V. Johnson
Private collections
Robert Rauschenberg Foundation, New York
Alexander W. Rutherfurd
Smithsonian Institution Libraries, Washington, D.C.
Smithsonian National Air and Space Museum, Washington, D.C.
Roberta W. Waddell
Stephen White Collection II
John G. Wolbach Library, Harvard College Observatory, Cambridge, Mass.
Yale Center for British Art, New Haven, Conn.

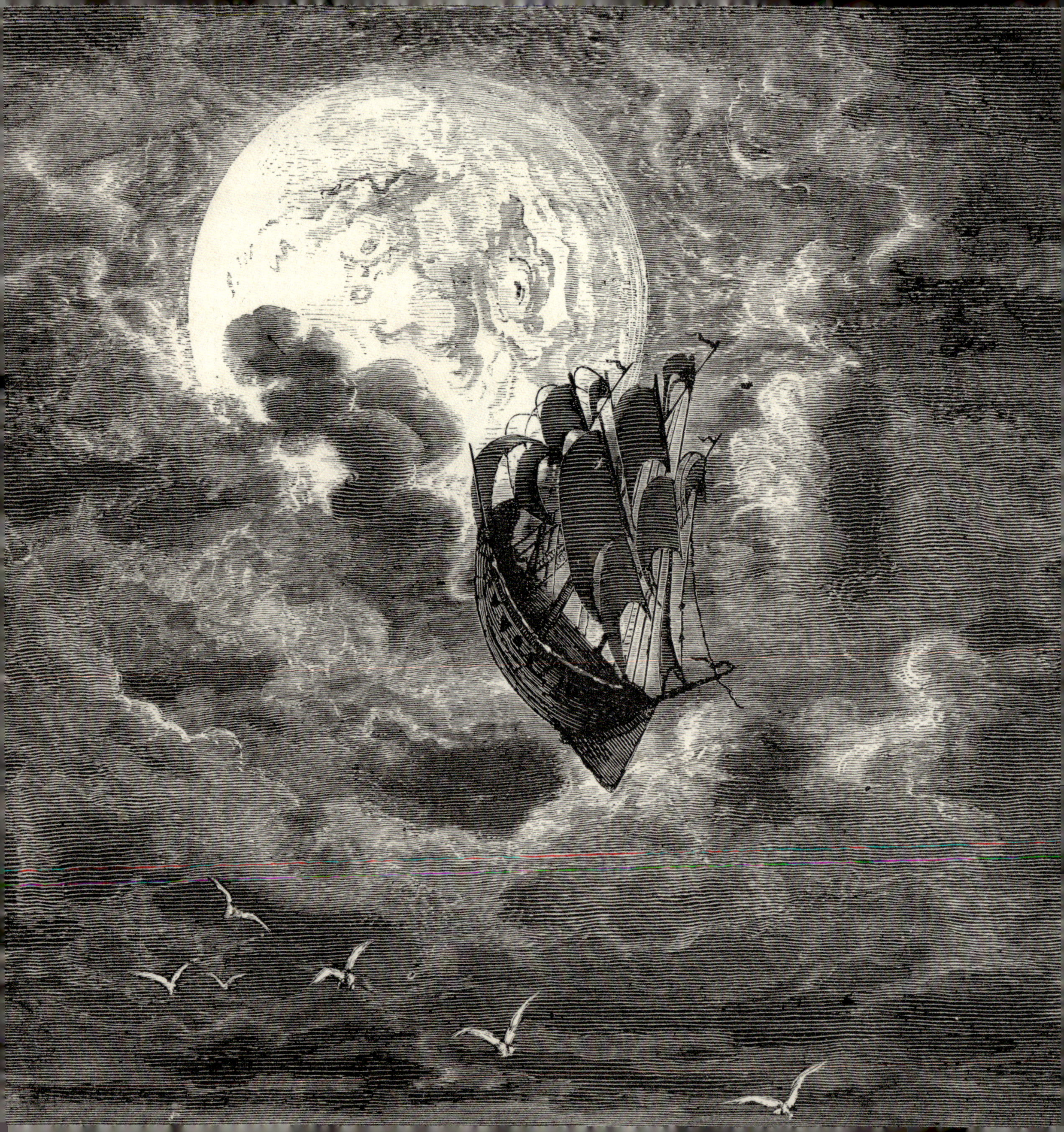

INTRODUCTION: THE MUSE OF APOLLO

TOM HANKS

FROM THE EARLIEST DAYS of human history, no matter where our kind was scattered, the moon was a mystery. Whether we lived in the equatorial zones or the latitudes of ice and snow, the moon hovered over us, perhaps the most enigmatic site in creation. Bright enough to illuminate the night, slimming to the narrowest of crescents before disappearing completely, the moon showed a reliable pattern. We learned that after going dark, the moon would return to us, once again growing full and glowing. Over time, the ancients saw that the moon returned to the same place, the same part of the sky overhead, and they were prescient enough to set in stone—literally, in some cultures—the timing of the moon's journey and her form (the feminine being the writer's choice of descriptive gender). The moon then became predictable, a constant in our lives, a heavenly body with a familiar appearance.

Those times she somehow blotted out the sun for a few minutes certainly caused no small amount of panic and prompted a plethora of wrong explanations, as did the occasions when she slowly turned bloodred in the nighttime sky.

Aha! The moon retained her mystery. She remained inexplicable. She surprised us often enough to keep us wondering just what was going on up there. We attempted to describe her in words and pictures—and yet questions remained. Was that a face looking down on us? Did the gods live there? Or a big rabbit? Or a divine woman? Maybe a race of cricket-men?

The more fanciful of us human beings began to imagine going there to see for ourselves. All we needed was the right set of wings, or a big enough catapult. Maybe we could rise up and meet her in a coal-powered balloon. Or the same fuel could power a locomotive to roar onto the moon's surface, right after the rail was laid from here to there. Better yet, and most likely, given the times, a hollow cannon shell would be fired from a God-size hunk of artillery; tucked inside would be intrepid voyagers with a few blankets and a picnic lunch. Alas, none of those theories of lunar travel would work.

Still, how delightful it was to ponder meeting a challenge that was thousands and thousands and thousands of years old.

When the manufacture of wings made it possible for us to soar, and rocket motors made it possible to loft radios, cameras, and dogs into the lifeless void of space, the moon suddenly came within our physical reach. The amount of time from Kitty Hawk to Sputnik was fabulously *short*—a mere wink of the eye of God and a bold testament to the enlightenment of humankind. The day, the hour, the moment finally arrived when the goal of landing a man on

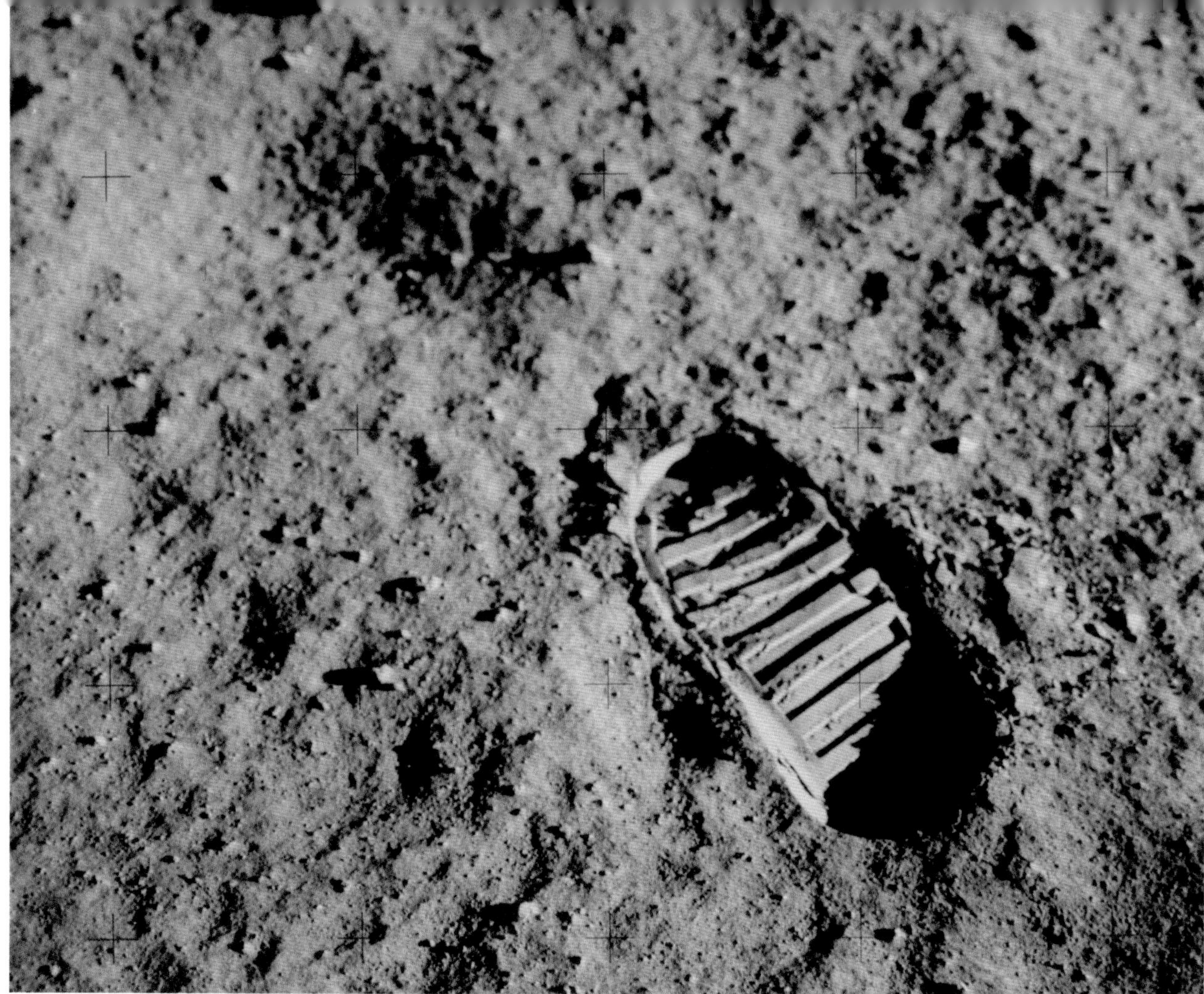

the moon was stated, *promised* (the descriptive gender of "man" is taken from the historical record). And the entire world accepted that not only was the voyage possible, it was *inevitable*.

There are two aspects of humankind's trips to the moon that are, I think, stunning.

First: The project was named for Apollo, the Greek god who drove the chariot of the sun across the sky (this proposition served as the rather cockamamy excuse for scholarship way back when; other theories abounded, of course). And the use of the ancient deities of mythology was excellent branding by NASA. Project Mercury hurled a single human being into space with all the daring and adventure of an old-time radio serial—*Captain Midnight*, or *Terry and the Pirates*. Project Gemini sent astronaut "twins" around and around the world with long lists of tasks to complete, gear to test, and physics to prove. These were flights of glamour. The banter could be witty. The photos were gorgeous! And the Gemini spacecraft itself was about the size of the sports car James Bond drove, or the Corvette that was driven across the country in *Route 66*.

But by the time Project Apollo began, there was no longer any trace of the cockamamy. Apollo knocked some sense into our promised voyage to the moon. The trip was dangerous. Three lives were lost, those of the first trio meant to fly Apollo. The men were killed in a horror—perhaps owing to hubris, but certainly as a result of a failure of imagination. How often in the history of our world has such a failure resulted in such a tragedy? Often. Alas.

We had to ask ourselves if the endeavor was going to be worth the cost. Did we have the wherewithal to risk more human lives? Were we willing to provide the money? Would we be able to go to the moon at all?

This is the second stunning fact of Apollo: that we chose to go to the moon on what was, up to just a few years earlier, truly a wing and a prayer. The science fiction of it all was fun. The science fact was so serious that the more foolish of us found we had disregarded that latter half of the challenge of putting a man on the moon: *returning him safely to the earth*.

Less than a decade before human eyes saw the earthrise from the far side of the moon, it was worried that, in the free fall of microgravity, an astronaut's eyeballs would float out of their sockets. The ability to swallow food in space was in question. (To allay that fear, John Glenn swallowed a pill on one of his three orbits.) It was theorized that a lunar walker would disappear into the moon's surface like a wayward hunter into quicksand, for how could that collection of dusty particles be solid enough to support human beings, much less the rocket ship that brought them?

In almost *no time at all* experience and example dismissed that nonsense, but what was never fully crossed was a threshold of safety. Anyone who flew into space, any member of an Apollo crew, would be cheating death. No one would fly from the earth to the moon with even odds of making it back.

And there you have the muse of Apollo. No matter how you ponder, explain, picture, or experience going to the moon, the fact is that humans were not made to survive there. And yet, we were made to want to go.

Apollo touched so many of the disciplines of humankind: the sciences, of course, but also politics (we had to beat the Soviets!), economics (industries all over the nation were engaged), and, as this volume details, the arts. Poets, journalists, painters, and filmmakers weighed in with their interpretations of Apollo. There was the inherent romance of the journey into the unknown (where no man had gone before). There was the national pride of the planting of the Stars and Stripes at six different lunar sites (though no territorial claims were made; the moon belongs to all humankind). And there was the moment in history when time was divided into two distinct eras: before Apollo took us to the moon, and after we had returned.

The big earthbound question of Apollo — was it worth it all? — is moot. One can look at the megaboot the program gave to science and technology and say, "You bet!" Or look at what going to the moon did for our collective consciousness: we saw our fragile little blue-green-white earth from a color television camera mounted on an electric car that had unfolded and then been driven across the moon to a place called Hadley and a valley named Taurus-Littrow. Likewise, it is moot that the trips were dangerous in the extreme. After all, but for astonishing good luck in timing and the maximum efforts of a few thousand people, a second trio of astronauts was nearly lost in horror, yet instead made history.

Even the ranting of idiots who can't comprehend the grand potential of our human abilities or accept the facts of our history cannot devalue the accomplishment of our going to our mysterious moon and returning safely to our good earth.

What remains for the people of earth is an even greater challenge now than that we faced before there were footprints in the Sea of Tranquility, the Ocean of Storms, and the Fra Mauro Highlands. Who will be the next to try? Who will dare to put in the work and accept the risks of flying to the moon? Who will attempt that next lunar landing with its so-very-slim margin for error? Who is going to go where no *woman* has gone before?

I hope it's you.

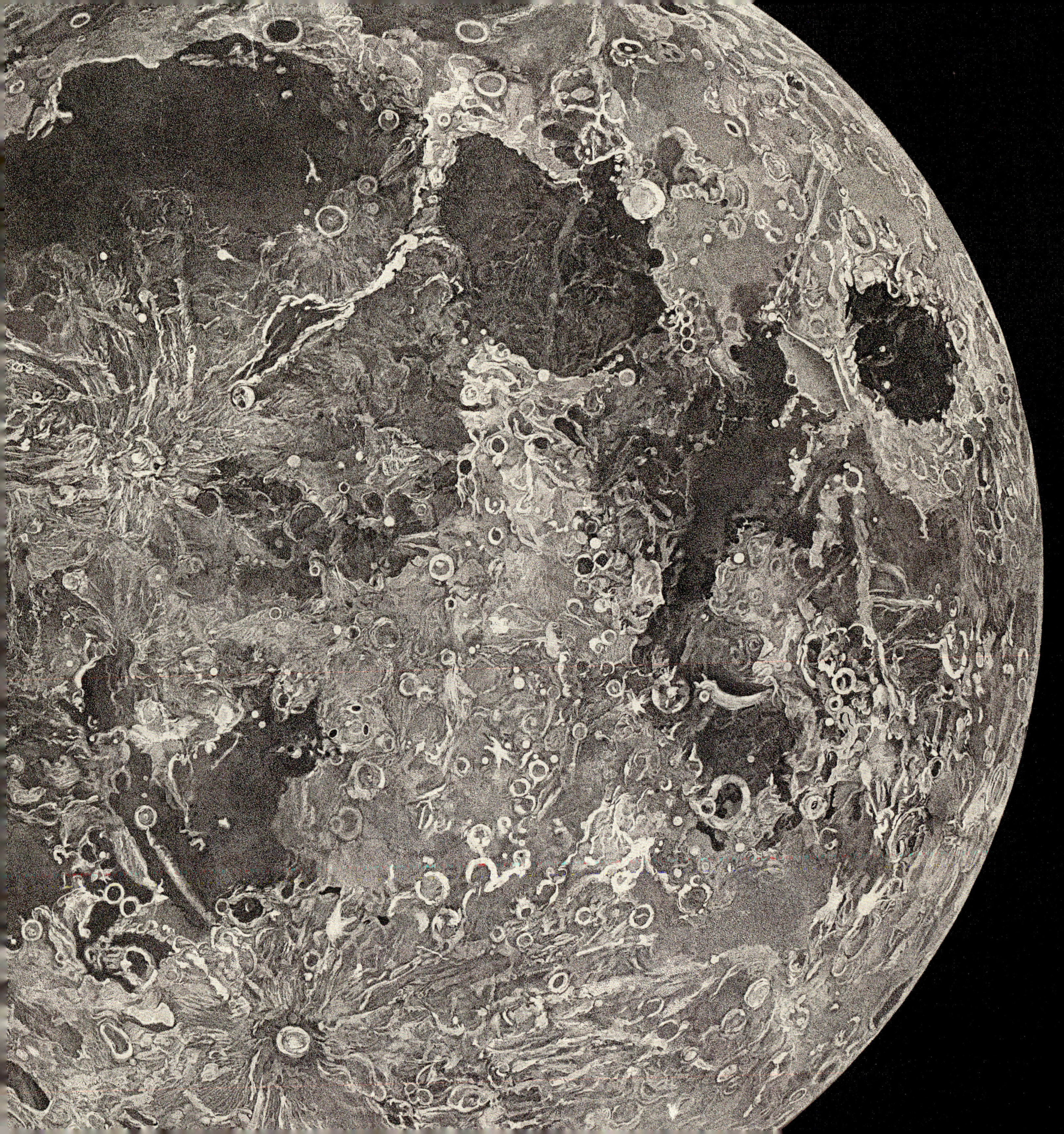

MAPPING THE MOON

BETH SAUNDERS

Nothing could be more interesting than [the moon's] appearance through that magnificent instrument, but to transfer it to the silver plate, to make something tangible of it was quite a different thing.

— JOHN ADAMS WHIPPLE

OVER THE COURSE OF nineteen nights in late 1609, the Tuscan astronomer Galileo Galilei occupied a tower of the Venetian church of San Giorgio Maggiore, training his homemade telescope on the heavens. He was not the first person to aim the newly invented (ca. 1608) instrument at the moon, but the impact of his action would be the most significant by far.[1] The observations recorded and disseminated in his 1610 book *Starry Messenger* (*Sidereus Nuncius*) transformed human understanding of the earth's satellite. While Renaissance science had inherited an Aristotelian view of the moon as a perfectly spherical and unblemished orb, Galileo's observations revealed its surface to be like that of our own planet: rugged and uneven, marked by valleys, craters, and mountains. His discovery changed the appearance of the moon in Western art, compelled new scientific inquiry into how and when the lunar surface formed, and inspired the question of whether life exists on the moon. Not satisfied with using written description alone to convey his astounding discoveries, Galileo included in his book illustrations based on his own drawings (pl. 1). This dissemination of visual evidence in support of his scientific observations established a new astronomical field concerned with describing and mapping the moon's surface, which became known as "selenography."

Following in the wake of Galileo's groundbreaking publication, astronomers and artists capitalized on technological improvements to telescopes to create ever more accurate visual records of the lunar surface. Over the course of the next two centuries, detailed maps of the moon chronicled newly identified topographical features and became instrumental evidence in scientific debates. Almost immediately upon the public announcement of the invention of photography in 1839, the medium became an important tool of astronomical research. During the nineteenth century, astronomers attempted to harness photography to render lunar maps of greater accuracy than the hand-drawn examples that preceded them. This turned out to be a significant technological challenge, and many observers concluded that photographic efforts paled in comparison to the great aesthetic and scientific achievements of earlier cartographic projects. However, because photography was understood as representing nature objectively, without the intervention of human subjectivity, photographs offered authoritative evidence, and their reproducibility enabled the dissemination of scientific knowledge to professionals and the public alike.[2]

A survey of hand-drawn lunar maps made in the seventeenth and eighteenth centuries demonstrates the visual field into which photographs of the moon entered and establishes the terms with which scientists evaluated them. Among the finest pieces of manual illustration is a series of three large engravings of the moon in different phases produced by the French artist and draftsman Claude Mellan in 1635 (pls. 3–5). Having trained in Paris and Rome under the painter Simon Vouet, Mellan worked for the wealthy French dilettante Nicolas-Claude Fabri Peiresc and the astronomer and mathematician Pierre Gassendi on their study of lunar eclipses.[3] The visually striking results of this collaboration remained unsurpassed in accuracy until the nineteenth century, when improvements to telescopes made it possible to see greater detail. Despite the incredible naturalism of Mellan's prints, they did not circulate widely and, lacking any identifying notations or scientific text, they were not especially useful maps. Instead, they dazzle the eye as a tour de force of artistic skill. The purported conflict between scientific utility and sheer aesthetic pleasure in images of the moon would remain a point of debate among astronomers well into the nineteenth century, suggesting that pictures of celestial bodies are seen as carrying an inherent sense of wonder somehow at odds with the requisite scientific objectivity.

For astronomers, maps accompanied by texts in scientific publications held greater practical value for communicating new discoveries and observations than did individual images. Nevertheless, the lavishness of early lunar atlases demonstrates the power of arresting images to engage the intellect. The amateur astronomer Johannes Hevelius's *Selenography, or the Description of the Moon* (*Selenographia, sive lunae descriptio*), considered the first book entirely dedicated to the moon, is a case in point.[4] Printed in Poland in 1647, it contains engravings detailing all forty phases of the moon, as well as three lunar maps. Framing a composite map of the moon are groups of putti holding banners, looking through telescopes, and making drawings (pl. 6). Symbols of divine agency, these Baroque elements present the cosmos as the heavenly domain of the gods; the field of selenography was after all named for Selene, mythical goddess of the moon. Hevelius's creations are the product of a worldview still bound to past traditions rather than rooted in tangible proof of direct observation. Despite the regard in which astronomers held *Selenography*—it served as the standard lunar atlas for over a century—it represents an early phase of discovery that would be superseded with the intellectual awakening of the Enlightenment.

Working in the second half of the eighteenth century, British pastel portraitist and amateur astronomer John Russell created numerous detailed sketches of the moon based on observations through his telescope (pl. 7), which provided the source material for several lunar projects.

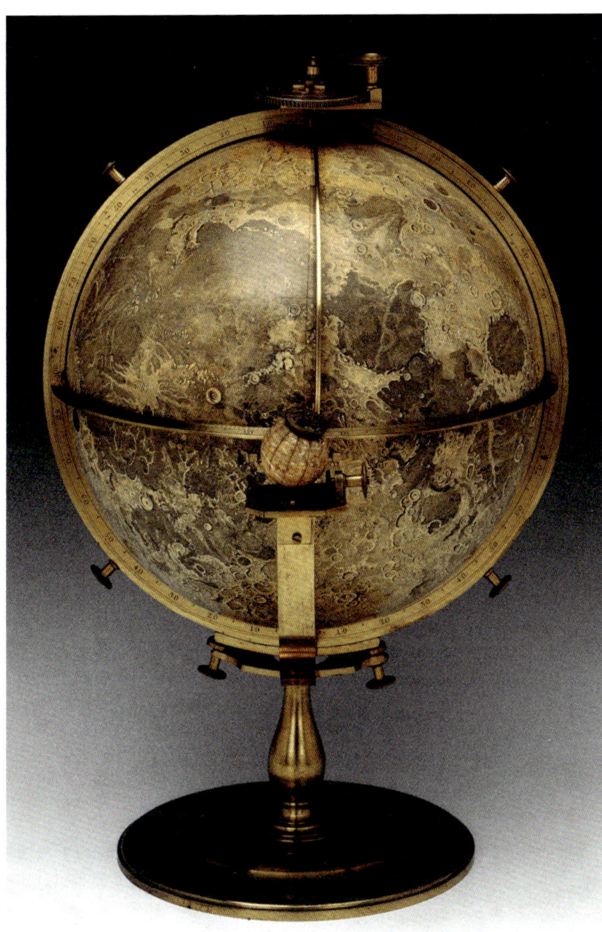

FIG. 1. John Russell (British, 1745–1806), *Selenographia*, 1797. Brass, papier-mâché, plaster, engraving and hand-coloring on paper, varnish; sphere diam. 11⅘ in. (30 cm). National Maritime Museum, Greenwich, London, Caird Fund

He produced a five-foot-diameter pastel rendering of the face of the moon, and he also sold globes made by affixing engravings based on his drawings to a papier-mâché sphere (fig. 1).[5] Notably, one side of the sphere is unembellished, its blank surface a reminder that the far side of the moon remained unseen and unknown. Although the commercial venture failed, Russell's son posthumously published two astonishing engravings made by his father, which scientists held in high esteem. An 1805 print of the moon depicted as though exposed to flat, even light (pl. 8) suggests that the "photographic" realism one might initially attribute to Russell's work was subservient to preconceived notions about how the moon should appear on the page—and about the very nature of representation. He aimed for seamless perfection, believing that although nature provided the wellspring for art, it was his duty to "correct" it in order to obtain an ideal result.[6]

The first half of the nineteenth century saw a fundamental shift in Western thought centered on the question of how knowledge was formed. Scientific positivism, the theory expounded by August Comte that understanding of the natural world should be based on sensory observation, privileged sight as the organ of perception. The camera and other optical devices such as the telescope functioned as extensions of the eye, making possible study of phenomena hitherto beyond the reach of human vision.[7] In a sense, images became more important than ever before, and photographs offered an unsurpassed verisimilitude, providing the raw visual data on which science was based.

A class of educated amateurs with ample time and financial resources dominated the field then known as "natural philosophy," and many of the earliest practitioners of photography came from its ranks. Perhaps the most representative figure linking the developments in photography and astronomy in this period is the British polymath John Herschel (fig. 2). Continuing work carried out by his father and aunt, the astronomers William and Caroline Herschel, he catalogued double stars and nebulae using the Great Forty-Foot Telescope at the family's famed observatory in Slough, England. In 1834 Herschel constructed a twenty-one-foot-long telescope at the Cape of Good Hope in South Africa from which he observed Halley's Comet and other astronomical phenomena. His work achieved popular acclaim, and even became fodder for a notorious hoax in 1835, when a series of sensational articles in the *New York Sun* claimed he had discovered life

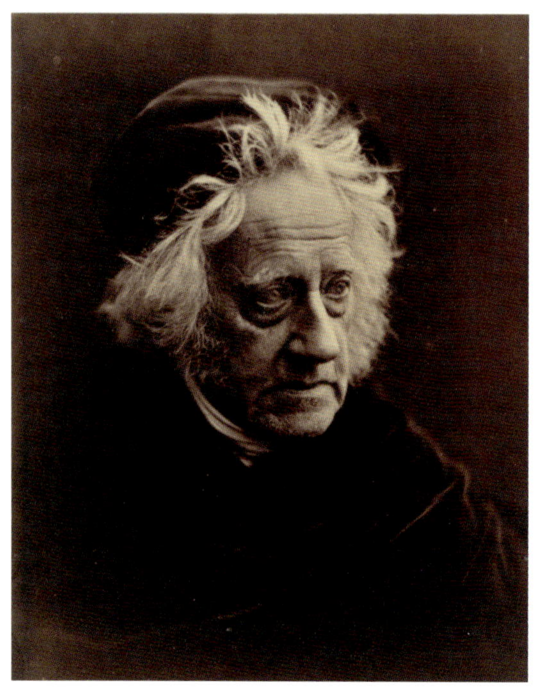

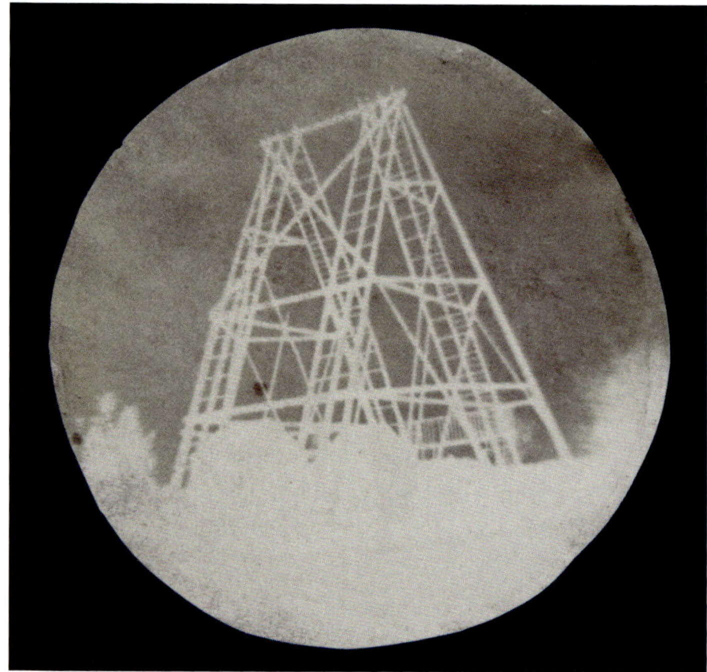

FIG. 2. Julia Margaret Cameron (British, b. India, 1815–1879), *Sir John Herschel*, 1867. Albumen silver print from glass negative, 12½ x 9 13/16 in. (31.8 x 24.9 cm). The Metropolitan Museum of Art, Gilman Collection, Purchase, Robert Rosenkranz Gift, 2005 (2005.100.25)

FIG. 3. Sir John Herschel (British, 1792–1871), *View of the Telescope at Slough*, 1839. Photogenic drawing, diam. 3 9/10 in. (9.8 cm). Museum of the History of Science, Oxford

on the moon.[8] Shortly after returning from South Africa, the astronomer learned that his friend William Henry Fox Talbot had developed a technique of capturing images of nature directly on paper coated with light-sensitive silver salts.[9] Herschel soon began his own experiments, producing an image of the framework for his father's telescope shortly before it was dismantled in late 1839 (fig. 3). In addition to making numerous contributions to the new medium—giving it the name "photography," discovering the fixing agent hypo, and inventing the red safety lamp and the cyanotype—he became a major promoter of the use of photography in the field of astronomy.

At roughly the same time in France, the astronomer François Arago, a close associate of Herschel, endorsed the invention of the daguerreotype, a photographic image captured on the surface of a polished metal plate coated in light-sensitive chemicals. In a speech to the French Chamber of Deputies aimed at securing a government annuity for Louis-Jacques-Mandé Daguerre, Arago enumerated the utility of the inventor's new medium to a variety of pursuits, among them mapping the moon: "We may hope to be able to make photographic maps of our satellite. In other words, it will be possible to accomplish within a few minutes one of the most protracted, difficult, and delicate tasks in astronomy."[10] Arago's statement implies that the camera could replace the labor of hours spent gazing into the telescope and recording observations by hand, and that the mechanization of the process would not only save time, but also produce more accurate representations of the moon through the elimination of human error. In the early days of photography, however, Arago's prophetic vision was far from an immediate reality. Urged by the Academy of Sciences in 1839, Daguerre attempted to record the moon using his invention, but it appeared as little more than a small smudge of light on the mirrored plate. Nevertheless, the mere possibility this pale reflection represented prompted the German naturalist Alexander von Humboldt to remark, "Even the face of the Moon leaves her portrait in Daguerre's mysterious substance."[11]

The promise of photography's application to astronomy quickly led others to conduct their own experiments. In the United States, the chemist John William Draper began his research with the daguerreotype almost immediately upon hearing the news of its invention, linking it with his study of the effects of light on chemical reactions. Draper's training in chemistry and optics—previously he had employed a camera obscura to observe how light passing through various liquids and chemicals affects the visible spectra—positioned him well for success with photography.[12] On March 16, 1840, he wrote in his laboratory notebook, "This evening I exposed a prepared plate to the moonbeams which had been conveyed by a double convex lens and fixed by a heliostat. Time half an hour on mercuralizing [*sic*] there was a strong image."[13] That March and April, he made a series of lunar daguerreotypes and presented his achievement to fellow scientists at the New York Lyceum of Natural History.[14] The moon is but one inch in diameter in a Draper daguerreotype from this era, yet the clarity and detail of the craters are astonishing, particularly where the sun's rays cast the longest shadows over the lunar surface, throwing the moon's topography into a breathtaking chiaroscuro (pl. 10). The resulting image is graphically striking in part due to the halo-like vignette encircling the moon, a crescent shape that evokes the lunar phases. Despite Draper's accomplishment—which surpassed Daguerre's—his efforts received only modest recognition from his contemporaries.

The failure of Draper's lunar daguerreotypes to draw great fanfare speaks to the difficulties of making and distributing photographs in this early period, and to the hurdles of astronomical photography in particular. An expensive and complicated pursuit, daguerreotypy required advanced knowledge of chemistry and optics, and it yielded a one-of-a-kind object; in order for an image to be reproduced and circulated, it had to be translated into another medium, typically lithography or wood engraving. Celestial bodies presented even greater problems for the photographer than earthbound subjects. Their distance meant that not only local weather but also the thermal state of the earth's atmosphere affected the quality of the resulting image. Exposure times were long and had to account for the apparent motion of celestial bodies across the sky, an effect actually owing to the earth's rotation. To solve this problem, the telescope required a clockwork mechanism, or clock drive, that moved the instrument in sync with the object of study. Much of the history of astronomical photography in the nineteenth century is a record of the incremental technological advances that made it possible to overcome these obstacles.

A daguerreotype made by the American portraitist Samuel Dwight Humphrey on September 1, 1849, in Canandaigua, New York, demonstrates the difficulty of manipulating the plate to account for celestial movement. It records nine exposures of the moon, each lasting between half a second and two minutes (pl. 12). Humphrey noted each exposure time on a slip of paper attached to the object and sent his result to Jared Sparks, the president of the Harvard College Observatory in Cambridge, Massachusetts. Although the photographer was not himself a scientist, his annotations connote a methodical experiment, the purpose of which was to demonstrate that a shorter exposure time produced a sharper image of the moon. The outcome is a stop-motion sequence of the satellite moving across the daguerreotype plate that presages later nineteenth-century developments in chronophotography by Jules Janssen, Etienne-Jules Marey, and

Eadweard Muybridge. No fewer than eighteen impressions of the moon dot the metallic surface of a multiple-exposure daguerreotype likely made in England by the Frenchman Antoine-François-Jean Claudet (pl. 13) around the same time as Humphrey's.[15] One of the first studio daguerreotypists in London, Claudet experimented with recording not only the moon but also electricity, clouds, and the sun viewed through fog. Mere millimeters across, the bright orbs and ellipses in these two daguerreotypes bear little resemblance to the moon, appearing instead as abstract renderings of light, space, and time.

During the 1840s, the problem of photographing celestial bodies was a major subject of research at the Harvard College Observatory.[16] Its first director, William Cranch Bond, purchased the largest telescope then known in the United States, an instrument more than twenty-two feet long with a fifteen-inch-diameter lens, known as the Great Refractor (pl. 11). Beginning in 1848, Bond partnered with the local daguerreotypist John Adams Whipple, who had already carried out his own trials with an amateur telescope. Using both the Great Refractor and a smaller telescope, the men performed a series of astrophotographic experiments that proved by far the most successful to date. In July 1850 they obtained a daguerreotype of the star Vega, and in March of the following year they accomplished the first in a series of successful lunar daguerreotypes using the Great Refractor (pl. 14). From these small images (the moon appears about two and a half inches in diameter), they produced copy daguerreotype enlargements, which they exhibited to acclaim in London at the Great Exhibition of 1851 and at the Royal Astronomical Society. Whipple, along with his commercial partner James Wallace Black, continued to collaborate with Harvard scientists throughout the 1850s, adapting innovations in photographic technology to their astronomical research (pl. 15).

The daguerreotypes Whipple and Bond displayed at the Great Exhibition had an enormous impact on astrophotography in Britain, where Liverpool became an important center of debate on the topic.[17] Members of the Photographic Society of Liverpool discussed the relative merits of lunar photography as compared to observational drawings and promoted the use of glass negatives instead of daguerreotypes. The major drawbacks of using daguerreotypy for this purpose were that the plates had to be small in order to focus enough light to capture the moon, and the resulting image could only be disseminated by rephotographing the one-of-a-kind object (hence the copy enlargements produced by Whipple and Bond). Frederick Scott Archer's 1851 introduction of wet-collodion glass negatives, much more sensitive than either daguerreotypes or paper negatives, thus marked an important development. An *Illustrated London News* column from 1854 reported on a lecture at the British Association in Liverpool, in which a glass positive of the moon was projected as a fifty-foot-diameter disk, allowing viewers to examine the lunar surface in unprecedented detail and with the spectacular effect of illumination (fig. 4).[18] The discussions surrounding lunar photography in Liverpool and reports in the popular press demonstrate the medium's important role in sharing scientific information with a wider public, which was critical not only to broadening popular understanding of fundamental natural phenomena, but also to securing financial backing of scientific research.[19]

From this Liverpool milieu, Warren De La Rue emerged as a key figure. Inspired by the Great Exhibition, he made his earliest trials with lunar photography in 1852 using a telescope of his own design and wet-collodion glass negatives.[20] Initially, his instrument was not fitted with a clockwork mechanism, so De La Rue employed an assistant to carefully move it in sync with the trajectory of the moon through the night sky. By 1856 he had designed a viable clock drive and entered a new phase of success, producing highly detailed albumen silver prints (pl. 17) and the first lunar stereoviews (pl. 19). By photographing the moon twice, a few hours apart on the same evening, De La Rue mimicked an effect that would otherwise require a specially designed bifocal camera. He believed that a three-dimensional view would assist astronomers in determining whether geological changes

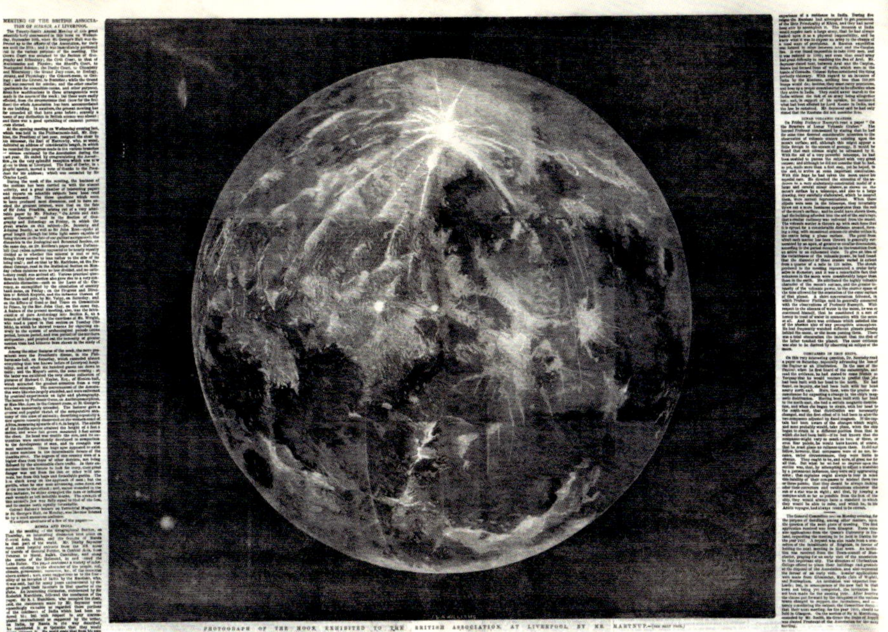
PHOTOGRAPH OF THE MOON EXHIBITED TO THE BRITISH ASSOCIATION AT LIVERPOOL BY MR. HARTNUP.

FIG. 4. J. and A. Williams, "Photograph of the Moon Exhibited to the British Association at Liverpool by Mr. Hartnup," *Illustrated London News*, September 30, 1854

were still actively occurring on the satellite, thus contributing to contemporary debates about whether its craters had been formed through volcanic activity or meteoric impact. Despite his achievements, De La Rue noted that the wet-collodion process had its drawbacks: because the photographic plate had to be exposed while wet, the resulting image often included debris and streaks that challenged its verisimilitude, and the chemicals tended to dry too quickly to accommodate the exposure times needed to capture celestial bodies. Photography had yet to eclipse drawing as a means of obtaining accurate astronomical records.

In addition to garnering the attention of the scientific community, De La Rue's photographs reached a wider public. In America, the studio photographer Austin Augustus Turner rephotographed De La Rue's lunar views, selling the pirated images as a boxed set of twelve cartes de visite (pl. 18). In England, De La Rue entered into an agreement with the highly regarded photographer Robert Howlett, who printed the lunar negatives and served as commercial distributor for the stereoviews.[21] De La Rue's images also contributed to Talbot's ongoing experiments with photographic printing. Talbot hoped to make the medium more viable for mass publication by producing engravings directly from photographs; this idea had important implications for the dissemination of scientific knowledge. Letters exchanged between 1857 and 1860 indicate that De La Rue sent the inventor examples of his lunar photographs, and in turn received samples of photographic engravings of plants.[22] De La Rue hoped that one of his lunar photographs might be reproduced by Talbot's method in the diary of the Royal Astronomical Society.[23] Although there is no evidence that this joint publication was executed, Talbot's photoglyphic engraving of the moon based on De La Rue's photograph is a tangible result of the collaboration (pl. 20). It may have been created as an initial proof for the project, which exemplifies the ways in which advancements in astronomy and the technology of photography propelled each other during the nineteenth century.[24]

Like De La Rue in England, the American photographer Lewis Morris Rutherfurd was a wealthy amateur who applied his personal resources to building his own observatory.[25] Inspired by the success of Harvard's Great Refractor, he constructed a fourteen-foot-long telescope in the backyard of his New York City home and obtained glass negatives of the moon, stars, and planets. Rutherfurd's greatest contribution to the field was to develop the first telescope specifically outfitted for photography. Because photographic plates are sensitive to a different spectrum of light than the naked eye, astronomers had to focus the instrument using trial and error; in 1864 Rutherfurd solved this problem by producing an achromatic lens specially corrected for the light-sensitivity of the photographic plate. By disregarding human sight in favor of the camera's eye, he was able to produce extraordinarily precise stellar maps as well as a series of stunning large-format photographs of the moon celebrated for their accuracy and beauty (pl. 21). Like De La Rue before him, Rutherfurd also produced stereoviews, traveling to various locations around the globe at different times of the year in order to obtain photographs of the full moon. Placed side by side in an optical viewer, the images had the effect of making accessible at once the entire surface of the moon visible from our planet under ideal circumstances. In other words, Rutherfurd produced a composite image, thus bringing to light an otherwise impossible view of the moon.

James Nasmyth, a successful industrialist and inventor—and a member of the same Liverpool-based circle of astronomers as De La Rue—pushed the limits of photographic truth even further.[26] His book *The Moon: Considered as a Planet, a World, and a Satellite* (1874), the product of decades of research, served as a compilation of available information on lunar geology and argued that the moon's mountainous and pockmarked surface was the result of volcanic activity.[27] Although he intended the publication's illustrations to persuade readers of his argument, the technique of their production calls into question the very concept of photography's evidentiary value. Because the difficulties

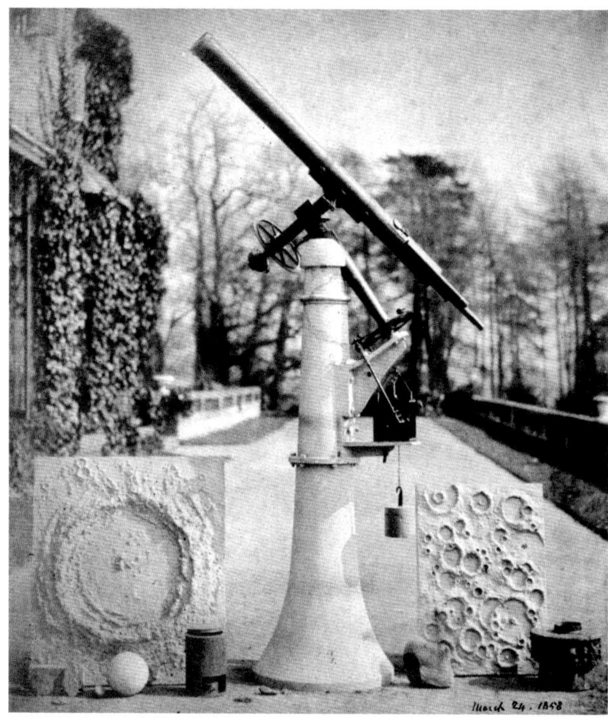

FIG. 5. James Nasmyth (British, 1808–1890), Lunar crater models and a refracting telescope outside the artist's home in England, 1858. Salted paper print, 8⅖ x 7½ in. (21.4 x 19 cm). National Media Museum, Bradford, U.K.

of astrophotography precluded the level of detail and three-dimensionality Nasmyth sought in his renderings of the moon's geographical features, he recorded what he saw through his telescope with black and white crayon on gray paper, then used his drawings as the basis for detailed plaster casts representing the lunar surface (fig. 5). He likely adopted the method from his father, a well-known Scottish landscape painter who used elaborate plaster models as studies for his paintings. Nasmyth then subjected his casts to specific lighting conditions and photographed them, producing an entirely fabricated but wholly convincing picture of the moon as it could not yet be seen by the naked eye (pls. 23, 24). In the image captioned "An ideal sketch of

'Pico,' an isolated lunar mountain 8000 feet high, as it would probably appear if seen by a spectator located on the Moon," the viewer confronts a magnificent mountain range from a position on the lunar surface — a vantage point humans would not achieve until the Apollo 11 landing in 1969. Nasmyth's vertiginous peaks are the stuff of fantasy, the product of a misinterpretation of shadows cast across the lunar surface. Far from decrying Nasmyth's methodology as unscientific, however, contemporaries such as De La Rue and Herschel lauded the visual effects and found his argument convincing. A review in the scientific journal *Nature* acclaimed, "No more truthful or striking representations of natural objects than those here presented have ever been laid before his readers by any student of science."[28] This statement suggests that the medium of photography itself imbued the images with a sense of veracity that gave them scientific authority.

In France, astronomer and writer Camille Flammarion blended fact and fantasy to an even greater degree. While employed at the Paris Observatory, he founded the French Astronomical Society, but he was also a spiritualist who believed in extraterrestrial life and the transmigration of souls after death. His *Astronomical Gallery* (*Galerie astronomique*), an 1867 series of cartes de visite, comprised ten photographs by Adolphe-Alexandre Martin of the illustrator Eugène Morieu's drawings of the planets and solar system, and two of the moon by De La Rue (pl. 22).[29] With explanatory texts by Flammarion printed on the backs of the cards, the resulting "gallery" serves as a succinct guide to the galaxy. Later, in *The Lands of the Sky* (*Les terres du ciel*, 1877), Flammarion transported his readers on a journey through the cosmos illustrated with imaginative engravings and reproductions of Rutherfurd's and Nasmyth's lunar photographs.[30] Flammarion's joyous amalgamations of scientific information and spectacular visuals helped promote public interest in astronomy.

The growing desire for accurate representations of the moon in this era challenged the commonly held association of photography with objective truth. In this field, the new medium was frequently pitted against drawing, which continued to be more reliable despite the subjectivity of its authorship.[31] In 1884, for example, the British Pre-Raphaelite painter John Brett produced a tenebrous rendering of Gassendi's crater in black chalk and gouache based on direct observation through the telescope that reveals his dedication to astronomy (pl. 25).[32] The work of French astronomer Etienne Léopold Trouvelot, an artist and scientist who joined the Harvard College Observatory in 1872, likewise demonstrates the persistence of manual astronomical illustration in the age of photography (pl. 26).[33] Working with a telescope outfitted with an etched-glass grid in the eyepiece, Trouvelot made sketches on gridded paper that became the basis for stunning pastels, fifteen of which he published as chromolithographs in *The Trouvelot Astronomical Drawings Manual* (1882, pls. 27, 28). In its introduction, the artist characterized photography as an aid at best, and one with serious limitations: "Although photography renders valuable assistance to the astronomer in the case of the Sun and Moon . . . for other subjects, its products are in general so blurred and indistinct that no details of any great value can be secured."[34] Trouvelot's pastels of the lunar Mare Humorum and a total solar eclipse show that he strove not merely to make a detailed record that might assist astronomers, but also to capture something of the sublimity of celestial bodies. In 1882 Trouvelot began working with photography at the Meudon Observatory just outside Paris under astronomer Jules Janssen, who famously declared that "the photographic plate will soon be the scientist's true retina."[35] Trouvelot's and Brett's technically accurate and aesthetically compelling drawings stand as a challenge to Janssen's opinion.

Trouvelot and Brett were among many in the nineteenth century to attempt a graphic interpretation of the sun's eclipse by the moon, which had proved to be a particularly awe-inspiring subject. Astronomers coordinated documentation across the globe, hoping to use these infrequent occurrences to gain insight into the composition of the sun's corona. As early as July 8, 1842, the physicist

Giovanni Alessandro Majocchi made the first successful daguerreotype of a solar eclipse in Italy, but he failed to capture the moment of totality.[36] Although the result of his effort does not survive, a number of daguerreotypes made during the eclipse of July 28, 1851 remain, including one created by Whipple at the Harvard College Observatory (pl. 29); a committee had formed earlier that year to advise astronomers on the best photographic practices to record the event.[37] In 1854 William and Frederick Langenheim of Philadelphia produced a remarkable series of seven daguerreotypes recording stages of the moon's transit across the sun during an eclipse (pl. 30).[38] Although in the Northern Hemisphere the moon would have passed from right to left, the trajectory appears laterally reversed on the uncorrected plates. In the resulting time-lapse sequence of photographs, another reversal appears to occur as well, with the crescents of sunlight taking on the look of the moon in its phases. Astronomers often traveled great distances to record the moment of totality from an optimal position. A photograph of the eclipse of May 6, 1883 (pl. 31) made by British astronomers H. A. Lawrence and Charles Ray Woods, stationed on remote Caroline Island in the Pacific, captures the moment of totality in an image that looks uncannily like a human eye staring out from space. Photographs of eclipses often border upon abstraction, and although their analysis requires specialist knowledge, they resonate with an almost elemental power.

In the late nineteenth century, the invention of dry-plate negatives greatly reduced exposure times, enabling photographers to more easily capture astronomical events like eclipses—and to create records that rivaled astronomical drawing. Their efforts were also aided by increasingly powerful telescopes constructed on a grandiose scale. Monumental instruments, from William Herschel's Great Forty-Foot Telescope to Harvard's Great Refractor to the Leviathan of Parsonstown in Ireland, sparked the public's imagination. The Leviathan—completed for the 3rd Earl of Rosse in 1845 and mentioned in Jules Verne's 1865 fantasy novel about space travel, *From the Earth to the Moon* (*De la terre à la lune*)—was so enormous its owner could stand in the mouth of the tube (fig. 6).[39] By the turn of the twentieth century, the optician and astronomer George Ritchey was constructing object glasses of an unprecedented size, including sixty- and one-hundred-inch mirrors for reflecting telescopes at the Mount Wilson Observatory in Los Angeles.[40] One photograph shows a Mount Wilson disk sitting atop the grinding machine in Ritchey's optical laboratory, its shimmering surface giving it the appearance of a celestial body brought to earth (pl. 33). These powerful telescopes enabled astronomers to turn their attention to the outer limits of the galaxy, documenting in photographs stars and nebulae invisible to the naked eye. The same instruments also made possible exquisite renderings of the moon so detailed they would continue to appear in astronomical publications for decades (pl. 34).

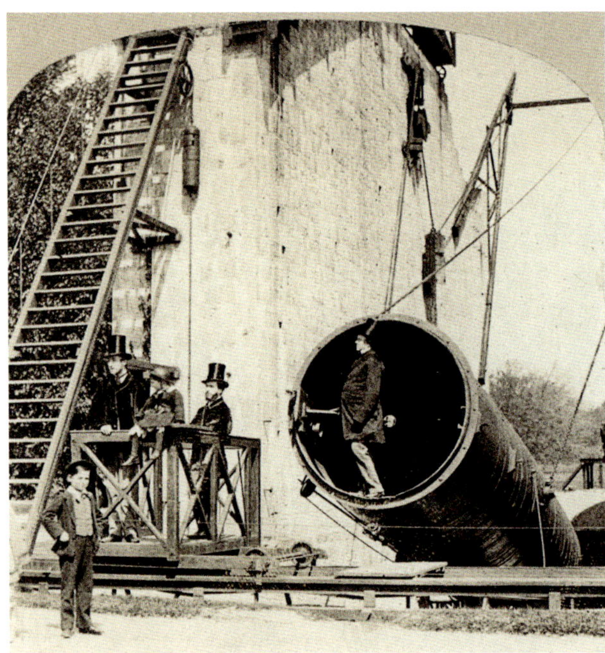

FIG. 6. Countess Mary of Rosse (British, 1813–1885), Lord Rosse's telescope at Birr Castle, Ireland, ca. 1857. Albumen silver print, 2 4/5 x 2 1/2 in. (7.1 x 6.4 cm). The Royal Society, London

In light of the era's rapid technological advancements, in 1889 Janssen, an expert in solar astronomy, called for the creation of a comprehensive lunar atlas.[41] Major observatories around the world responded to the challenge, competing to be first to achieve this milestone in selenography. In 1903 the director of the Harvard College Observatory, William H. Pickering, published the first complete photographic lunar atlas, *The Moon: A Summary of the Existing Knowledge of Our Satellite, with a Complete Photographic Atlas* (pl. 35). Aimed at scholars and laymen alike, the book became the standard reference on the subject. It reproduced in halftone photographs made at a field observatory in Mandeville, Jamaica, a site that offered more favorable environmental conditions for astronomical photography than those of Cambridge. Due to the small size and lack of detail in the illustrations, however, Pickering's atlas suffered in comparison to other contemporaneous efforts. One reviewer suggested that the poor reproductions had led Pickering to jump to false conclusions (he interpreted bright patches on the photographic plate as lunar ice caps); high-quality photographs had become essential to convincing scientific arguments.[42]

During the same period, astronomers at the University of California's Lick Observatory in Mount Hamilton were hard at work on their own lunar atlas. Unlike Pickering, Lick astronomer Edward Holden turned to skilled engravers to translate his photographic images into detailed and accurate prints. For this delicate work, he employed the New York Photogravure and Color Company, supplying them with enlarged glass positives made from original glass negatives (similar to the transparency reproduced here, as pl. 36). Such collaborations between astronomer and draftsman were sometimes fraught by competing motivations and visual lexicons. Too much invention on the part of the engraver could be detrimental to the scientific authority of the image; a lack of artistry, however, would result in inferior images that risked losing their interpretive value.[43] Although contemporaries deemed the *Lick Observatory Atlas* (1896–97) a noteworthy scientific accomplishment, it ultimately failed to compete with the ambition and aesthetic virtuosity of a significant French effort.

On July 9, 1894, Maurice Loewy and Pierre Puiseux of the Paris Observatory had announced to the French Academy of Sciences their intention to produce an authoritative lunar map on a grand scale using the institution's powerful telescope.[44] The equatorial telescope, completed in 1883, used a clockwork mechanism adjustable to thirty-six hundred speeds that allowed for exposures as short as half a second.[45] Loewy and Puiseux employed mammoth dry-plate glass negatives that recorded the moon in images six to seven inches in diameter. The partners created photographic enlargements that were then used as the basis for detailed photogravures of an unprecedented size (pls. 37, 38), produced by the French state printing house, the Imprimerie Nationale. Issued in twelve fascicles, the eighty-three photogravures were accompanied by explanatory text and by tissue overlays that delineated in labeled contours the moon's topographical features. The monumental mapping project concluded in 1910 with the publication of the final fascicle. A capstone of nineteenth-century astronomical photography, Loewy and Puiseux's atlas would remain the most comprehensive set of lunar maps for more than fifty years. In 1914 their assistant at the Paris Observatory, Charles Le Morvan, published a selection of the images as *Systematic Photographic Map of the Moon* (pl. 39). When placed in a grid, the gravures of his portfolio form two spectacular views of the waxing and waning full moon, testaments to the aesthetic achievement of the project. Despite impressive advances over the course of the long nineteenth century—from Russell's keenly observed half-globe to the photographic impressions of Loewy and Puiseux—there was yet more moon to be mapped. For much of the twentieth century, until the height of the space race, the far side of the moon would remain unknown and uncharted territory, a blank canvas upon which fantasies of extraterrestrial life and space travel could be projected.

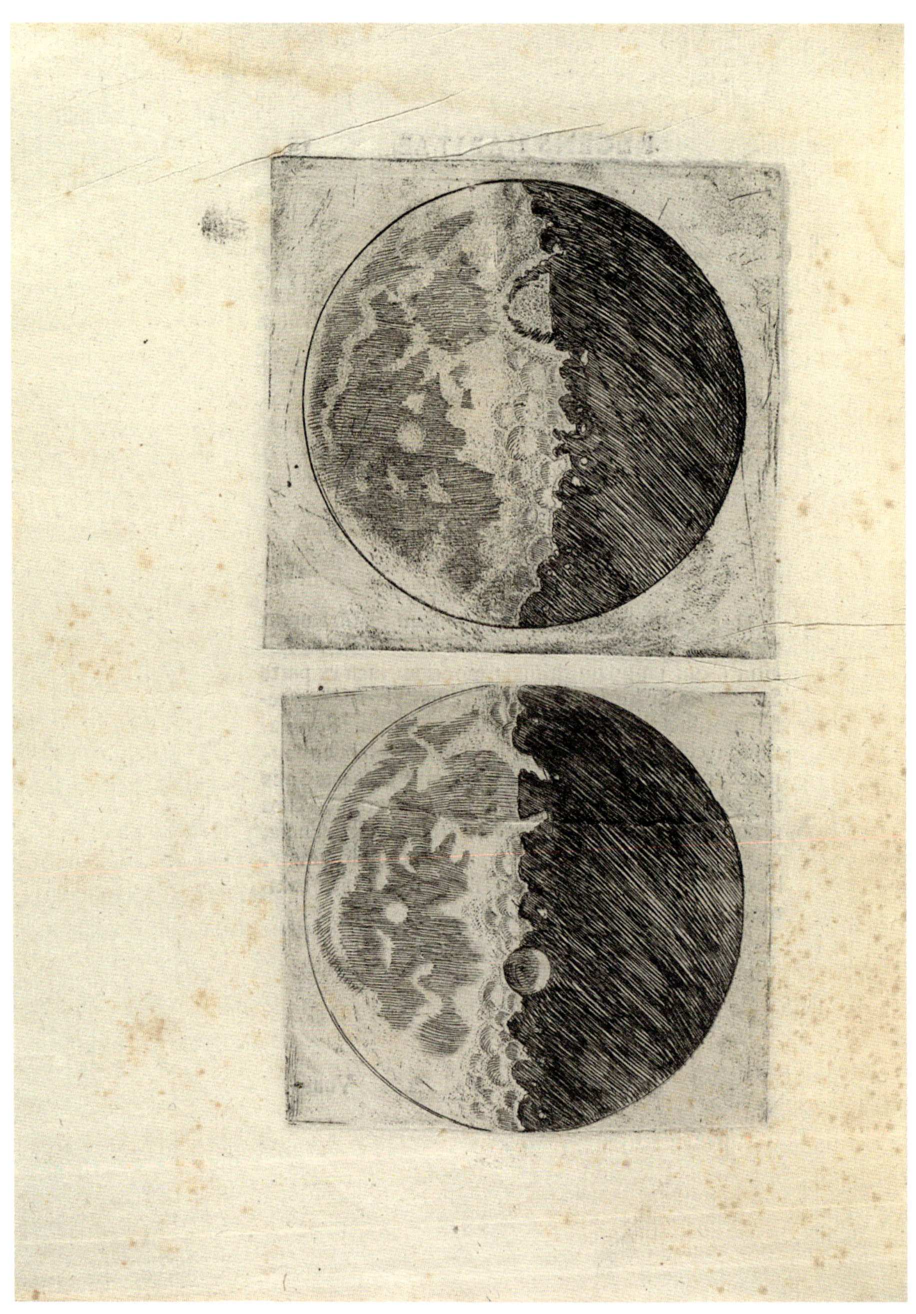

1 | GALILEO GALILEI | *Two Drawings of Waxing Moon*, in *Starry Messenger* (*Sidereus Nuncius*), 1610

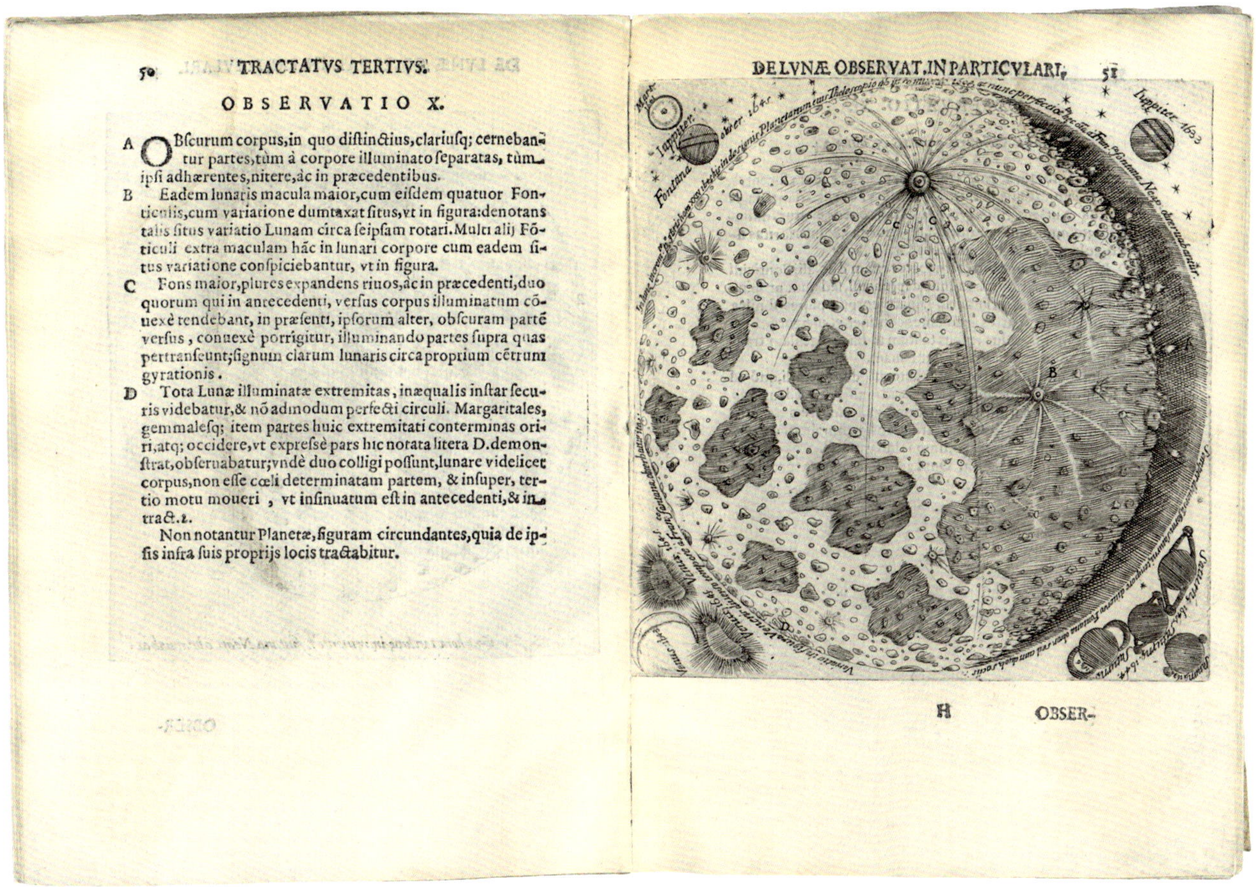

50 TRACTATVS TERTIVS.

OBSERVATIO X.

A OBſcurum corpus,in quo diſtinctius,clariuſq; cernebantur partes,tùm à corpore illuminato ſeparatas, tùm ipſi adhærentes,nitere,àc in præcedentibus.

B Eadem lunaris macula maior,cum eiſdem quatuor Fonticulis,cum variatione dumtaxat ſitus,vt in figura:denotans talis ſitus variatio Lunam circa ſeipſam rotari.Multi alij Fõticuli extra maculam hãc in lunari corpore cum eadem ſitus variatione conſpiciebantur, vt in figura.

C Fons maior,plures expandens riuos,àc in præcedenti,duo quorum qui in antecedenti, verſus corpus illuminatum cõuexè tendebant, in præſenti, ipſorum alter, obſcuram partẽ verſus , conuexè porrigitur, illuminando partes ſupra quas pertranſeunt;ſignum clarum lunaris circa proprium cẽtrum gyrationis .

D Tota Lunæ illuminatæ extremitas , inæqualis inſtar ſecuris videbatur,& nõ admodum perfecti circuli. Margaritales, gemmaleſq; item partes huic extremitati conterminas oriri,atq; occidere,vt expreſsè pars hic notata litera D.demonſtrat,obſeruabatur;vndè duo colligi poſſunt,lunare videlicet corpus,non eſſe cœli determinatam partem, & inſuper, tertio motu moueri , vt inſinuatum eſt in antecedenti,& in tract.2.

Non notantur Planetæ,figuram circundantes,quia de ipſis infra ſuis proprijs locis tractabitur.

DE LVNÆ OBSERVAT. IN PARTICVLARI. 51

H OBSER-

2 | FRANCESCO FONTANA | *Lunar Map*, in *New Observations of Heavenly and Earthly Objects (Novae coelestium, terrestriumque rerum observationes)*, 1646

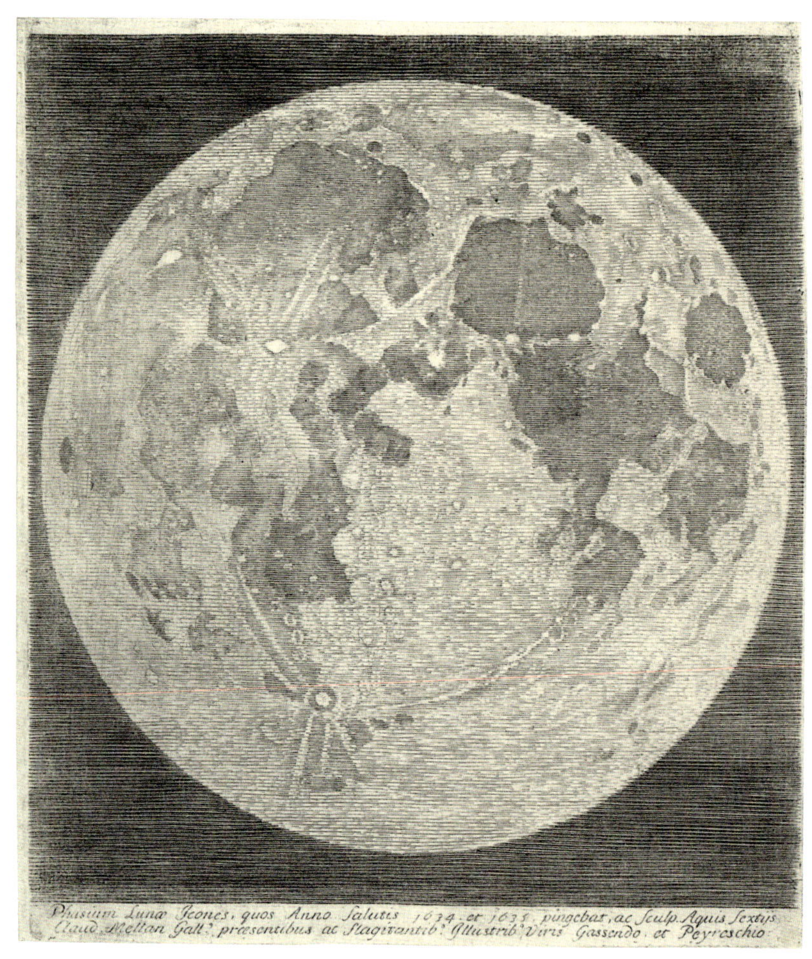

3–5 | CLAUDE MELLAN | *Full Moon, The Moon in Its First Quarter,* and *The Moon in Its Final Quarter,* 1635

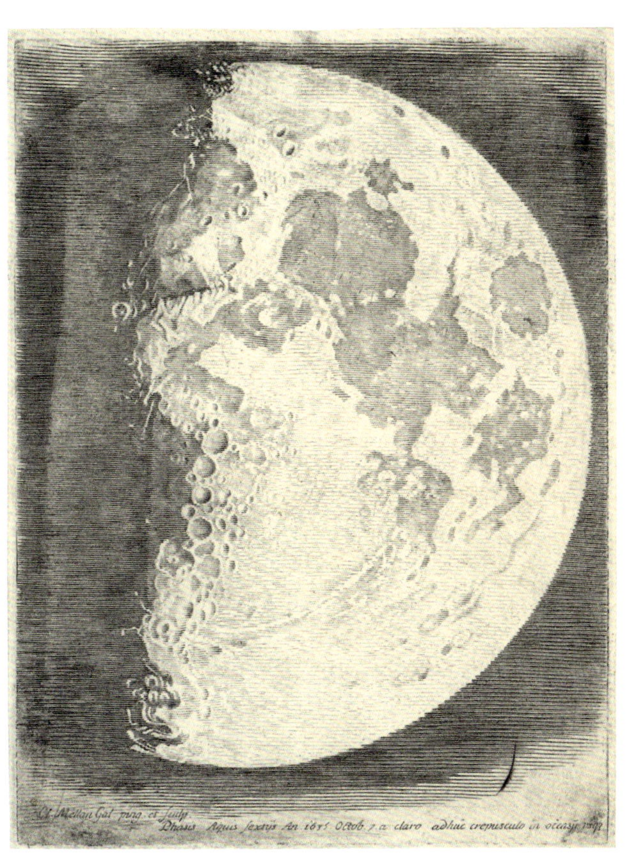
Cl. Mellan Gal. ping. et sculp.
Phasis Aquis Sextiis An 1635 Octob 7 a claro adhuc crepusculo in

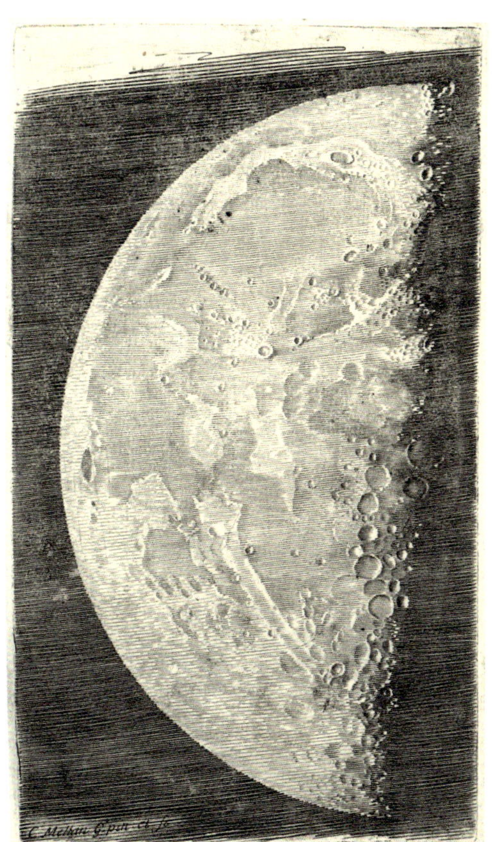
C. Mellan G. pin. et sc.

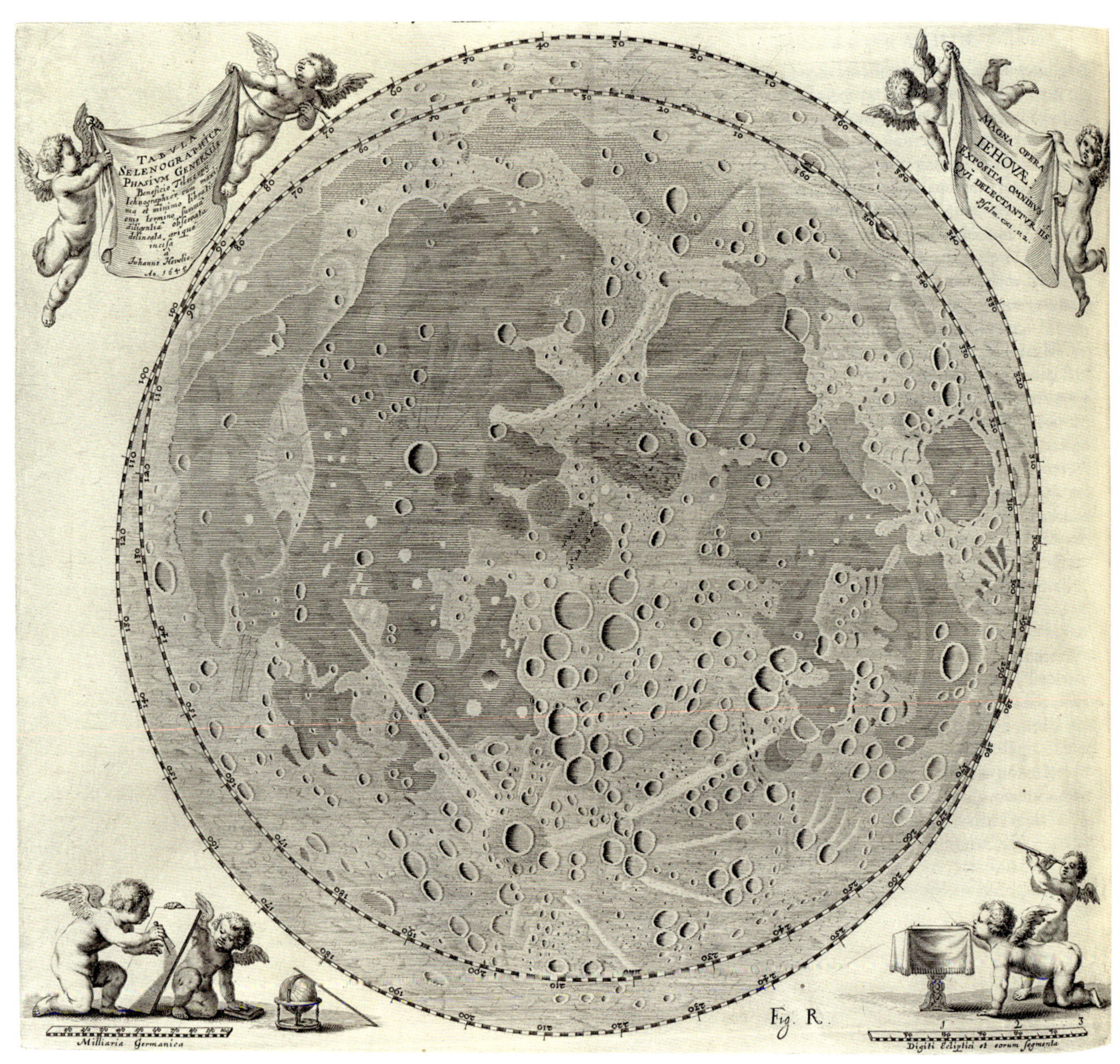

6 | JOHANNES HEVELIUS | *Lunar Map*, in *Selenography, or the Description of the Moon* (*Selenographia, sive lunae descriptio*), 1647

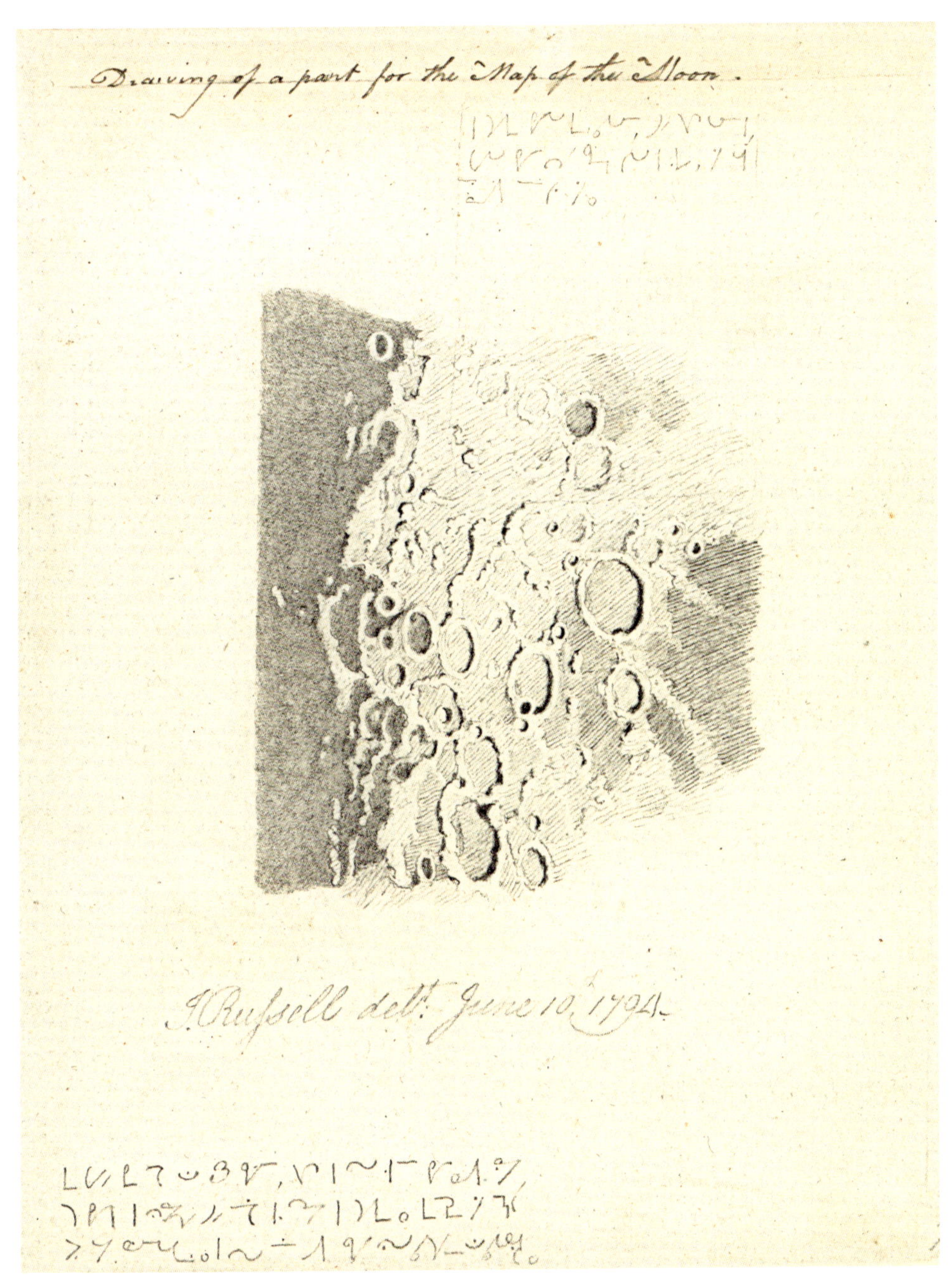

7 | JOHN RUSSELL | *A Drawing of a Part for the Map of the Moon*, 1794

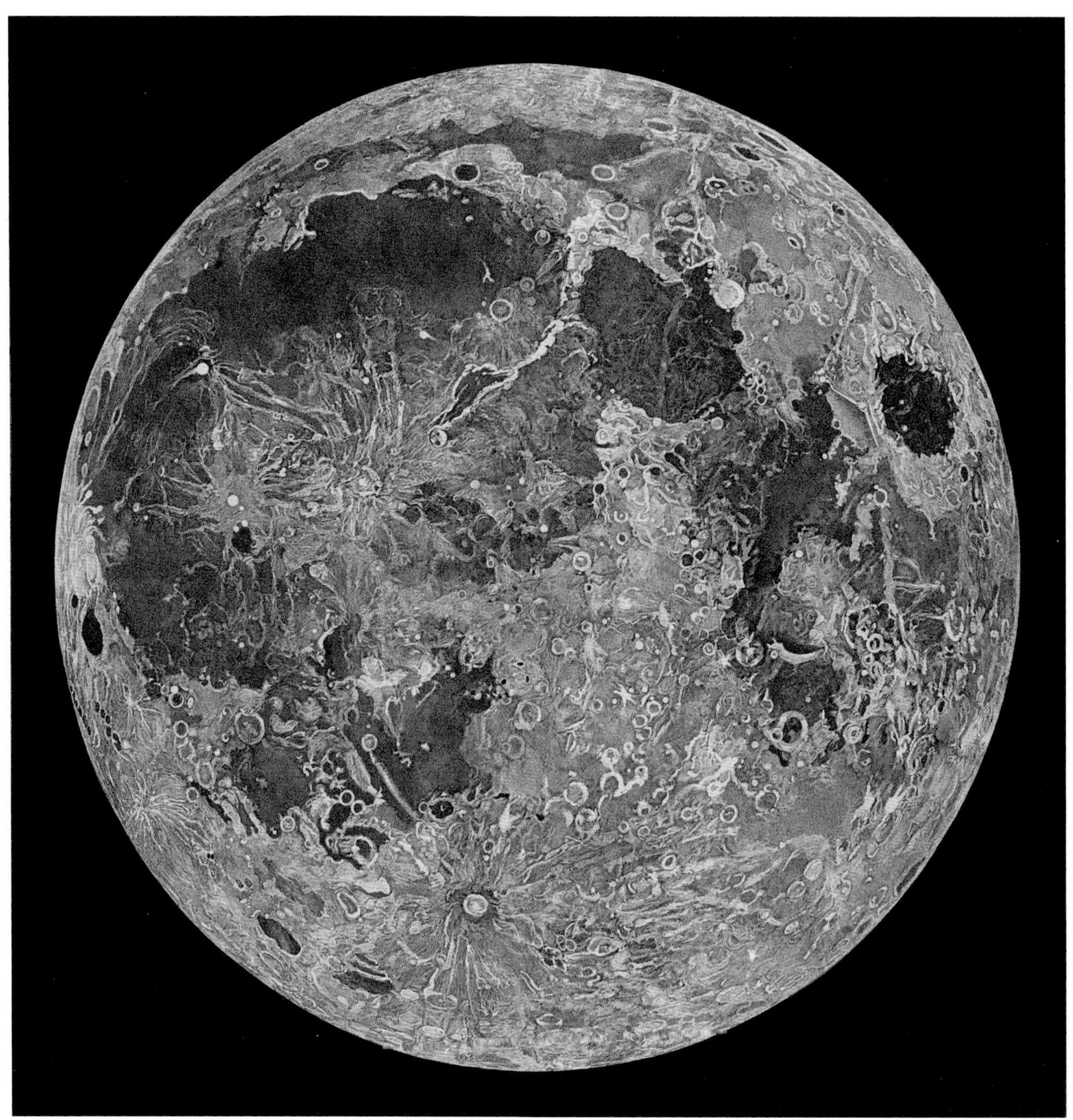

8 | JOHN RUSSELL | *Lunar Planisphere, Flat Light*, 1805

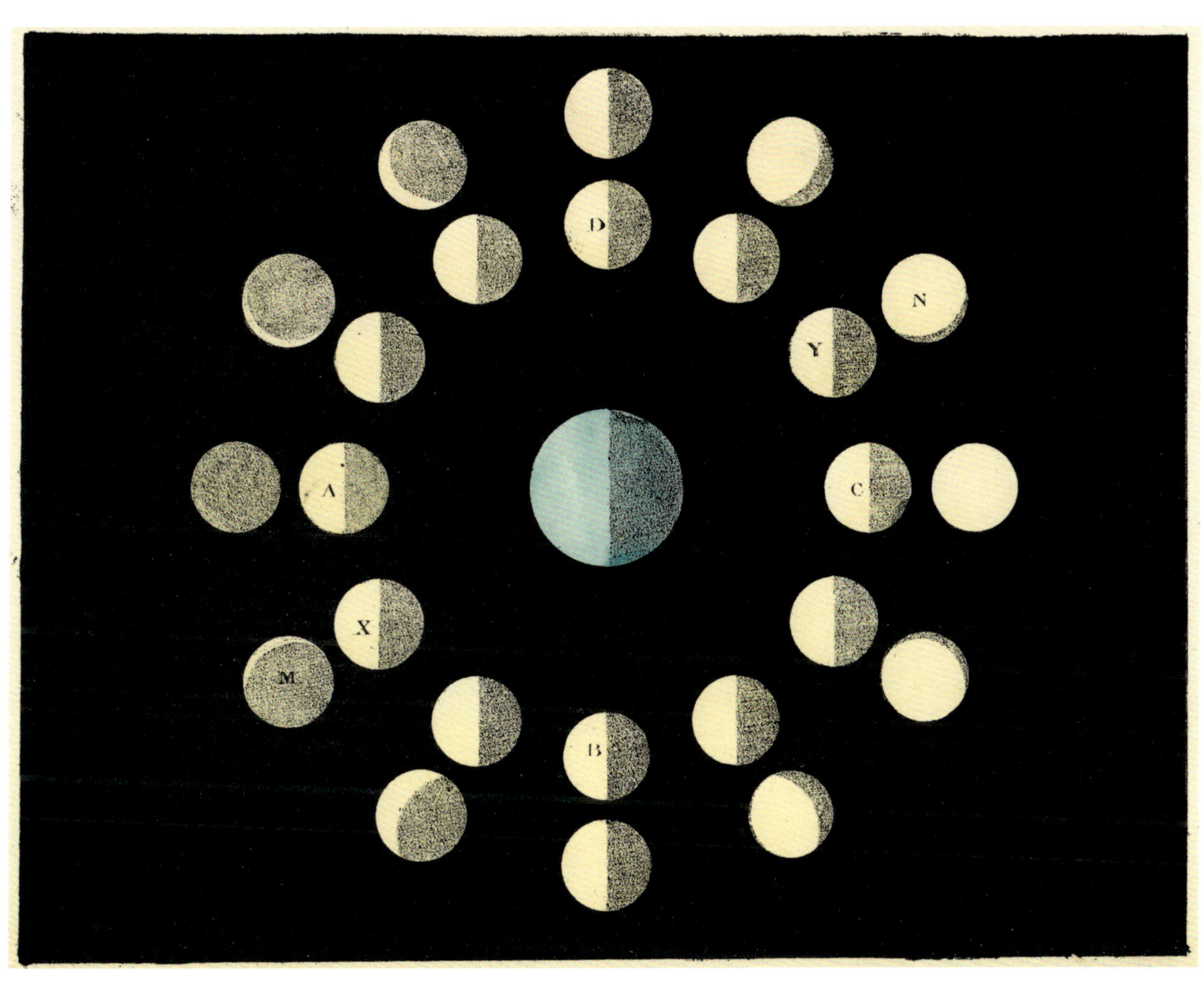

9 | CHARLES F. BLUNT | *The Moon's Phases*, in *Lecture on Astronomy: Beauty of the Heavens; a Pictorial Display of the Astronomical Phenomena of the Universe*, 1842

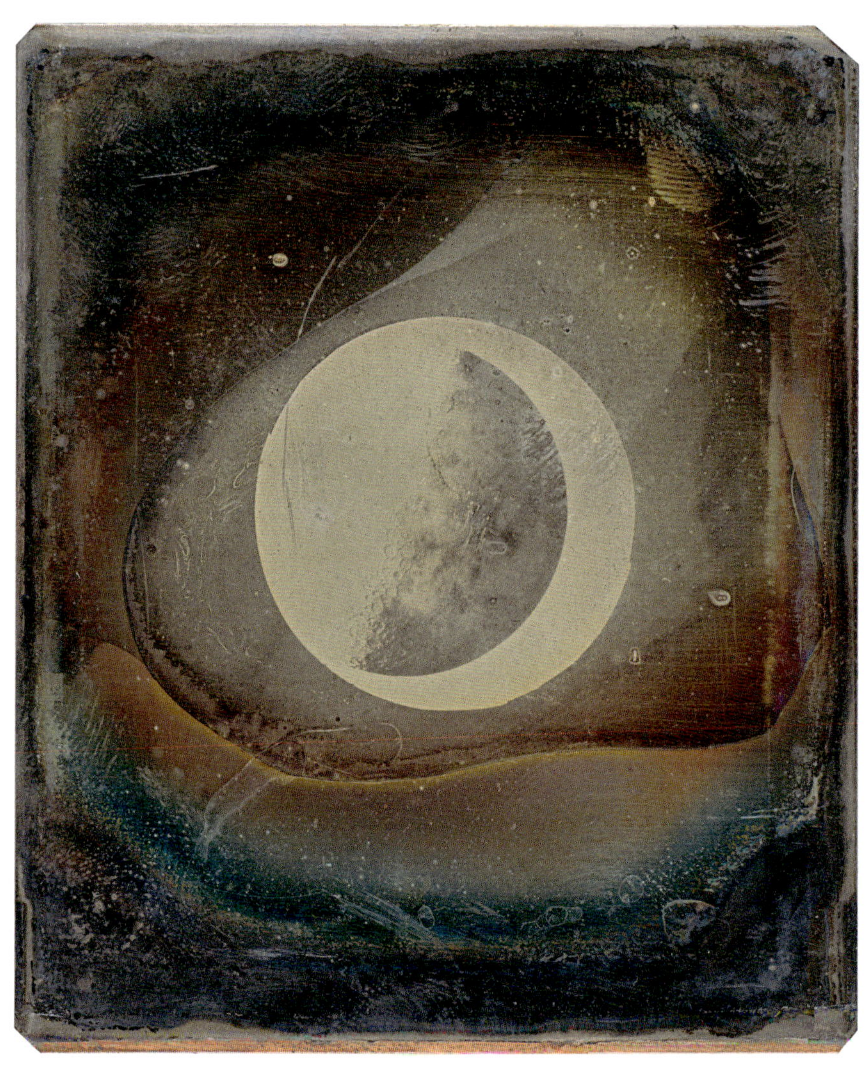

10 | JOHN WILLIAM DRAPER | *Moon*, 1840s

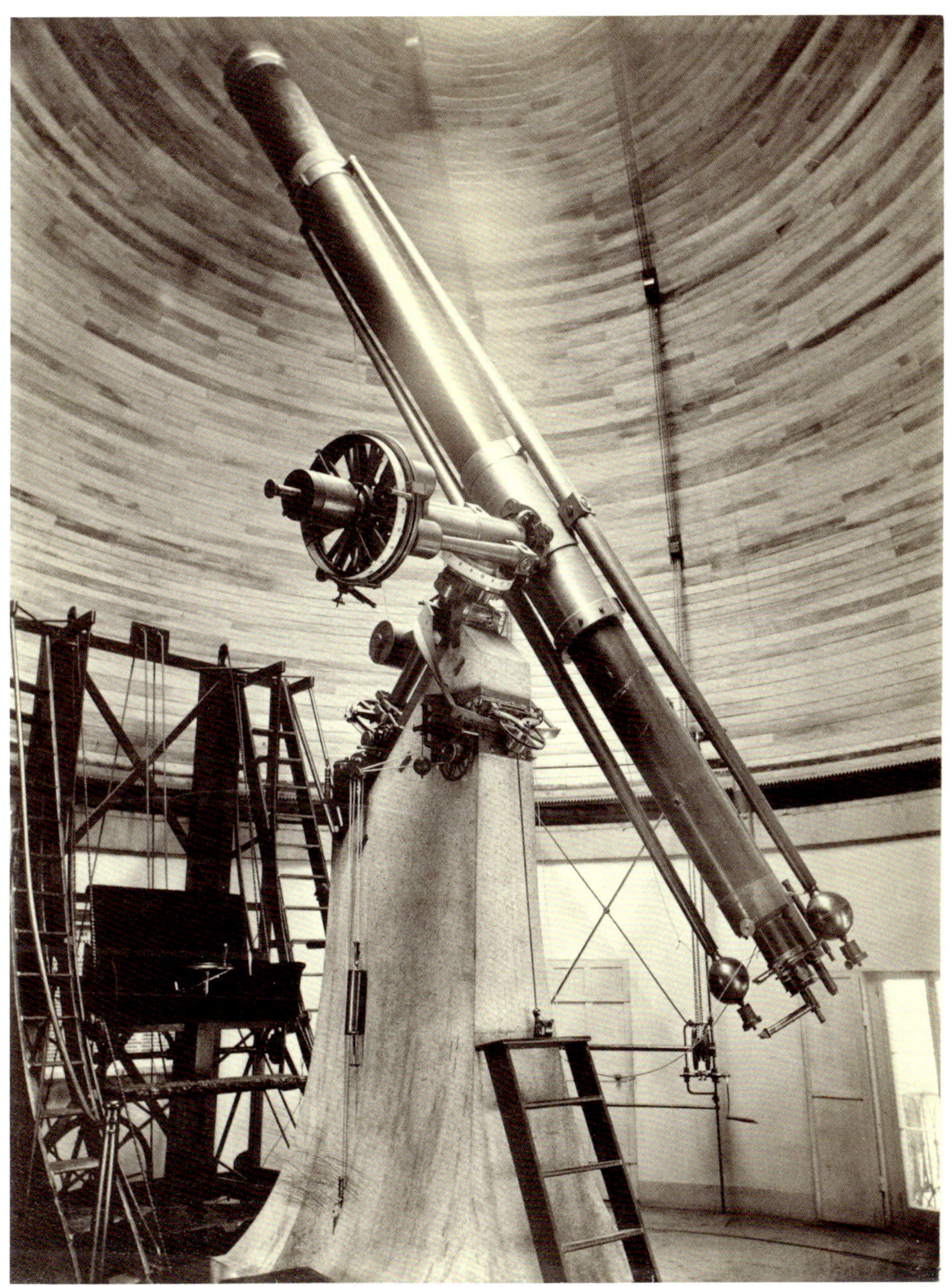

11 | GUSTAVUS W. PACH | *The Great Refractor*, ca. 1880

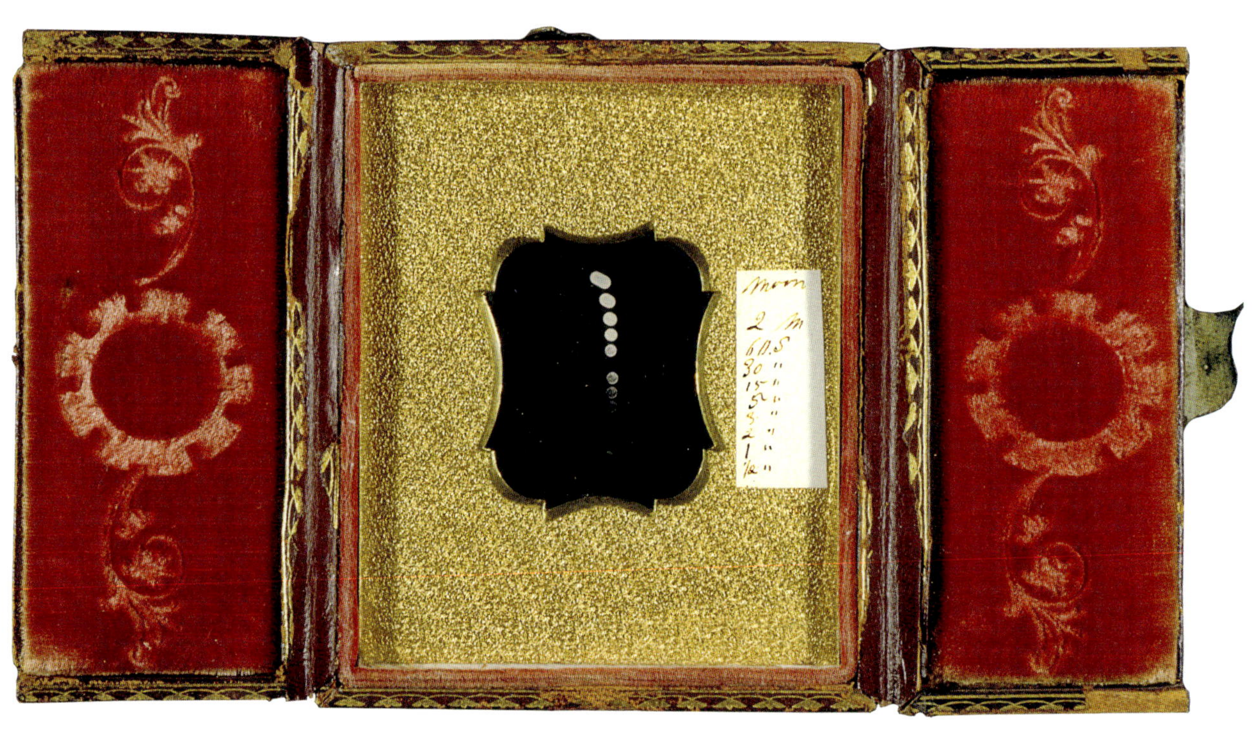

12 | SAMUEL DWIGHT HUMPHREY | *Multiple Exposures of the Moon: Nine Exposures Ranging from Two Minutes to Half a Second*, 1849

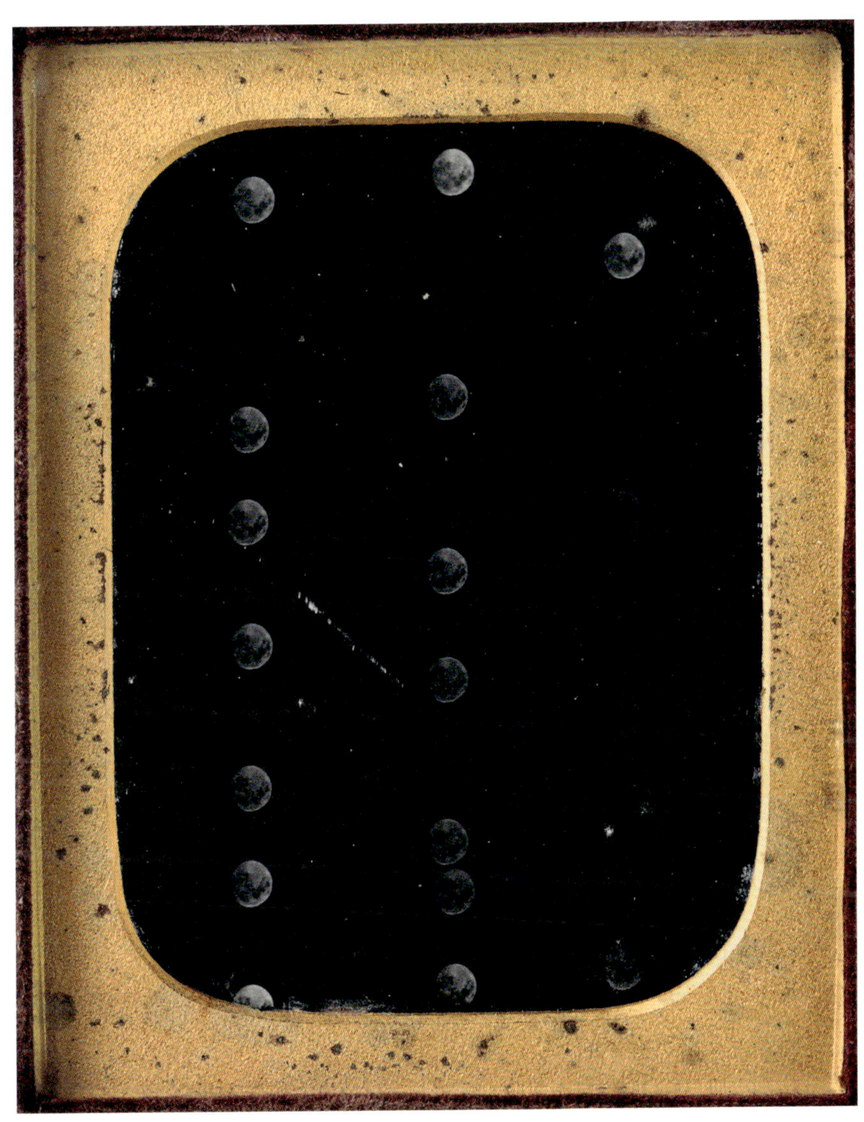

13 | ANTOINE-FRANÇOIS-JEAN CLAUDET | *Multiple Exposures of the Moon*, 1846–52

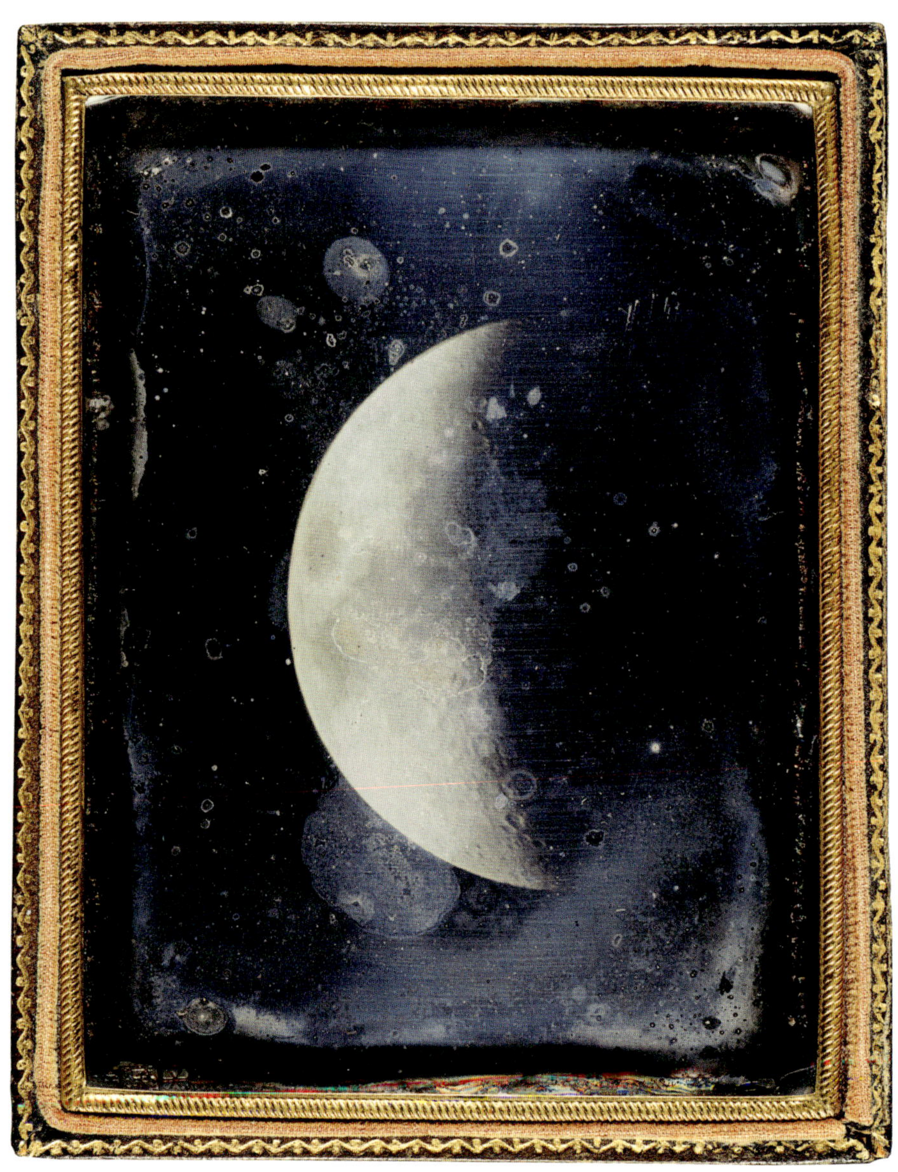

14 | JOHN ADAMS WHIPPLE | *View of the Moon*, 1852

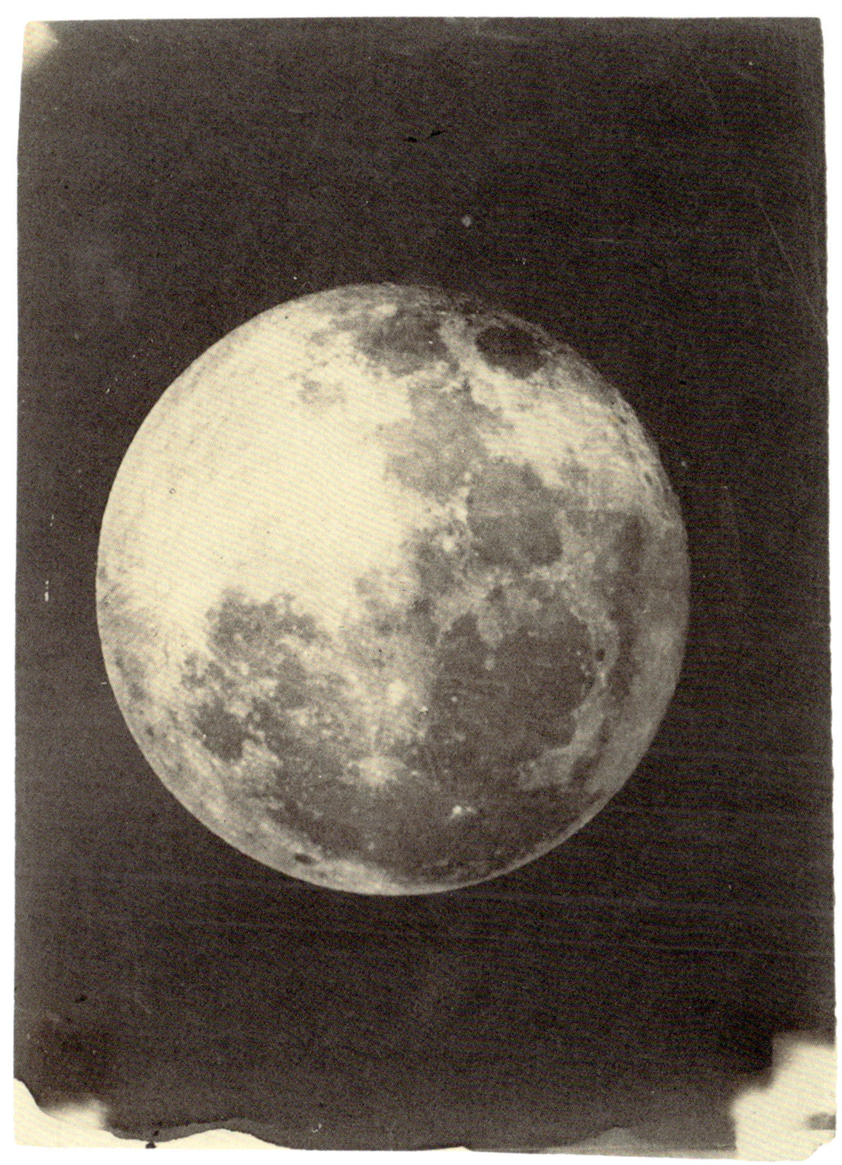

15 | JOHN ADAMS WHIPPLE AND JAMES WALLACE BLACK | *The Moon*, 1857–60

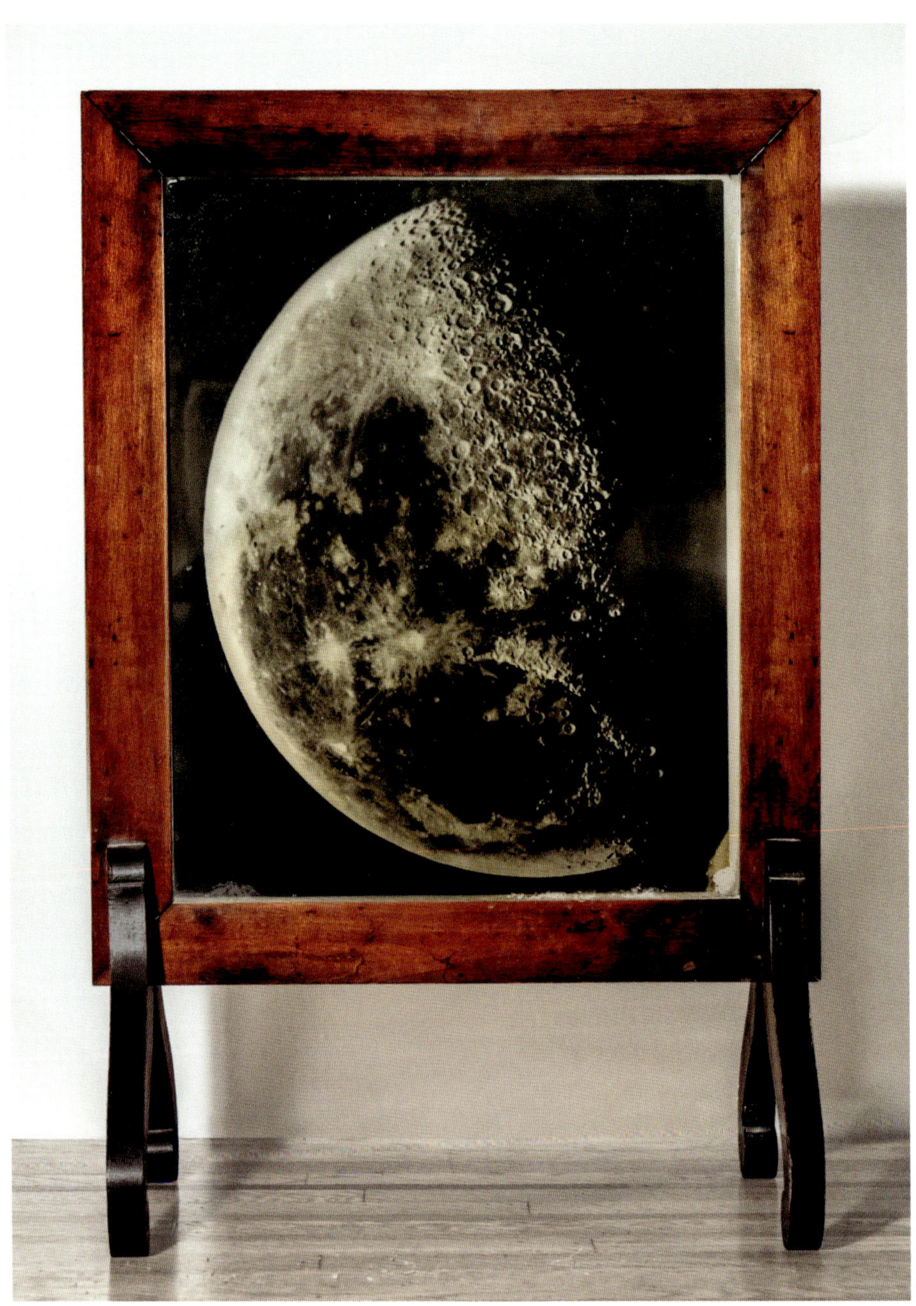

16 | HENRY DRAPER | *Lunar Transparency*, 1863

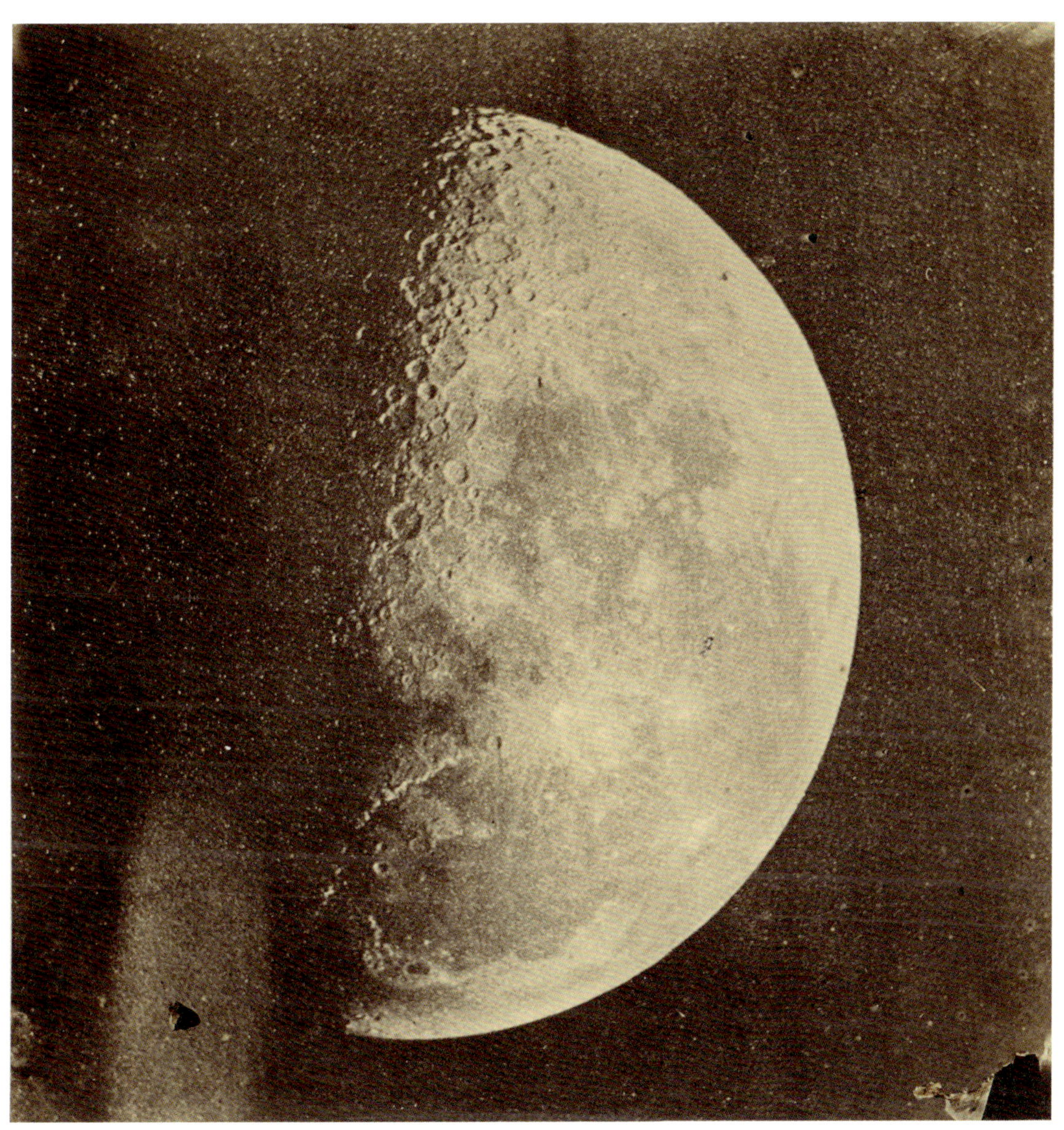

17 | WARREN DE LA RUE | *The Moon*, ca. 1856

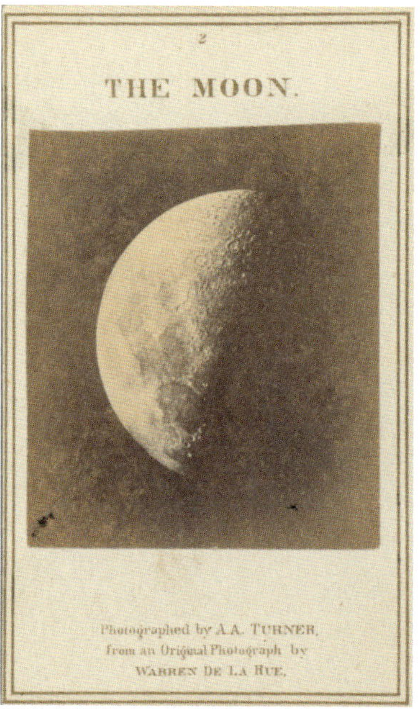

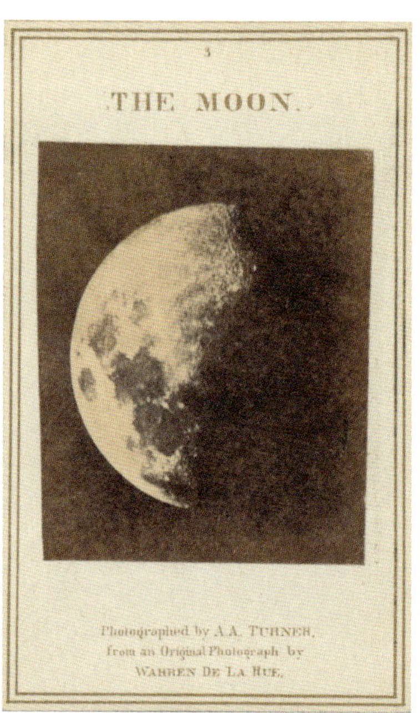

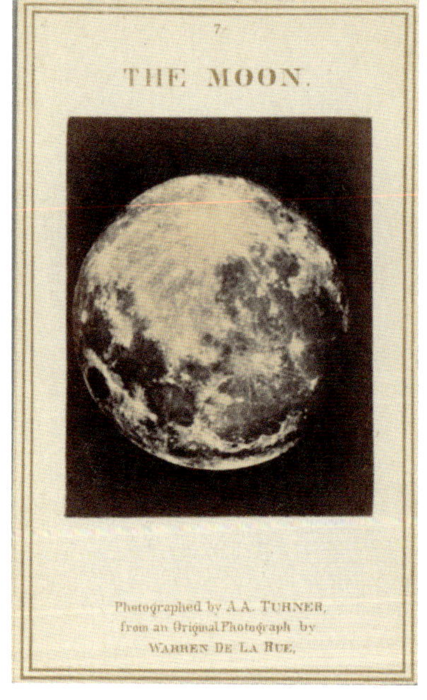

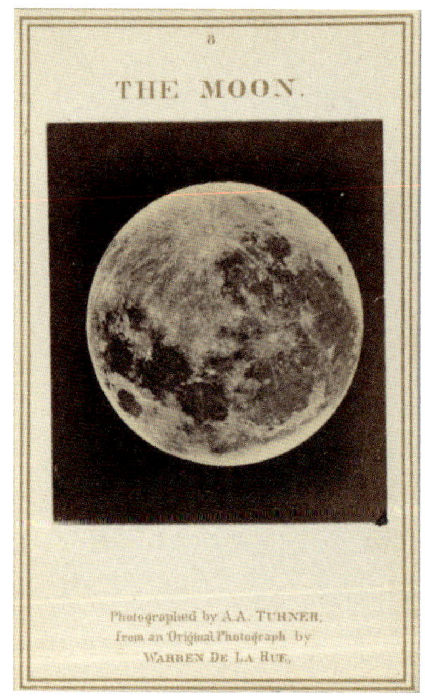

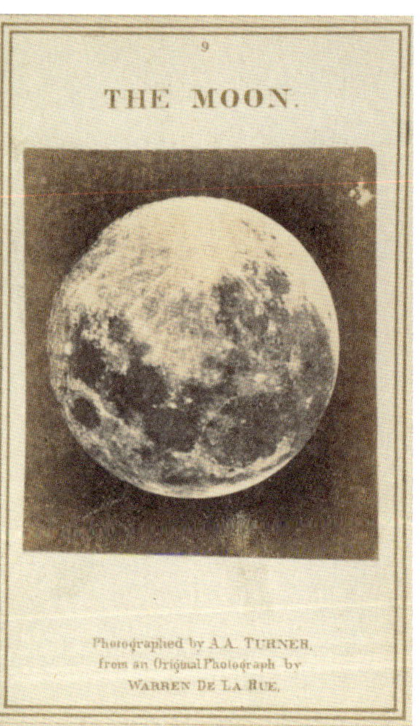

18 | AUSTIN AUGUSTUS TURNER, AFTER WARREN DE LA RUE | *Twelve Photographs of the Moon*, 1863

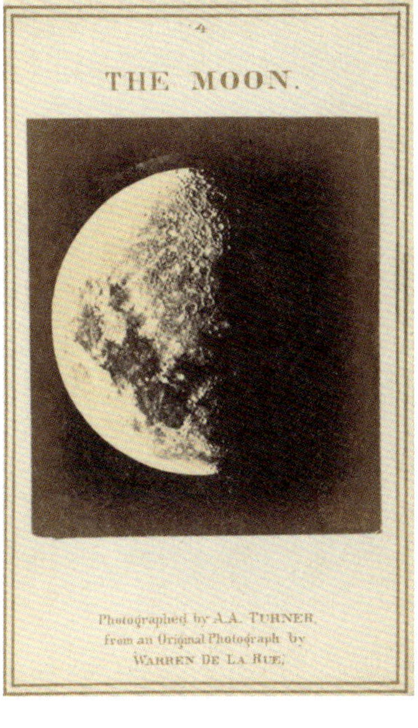
4
THE MOON.
Photographed by A.A. TURNER,
from an Original Photograph by
WARREN DE LA RUE.

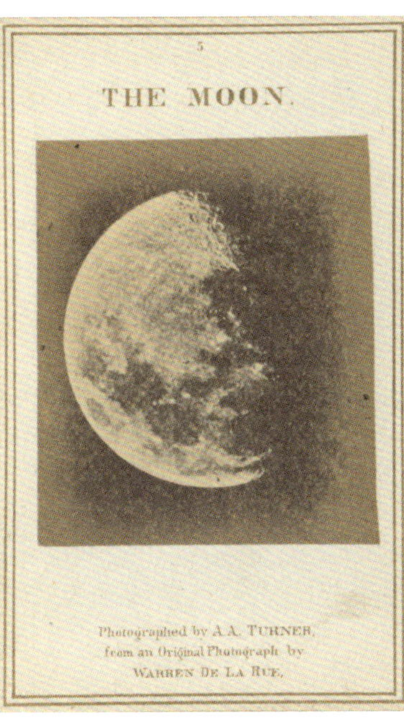
5
THE MOON.
Photographed by A.A. TURNER,
from an Original Photograph by
WARREN DE LA RUE.

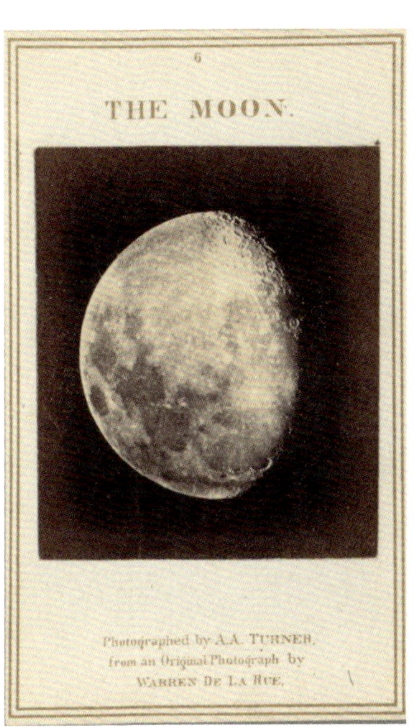
6
THE MOON.
Photographed by A.A. TURNER,
from an Original Photograph by
WARREN DE LA RUE.

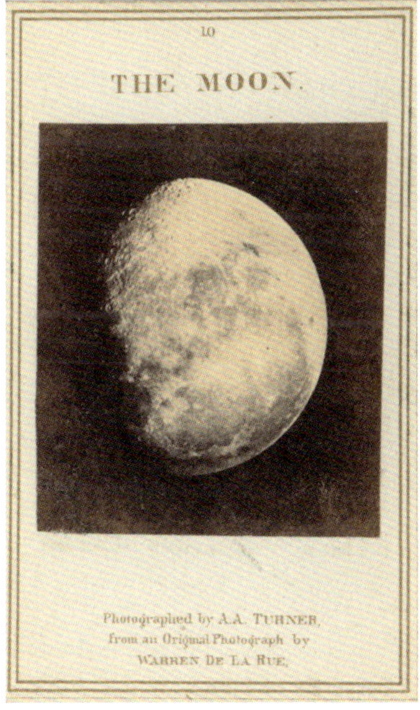
10
THE MOON.
Photographed by A.A. TURNER,
from an Original Photograph by
WARREN DE LA RUE.

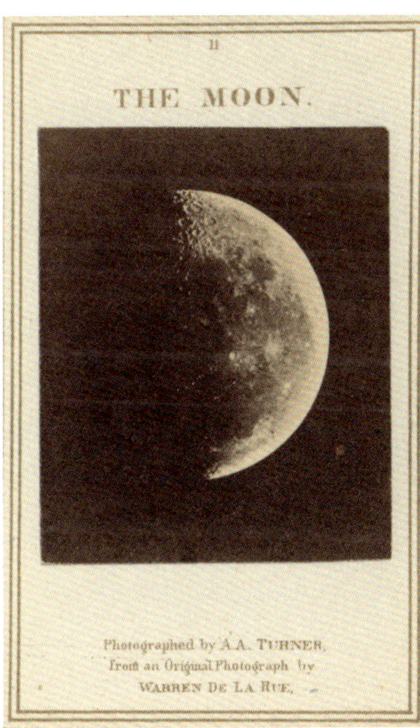
11
THE MOON.
Photographed by A.A. TURNER,
from an Original Photograph by
WARREN DE LA RUE.

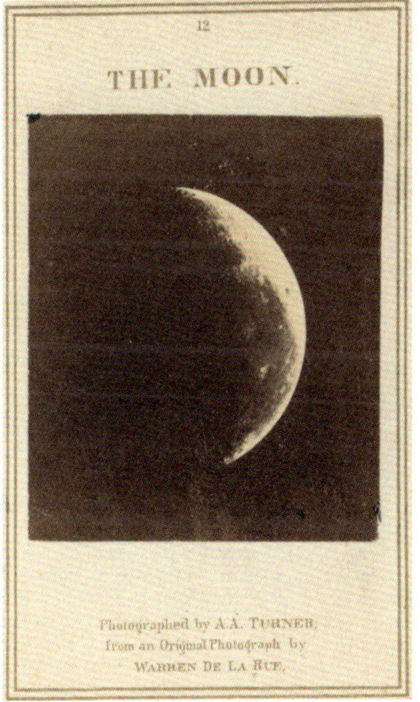
12
THE MOON.
Photographed by A.A. TURNER,
from an Original Photograph by
WARREN DE LA RUE.

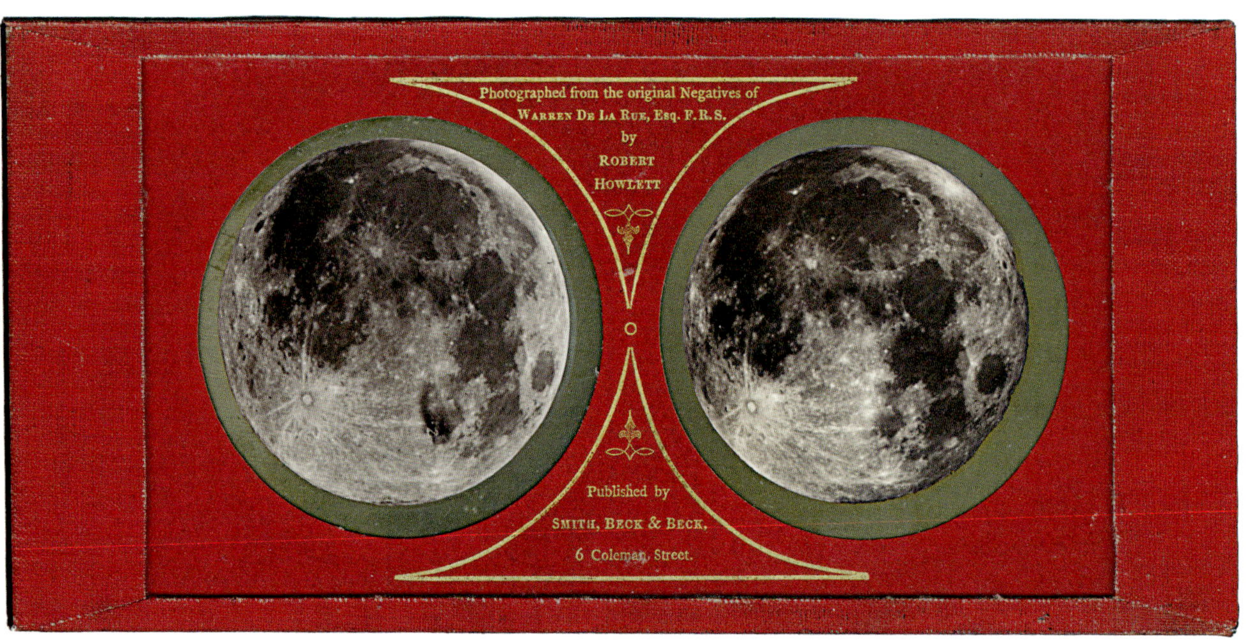

19 | WARREN DE LA RUE AND ROBERT HOWLETT | *The Moon*, ca. 1858

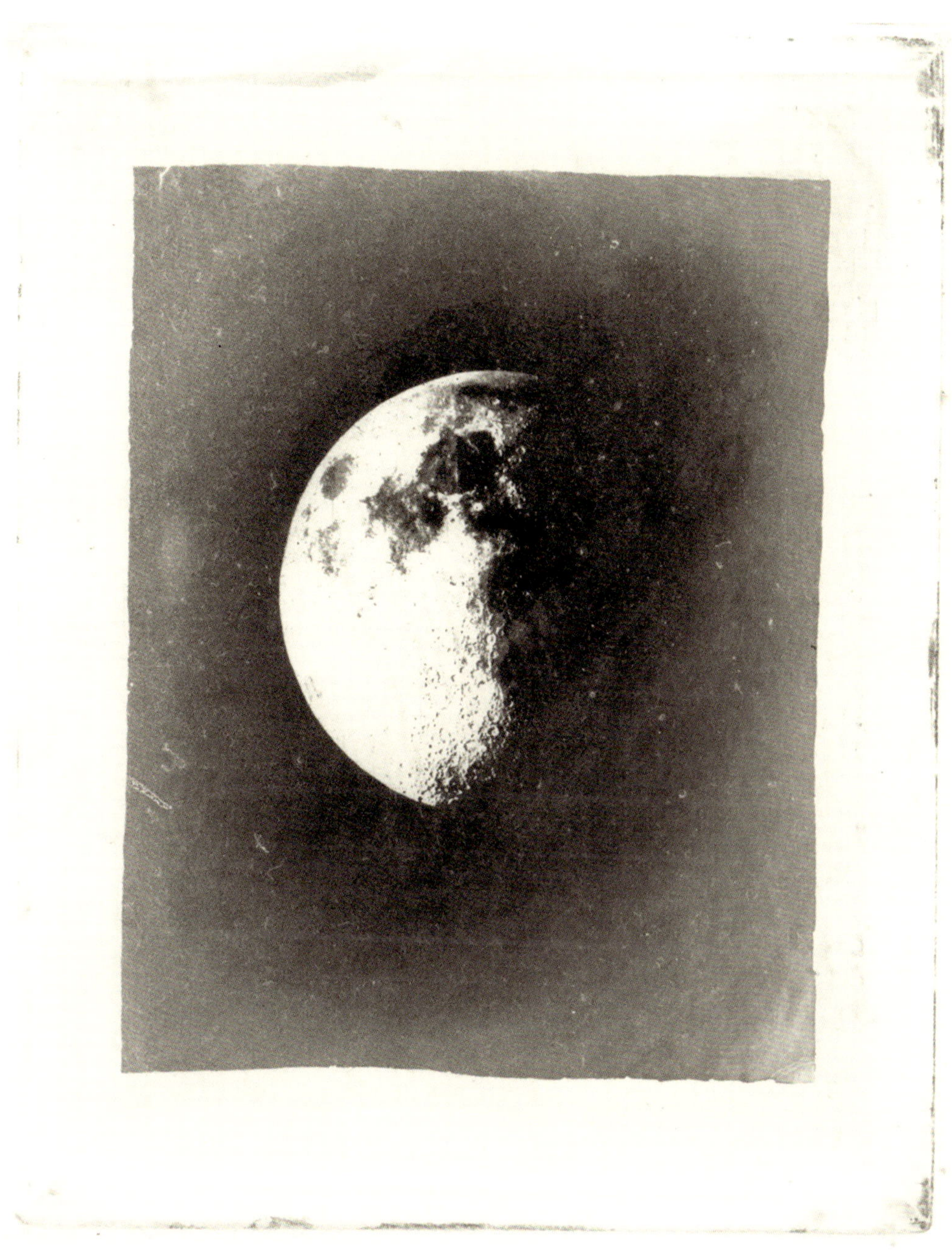

20 | WILLIAM HENRY FOX TALBOT AND WARREN DE LA RUE | *The Half Moon*, ca. 1864

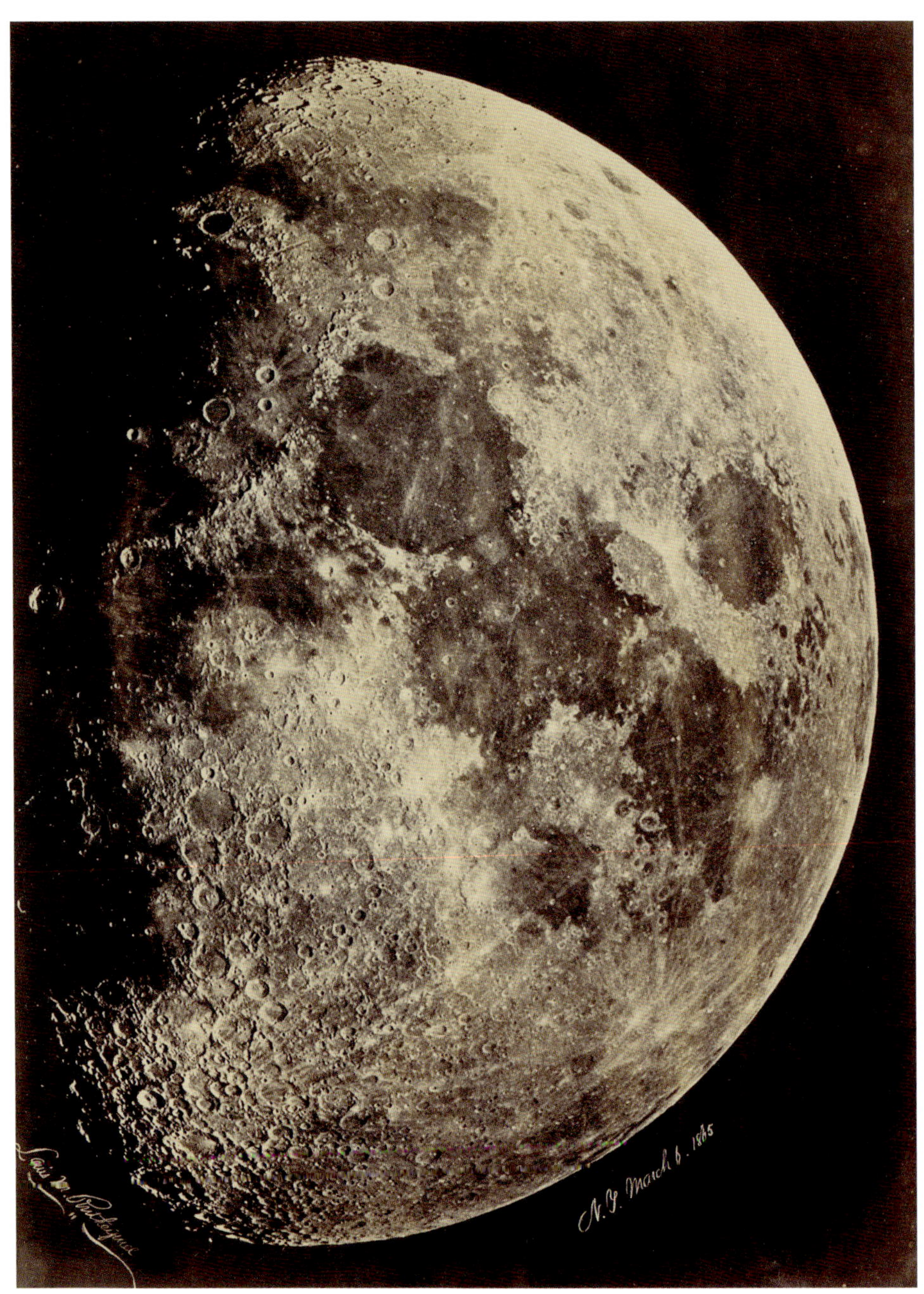

21 | LEWIS MORRIS RUTHERFURD | *The Moon, New York*, 1865

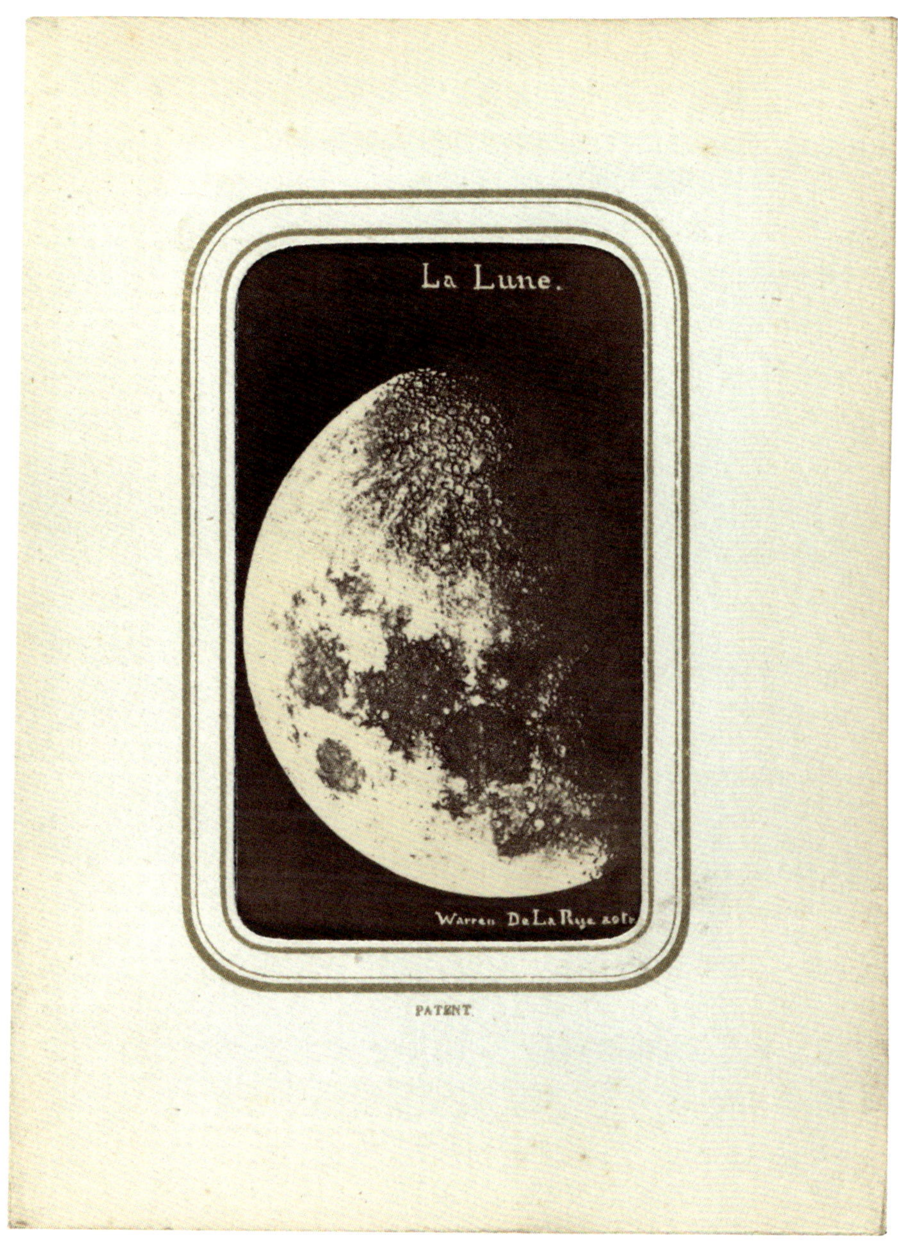

22 | CAMILLE FLAMMARION AND ADOLPHE-ALEXANDRE MARTIN, AFTER WARREN DE LA RUE |
The Moon (*La Lune*), in *Astronomical Gallery, Photographs of Science and Art* (*Galerie astronomique, photographie des sciences et des art*), 1867

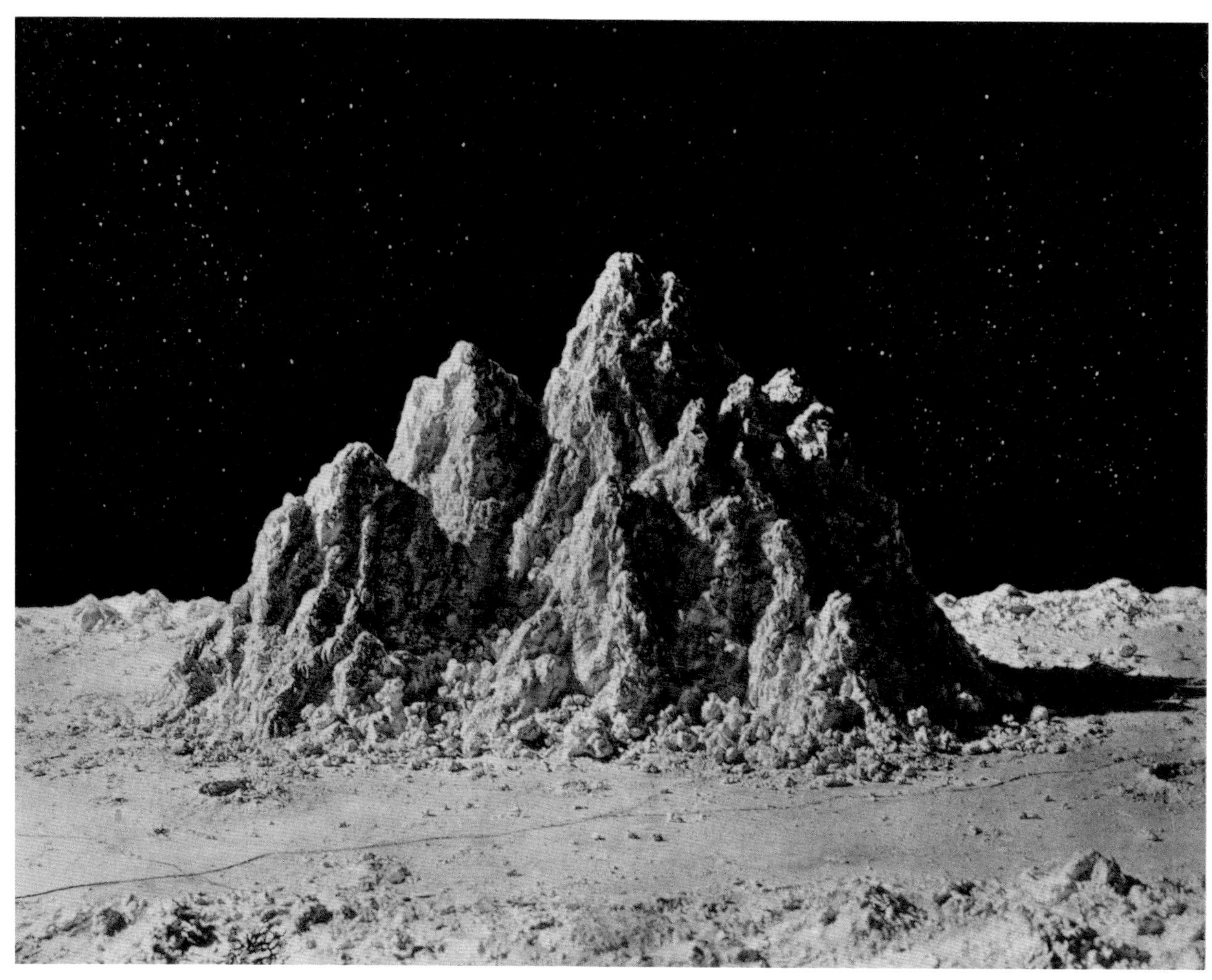

23 | JAMES NASMYTH | *An Ideal Sketch of "Pico,"*
in *The Moon: Considered as a Planet, a World, and a Satellite*, by Nasmyth and James Carpenter, 3rd edition, 1885

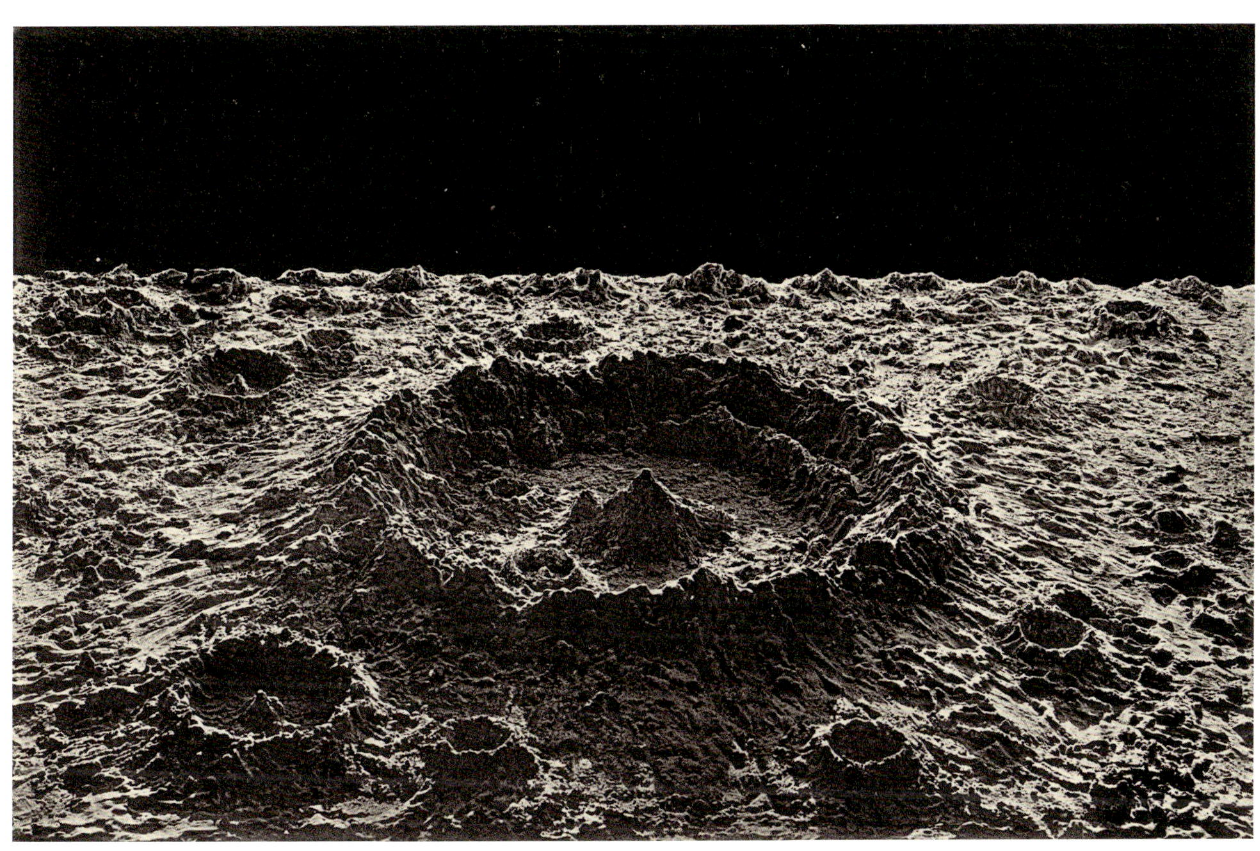

24 | JAMES NASMYTH | *Normal Lunar Crater,*

in *The Moon: Considered as a Planet, a World, and a Satellite*, by Nasmyth and James Carpenter, 1st edition, 1874

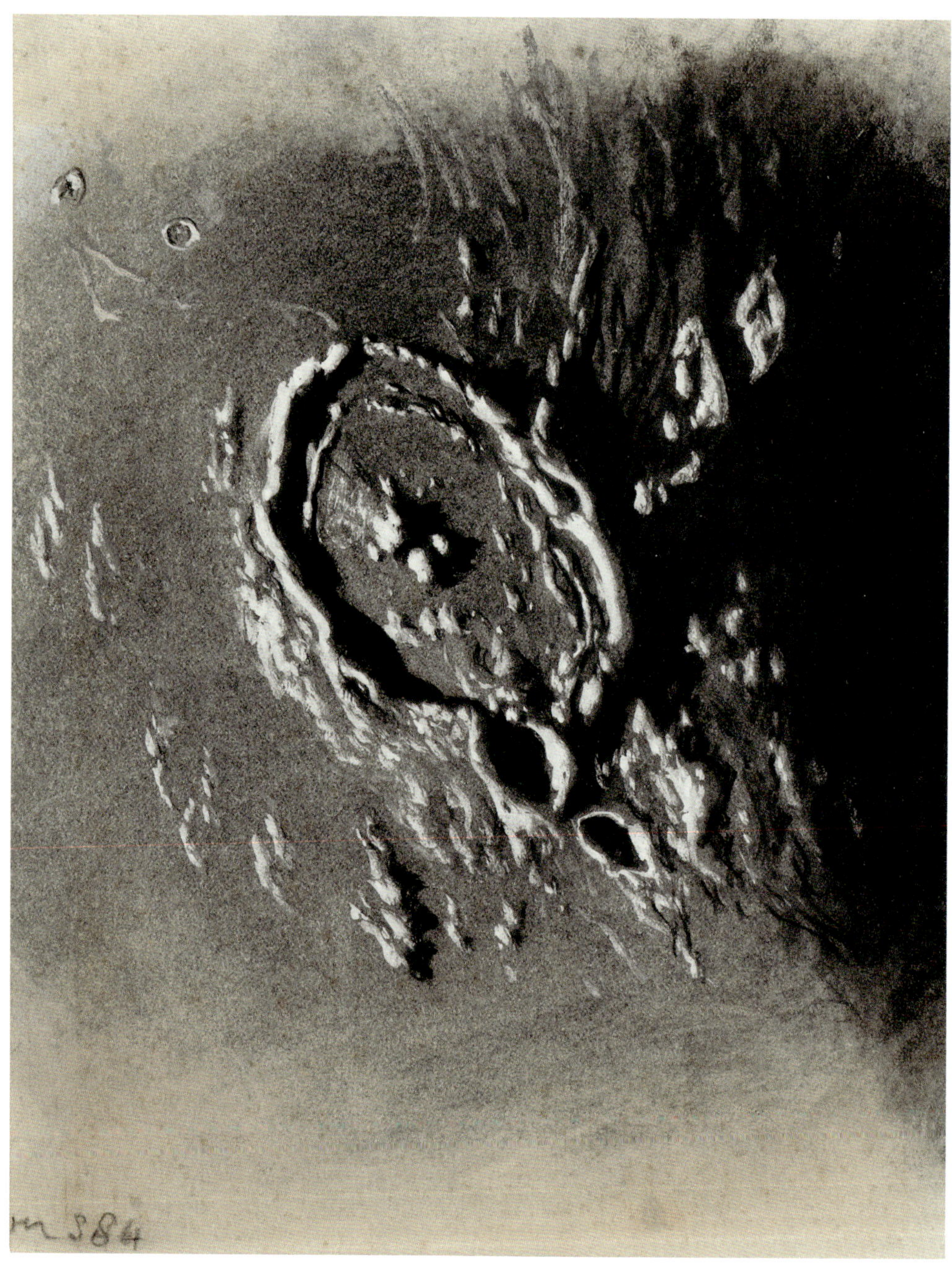

25 | JOHN BRETT | *Gassendi's Crater on the Moon*, 1884

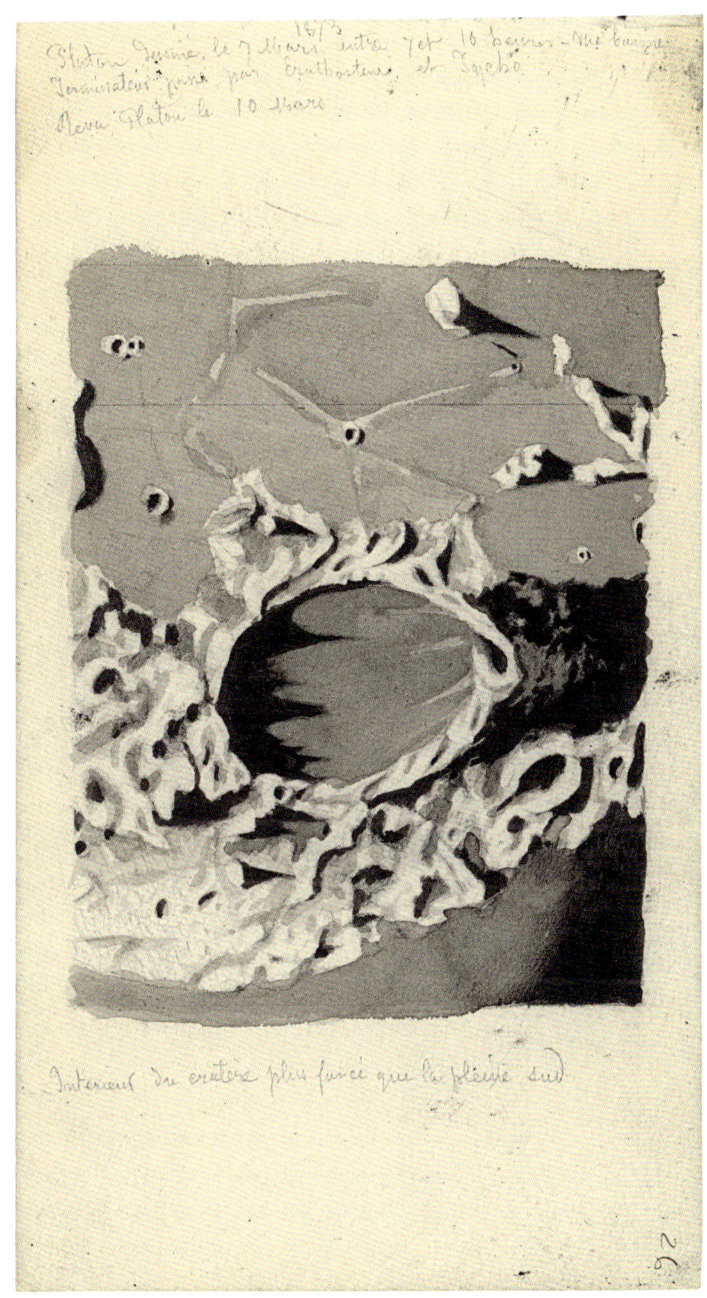

26 | ETIENNE LÉOPOLD TROUVELOT | *Interior of Crater*, in *Astronomical Notebook*, 1872–73

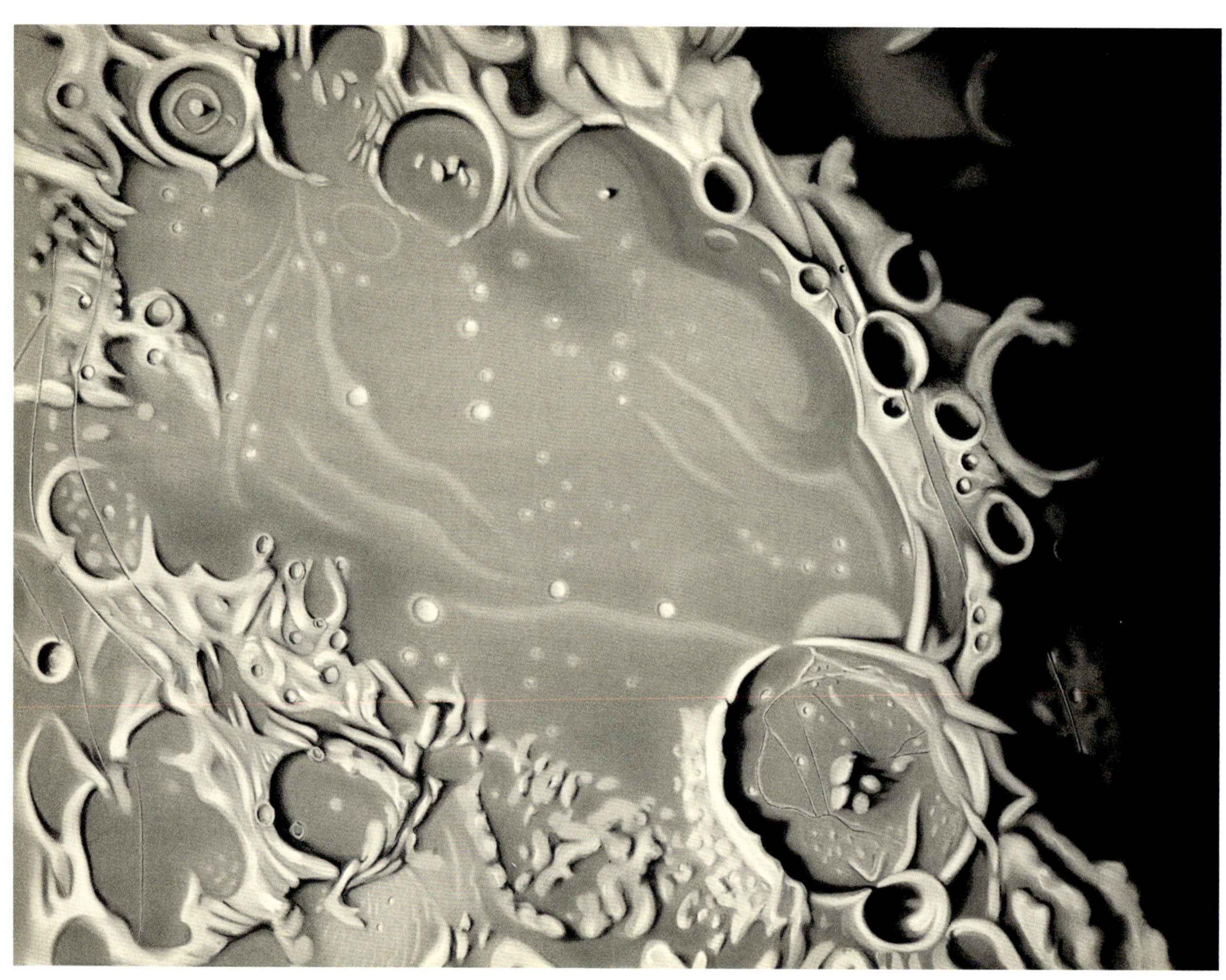

27 | ETIENNE LÉOPOLD TROUVELOT | *Mare Humorum*, from *The Trouvelot Astronomical Drawings Manual*, 1882

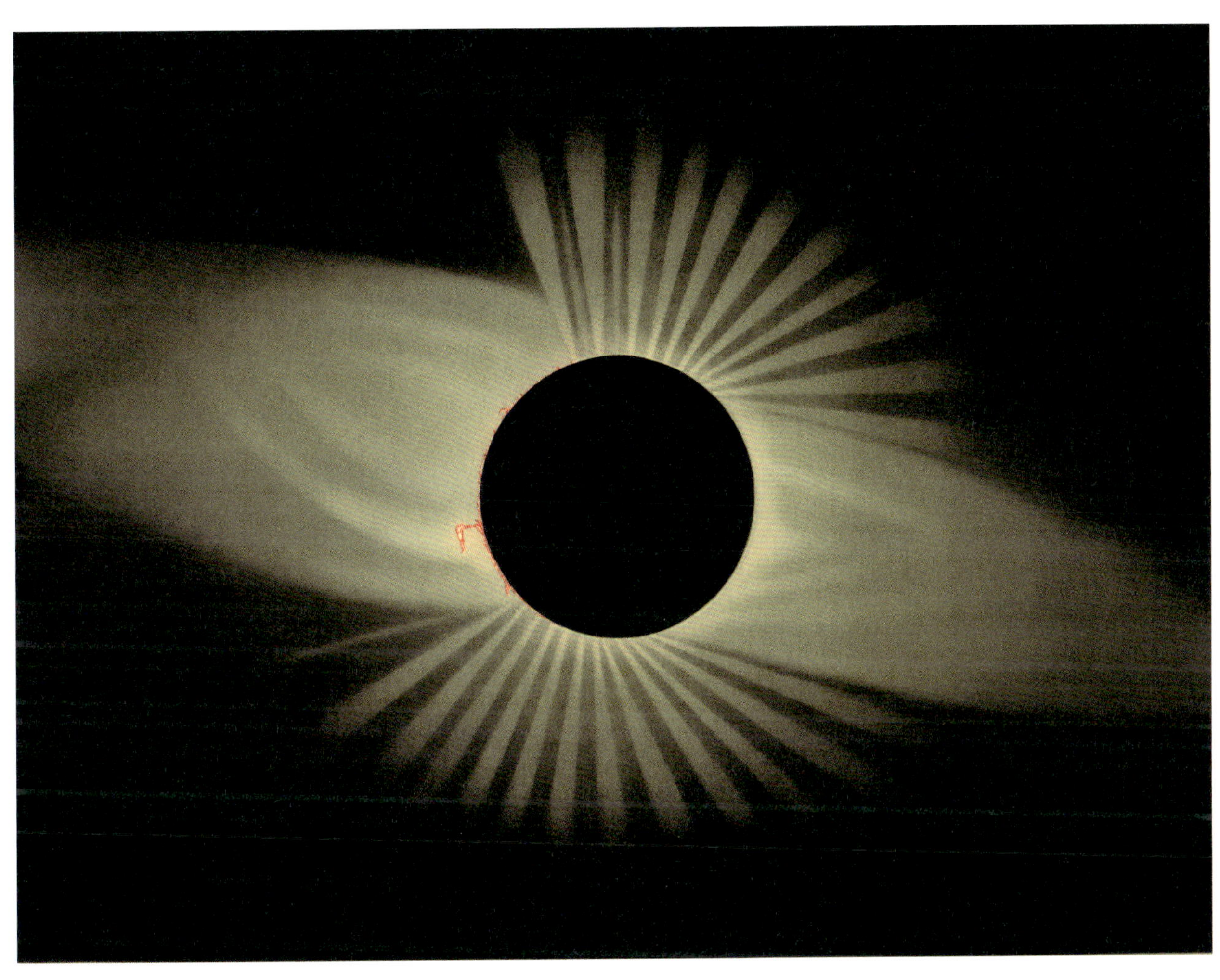

28 | ETIENNE LÉOPOLD TROUVELOT | *Total Eclipse of the Sun, Observed July 29, 1878, at Creston, Wyoming Territory*, from *The Trouvelot Astronomical Drawings Manual*, 1882

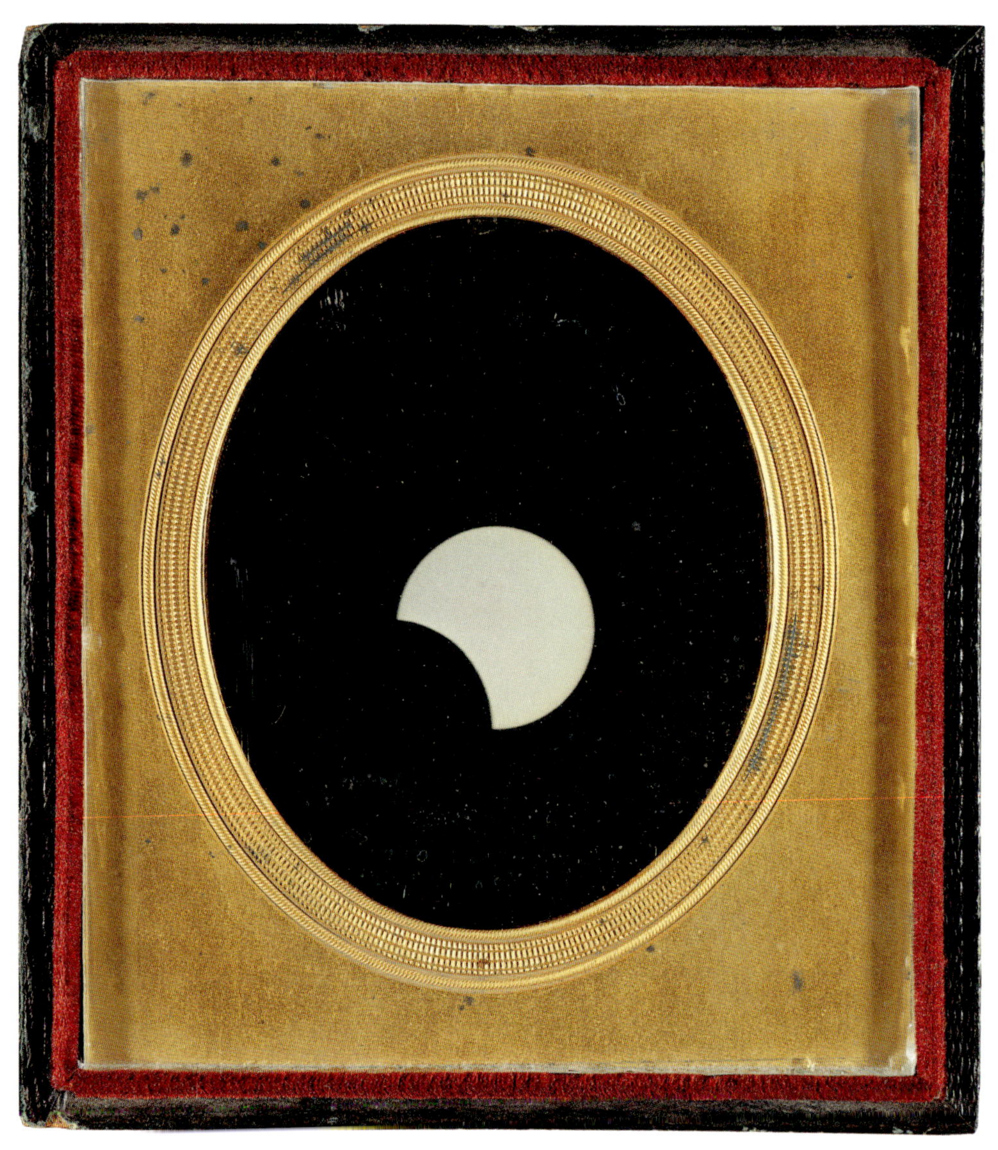

29 | JOHN ADAMS WHIPPLE | *Partial Eclipse of the Sun*, 1851

30 | WILLIAM AND FREDERICK LANGENHEIM | *Eclipse of the Sun*, 1854

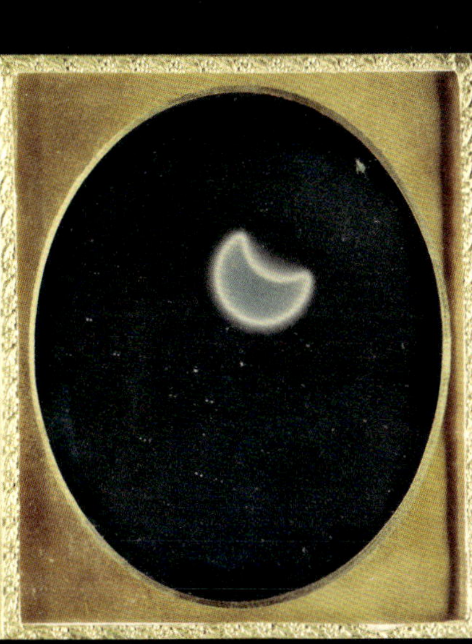
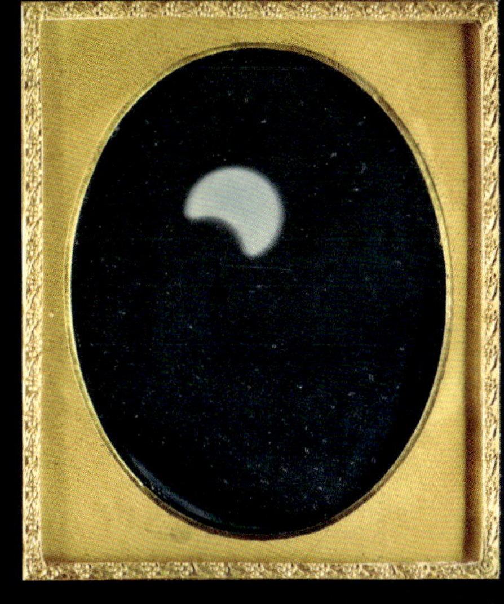

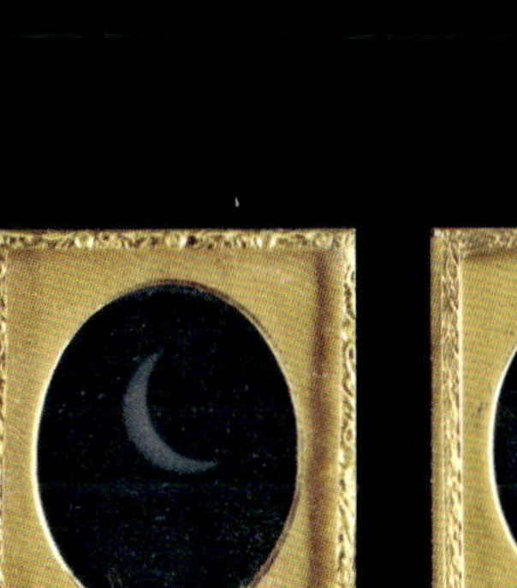
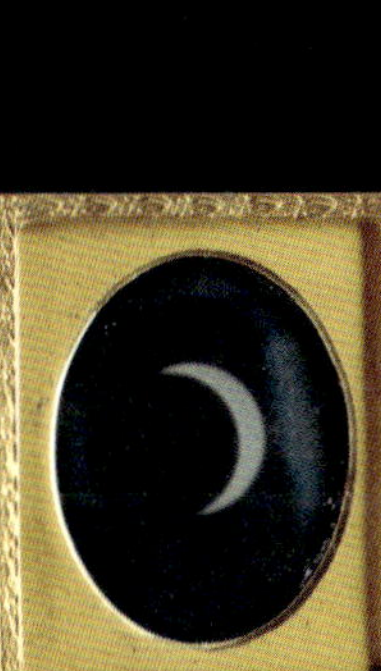
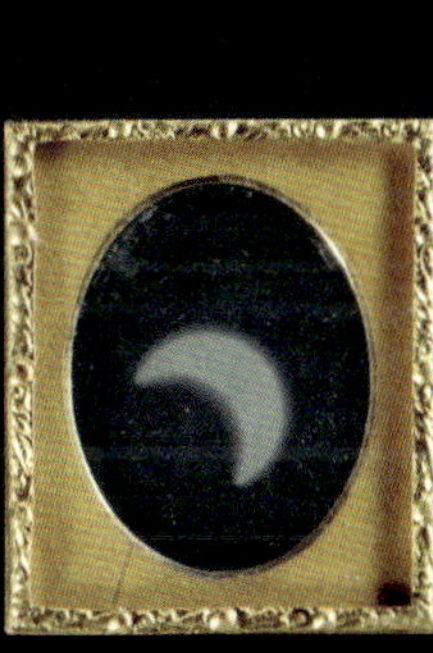

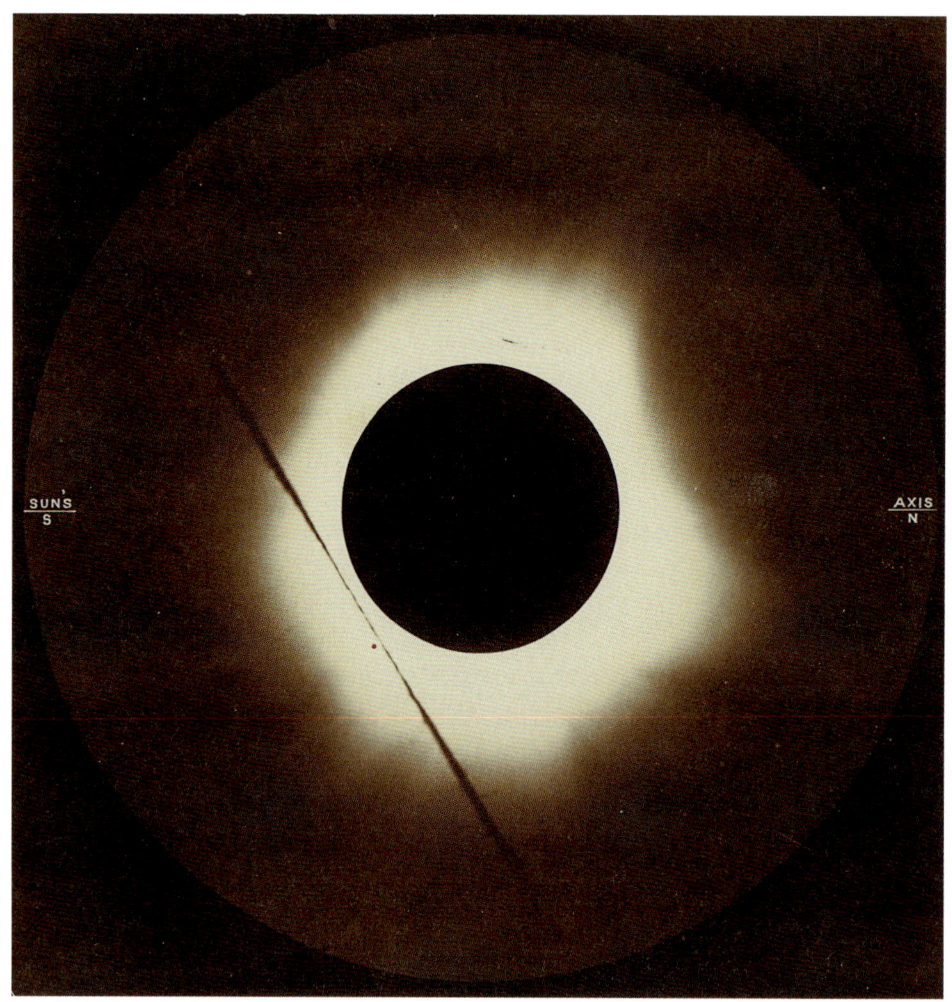

31 | H. A. LAWRENCE AND CHARLES RAY WOODS | *Solar Eclipse from Caroline Island*, 1883

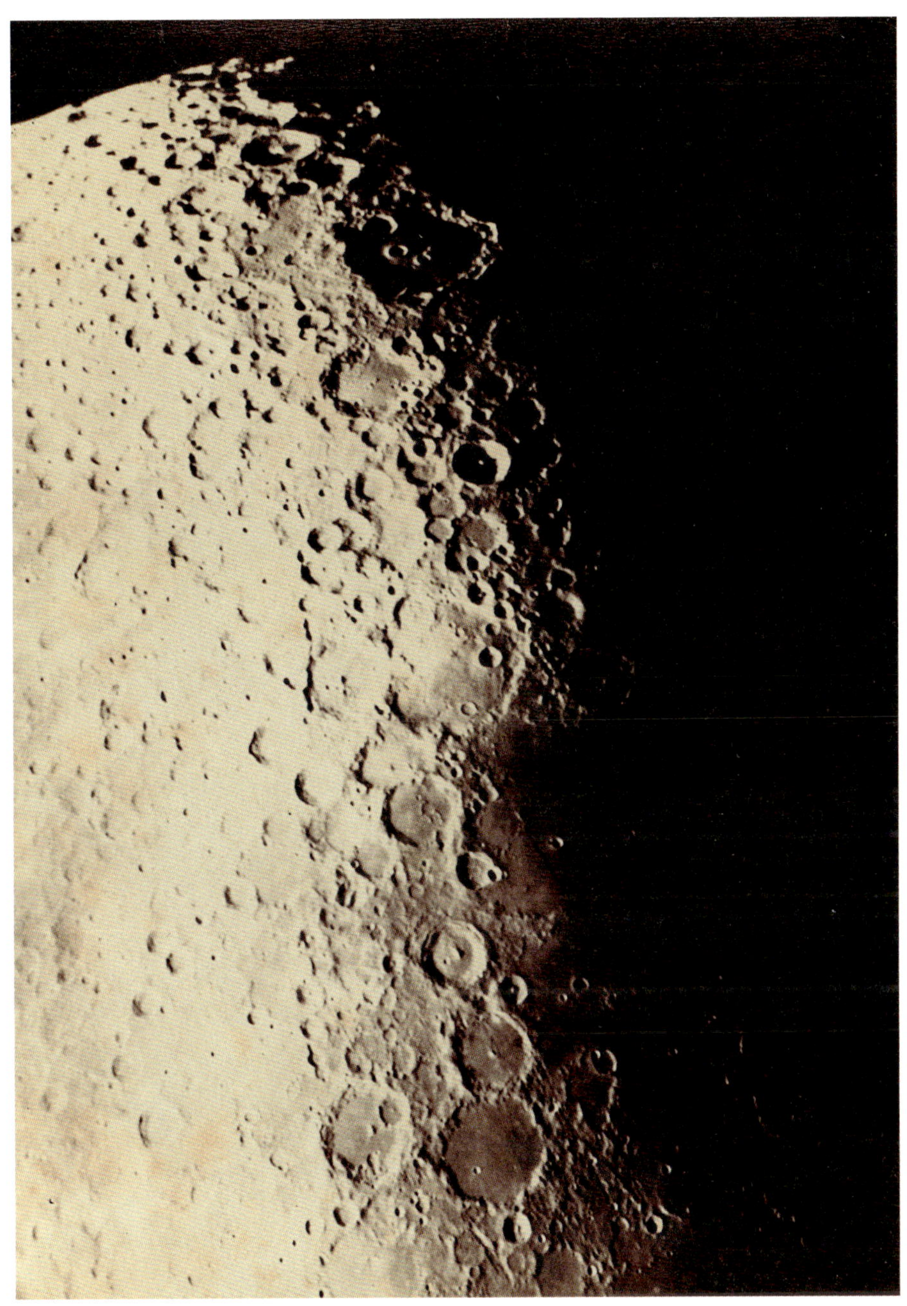

32 | PAUL AND PROSPER HENRY | *Lunar Photograph, South Pole*, 1890

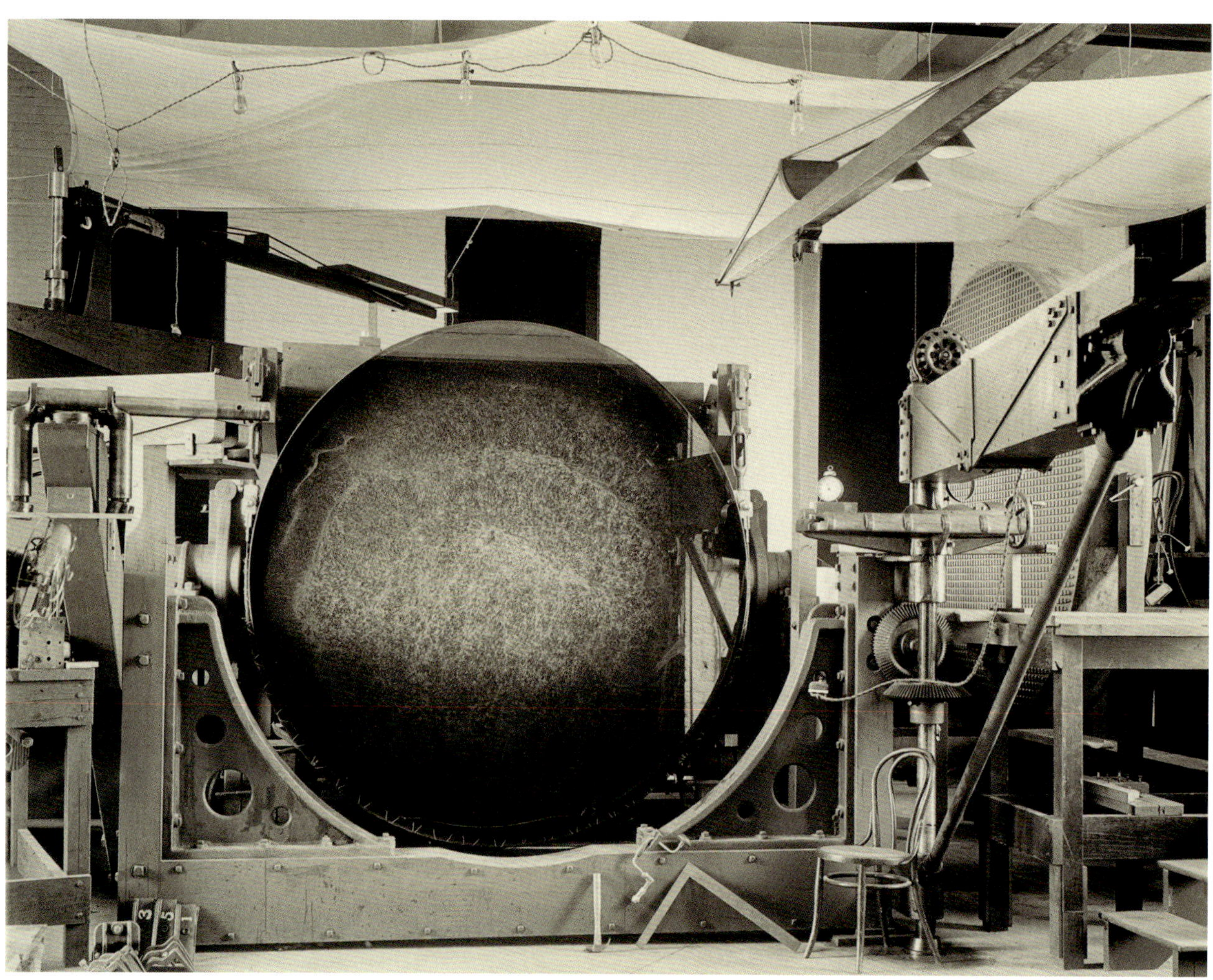

33 | GEORGE W. RITCHEY | *Mirror for Mount Wilson Observatory Telescope*, early 20th century

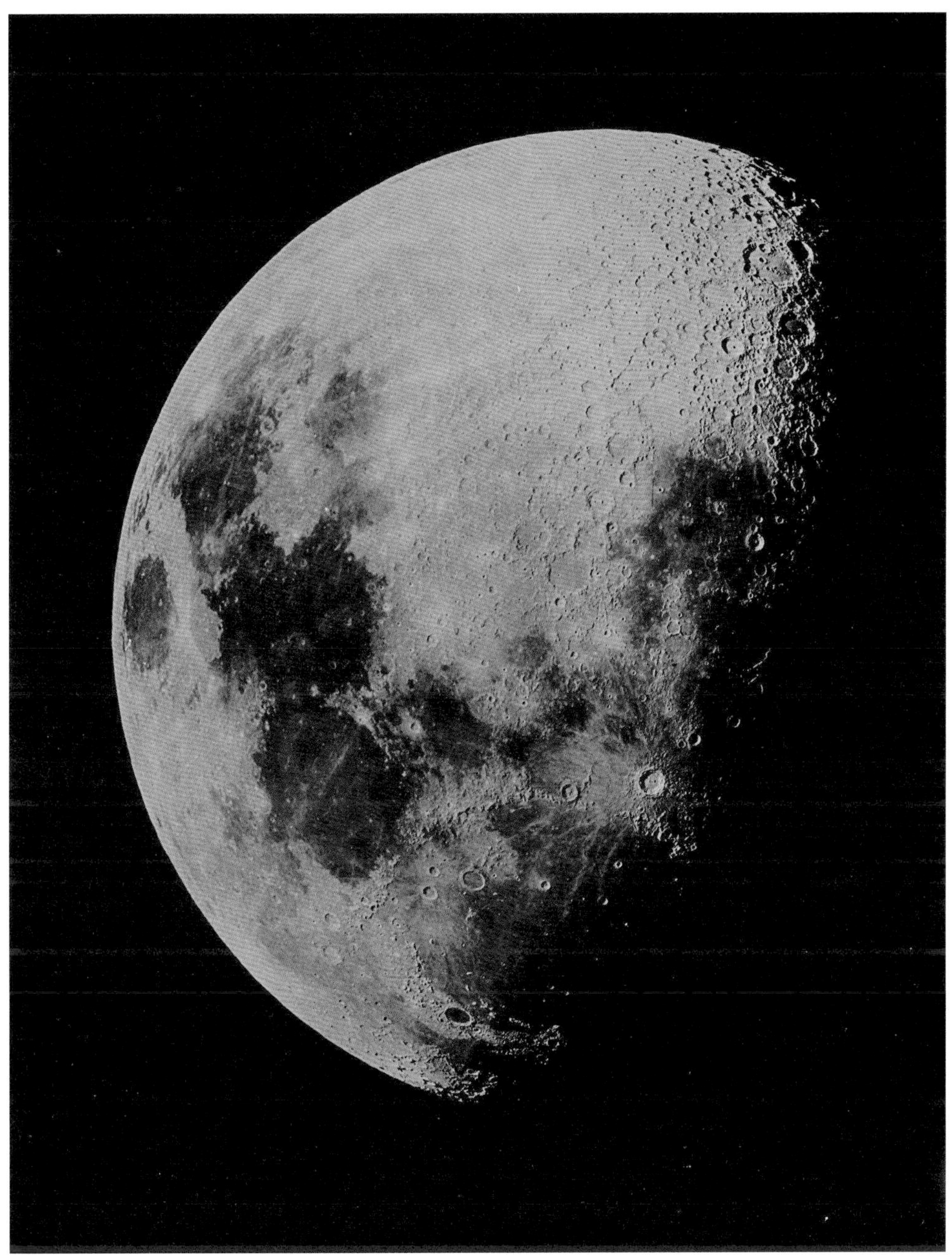

34 | GEORGE W. RITCHEY | *The Moon (10 Days Old)*, ca. 1901

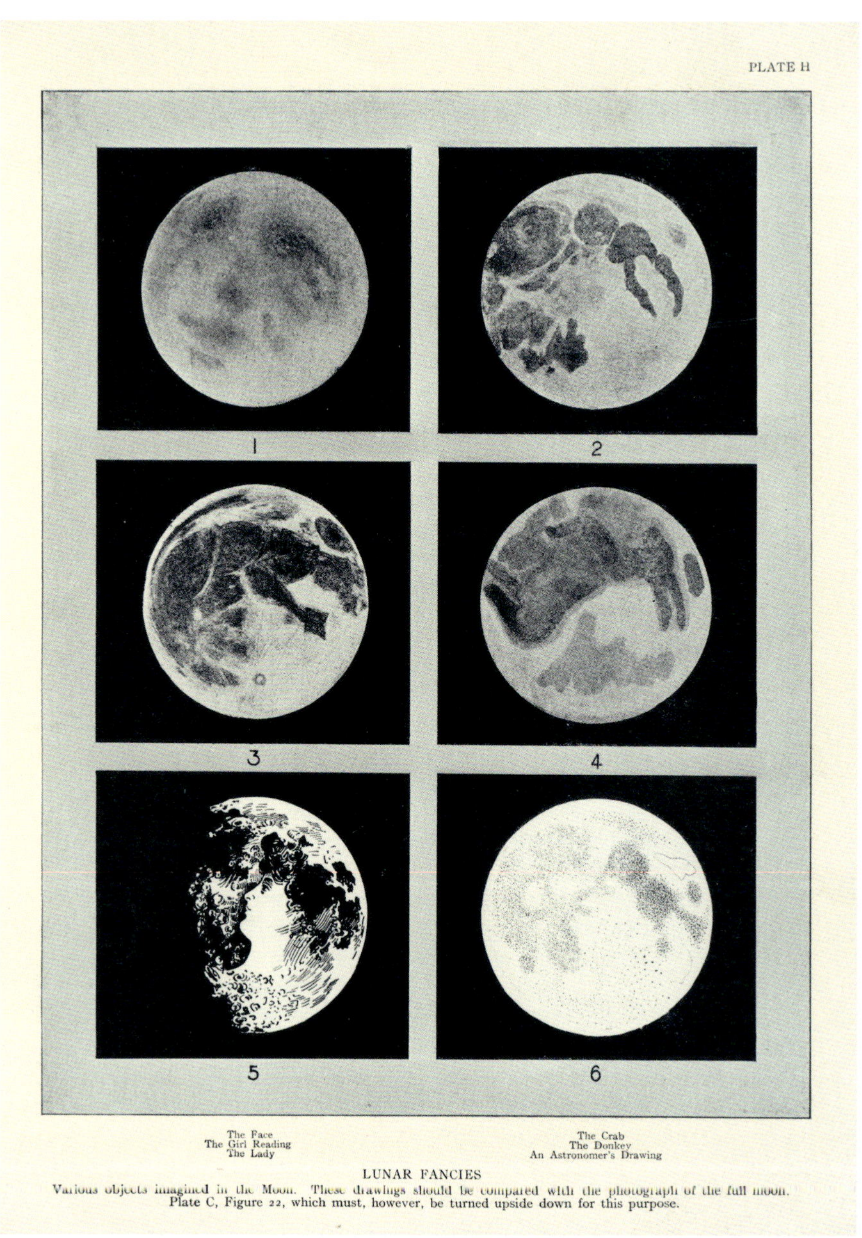

35 | WILLIAM H. PICKERING | *Lunar Fancies*,

in *The Moon: A Summary of the Existing Knowledge of Our Satellite, with a Complete Photographic Atlas*, 1903

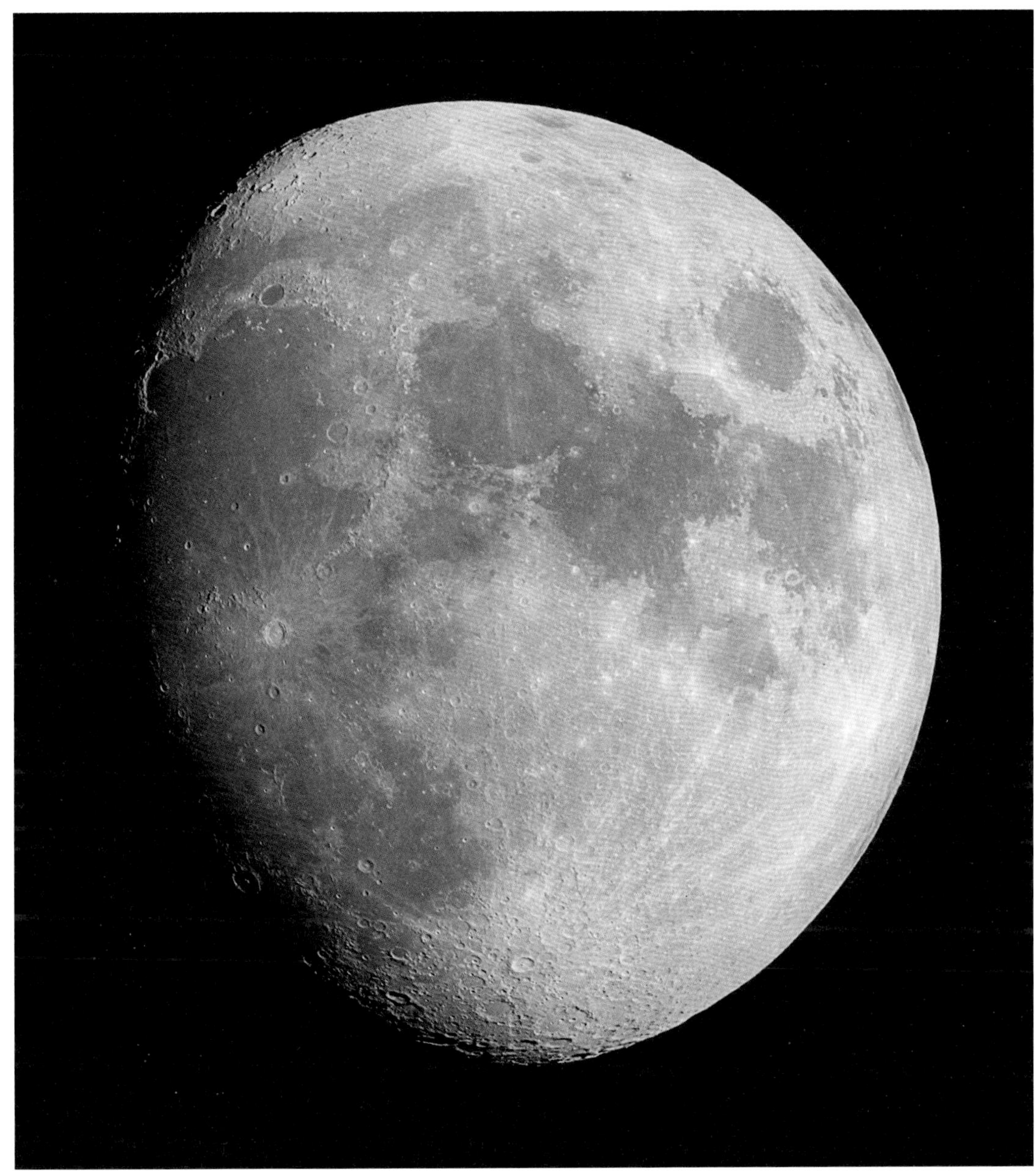

36 | *Transparency of the Moon from Negatives Made at the Lick Observatory, Mount Hamilton, California*, ca. 1896

OBSERVATOIRE DE PARIS

GRAND EQUATORIAL COUDE

ATLAS PHOTOGRAPHIQUE DE LA LUNE

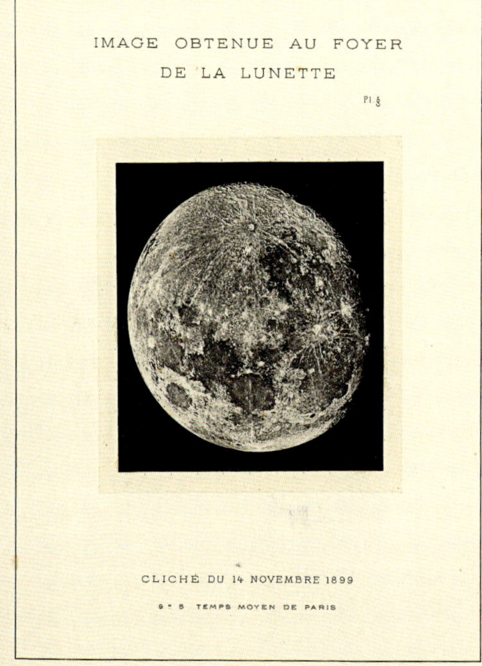

HÉLIOGRAVURES D'APRÈS LES CLICHÉS ET LES AGRANDISSEMENTS

EXÉCUTÉS PAR MM. LŒWY ET PUISEUX

ASSISTÉS DE M. LE MORVAN

PARIS 1903

37 | MAURICE LOEWY AND PIERRE PUISEUX |

Fascicle 7 (plates 36–41), from *Photographic Atlas of the Moon* (*Atlas photographique de la lune*), 1896–1910

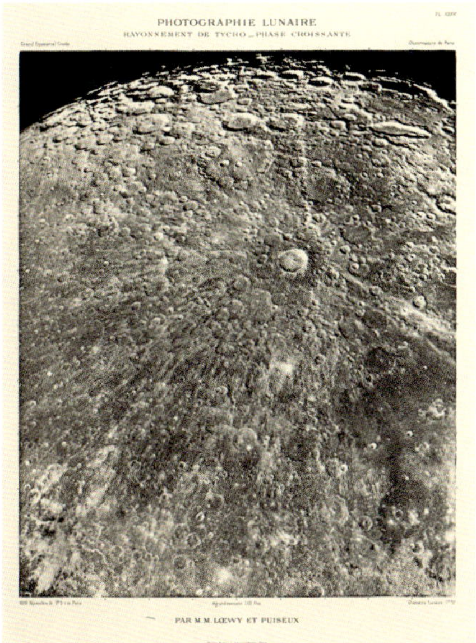
PHOTOGRAPHIE LUNAIRE
RAYONNEMENT DE TYCHO _ PHASE CROISSANTE
PAR M.M. LŒWY ET PUISEUX

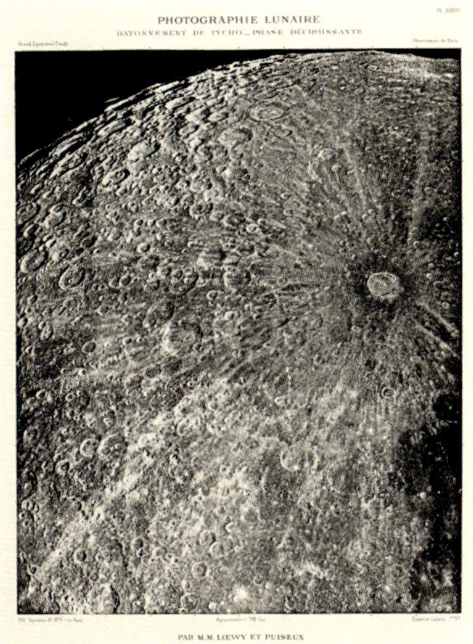
PHOTOGRAPHIE LUNAIRE
RAYONNEMENT DE TYCHO _ PHASE DÉCROISSANTE
PAR M.M. LŒWY ET PUISEUX

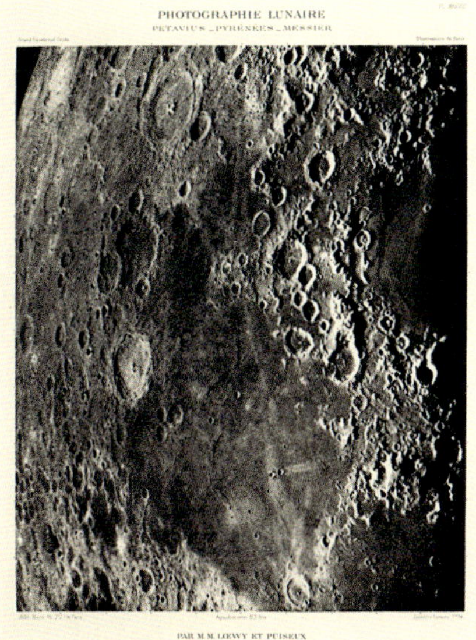
PHOTOGRAPHIE LUNAIRE
PETAVIUS _ PYRÉNÉES _ MESSIER
PAR M.M. LŒWY ET PUISEUX

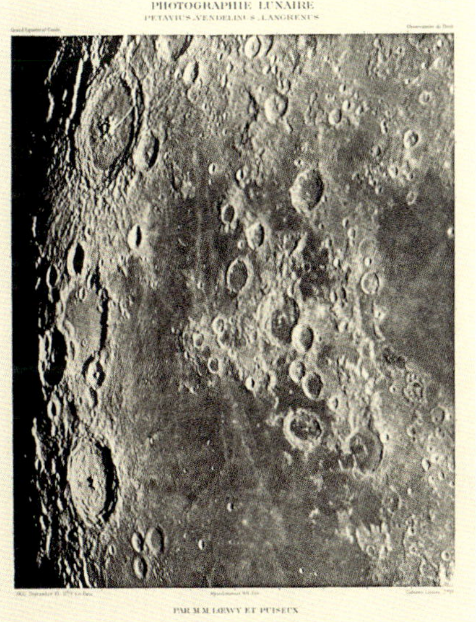
PHOTOGRAPHIE LUNAIRE
PETAVIUS _ VENDELINUS _ LANGRENUS
PAR M.M. LŒWY ET PUISEUX

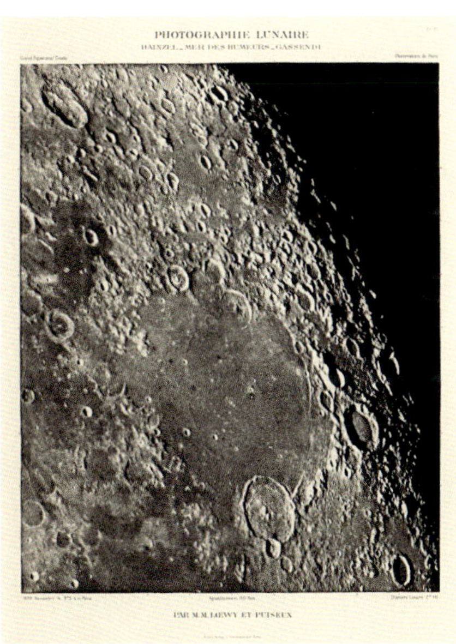
PHOTOGRAPHIE LUNAIRE
HAINZEL _ MER DES HUMEURS _ GASSENDI
PAR M.M. LŒWY ET PUISEUX

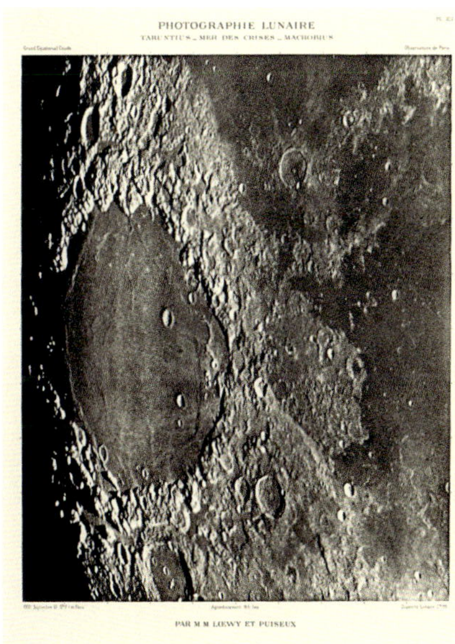
PHOTOGRAPHIE LUNAIRE
TARUNTIUS _ MER DES CRISES _ MACROBIUS
PAR M.M. LŒWY ET PUISEUX

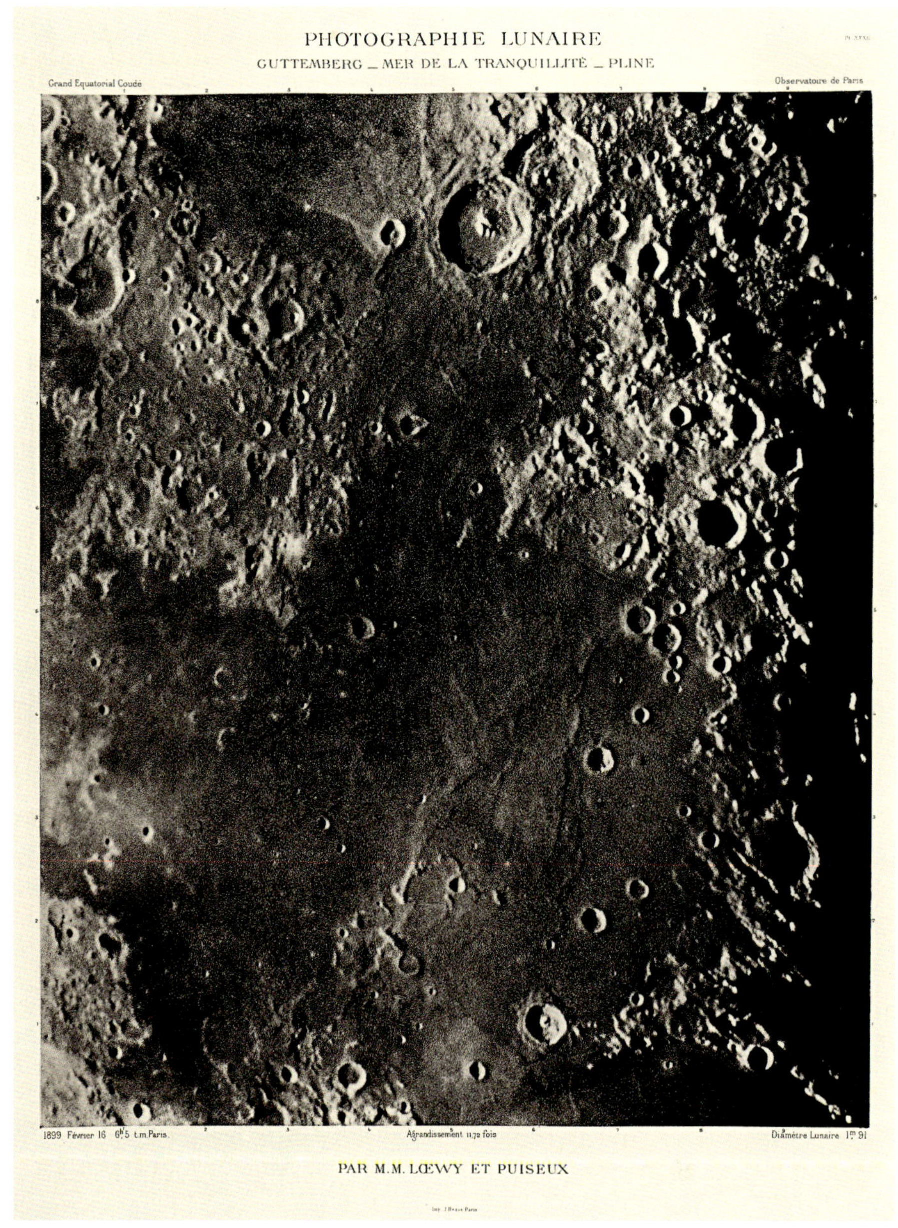

38 | MAURICE LOEWY AND PIERRE PUISEUX |

Plate 32 (without and with tissue overlay), from *Photographic Atlas of the Moon* (*Atlas photographique de la lune*), 1896–1910

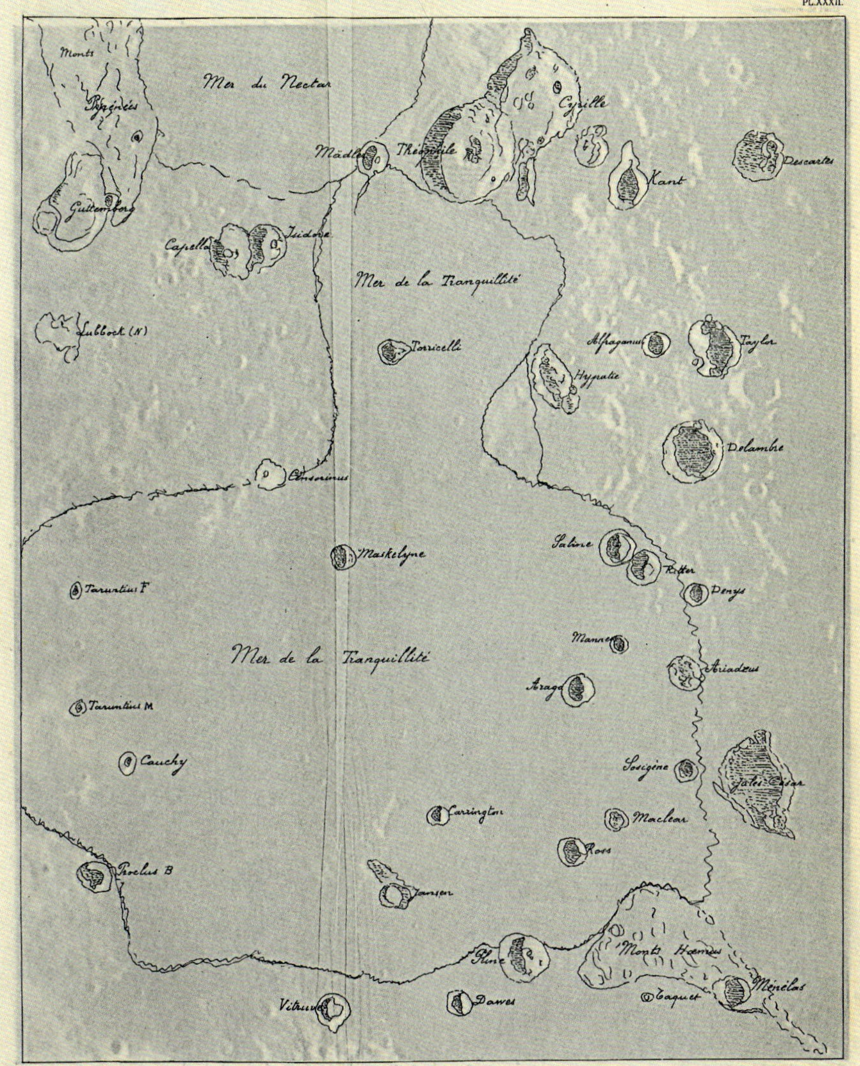
PHOTOGRAPHIE LUNAIRE
GUTTEMBERG _ MER DE LA TRANQUILLITÉ _ PLINE
PL.XXXII.
Monts Pyrénées
Mer du Nectar
Cyrille
Mädler
Théophile
Kant
Descartes
Guttemberg
Capella
Isidore
Mer de la Tranquillité
Lubbock (N)
Torricelli
Alfraganus
Taylor
Hypatie
Delambre
Censorinus
Maskelyne
Sabine
Ritter
Taruntius F
Denys
Mer de la Tranquillité
Manners
Ariadæus
Arago
Taruntius M
Cauchy
Sosigène
Jules César
Carrington
Maclear
Ross
Proclus B
Jansen
Monts Hæmus
Pline
Ménélas
Vitruve
Dawes
Taquet
PAR M.M LŒWY ET PUISEUX

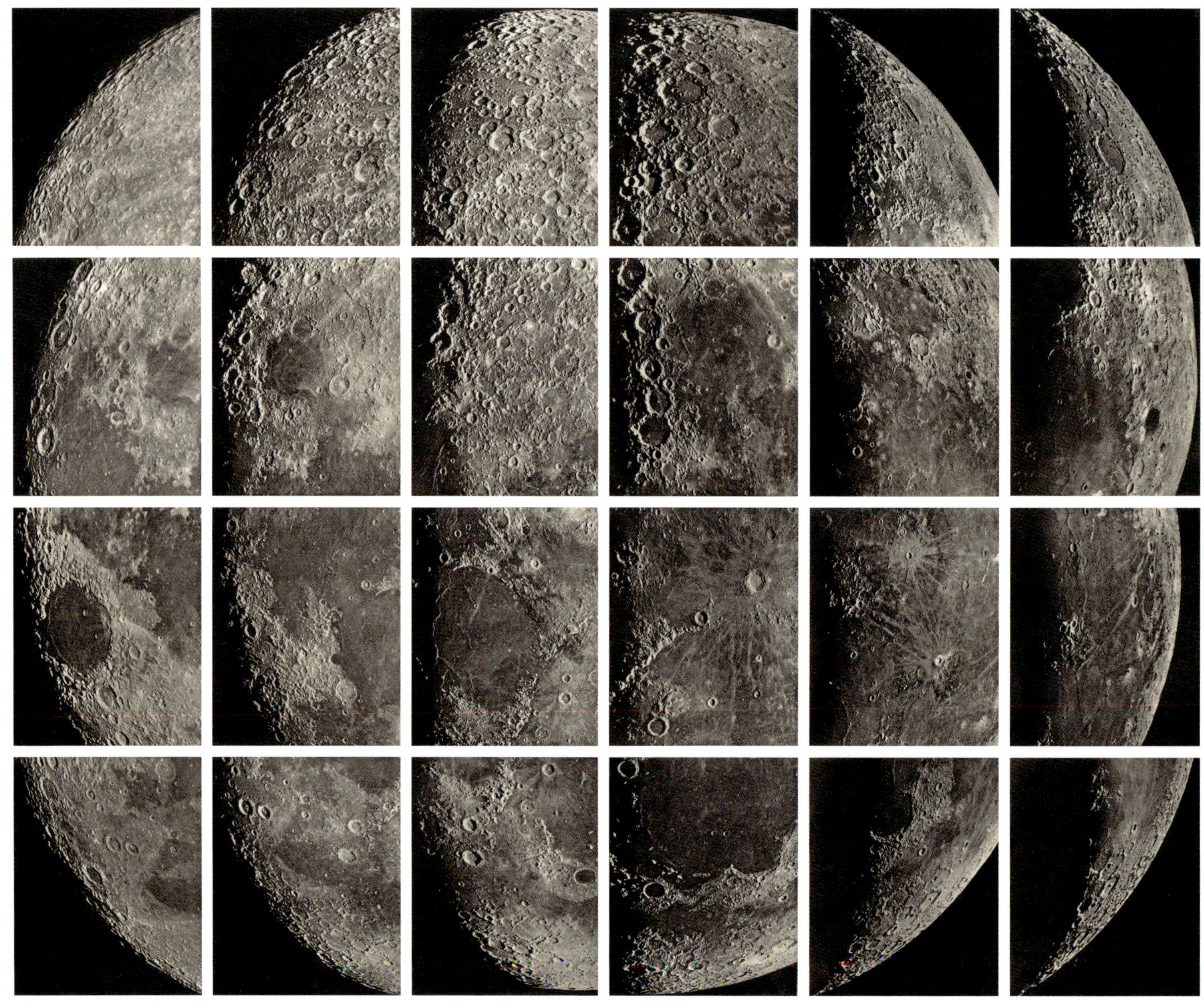

39 | CHARLES LE MORVAN | *Systematic Photographic Map of the Moon, Increasing and Decreasing Phases*, 1899–1909, published 1914

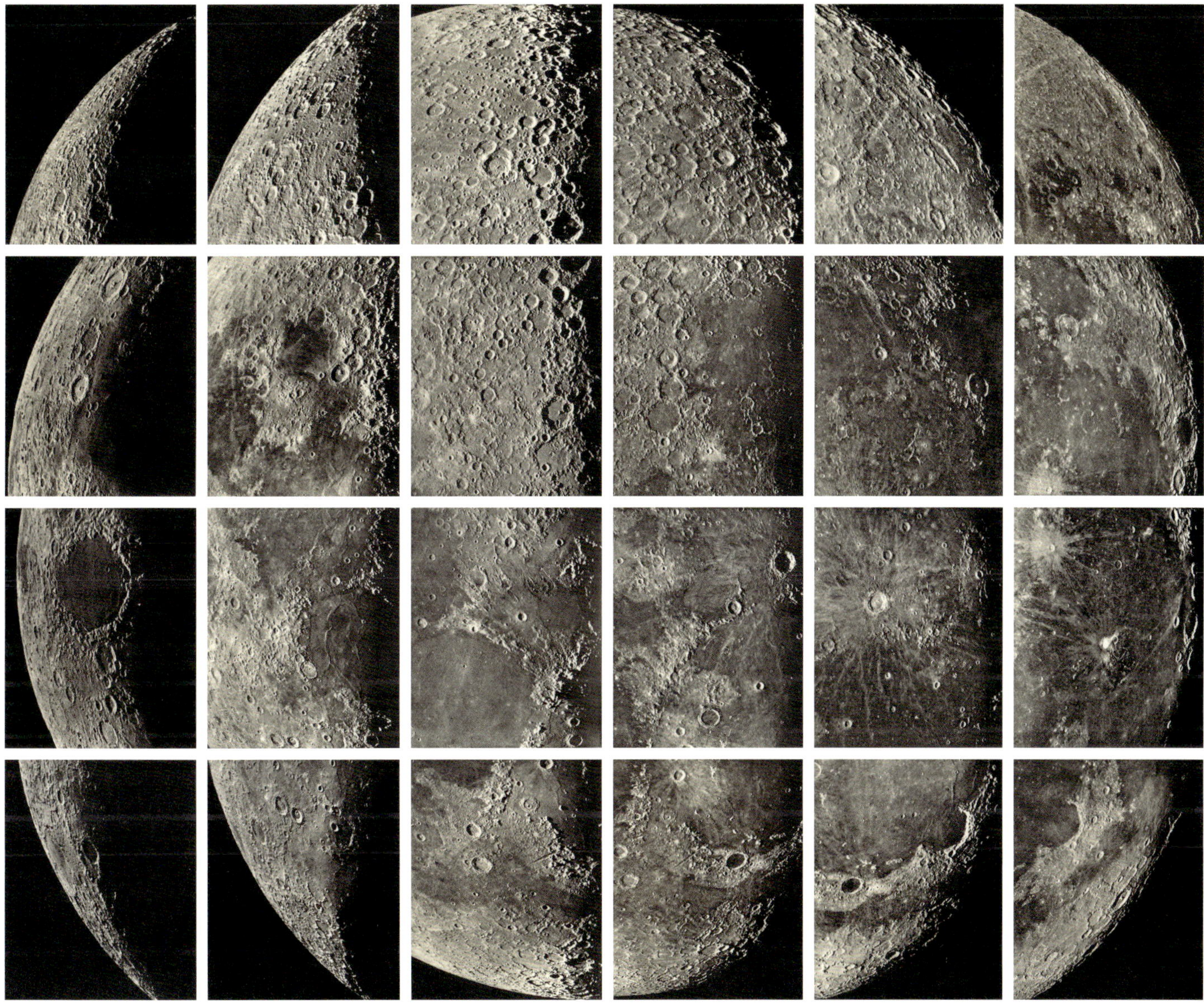

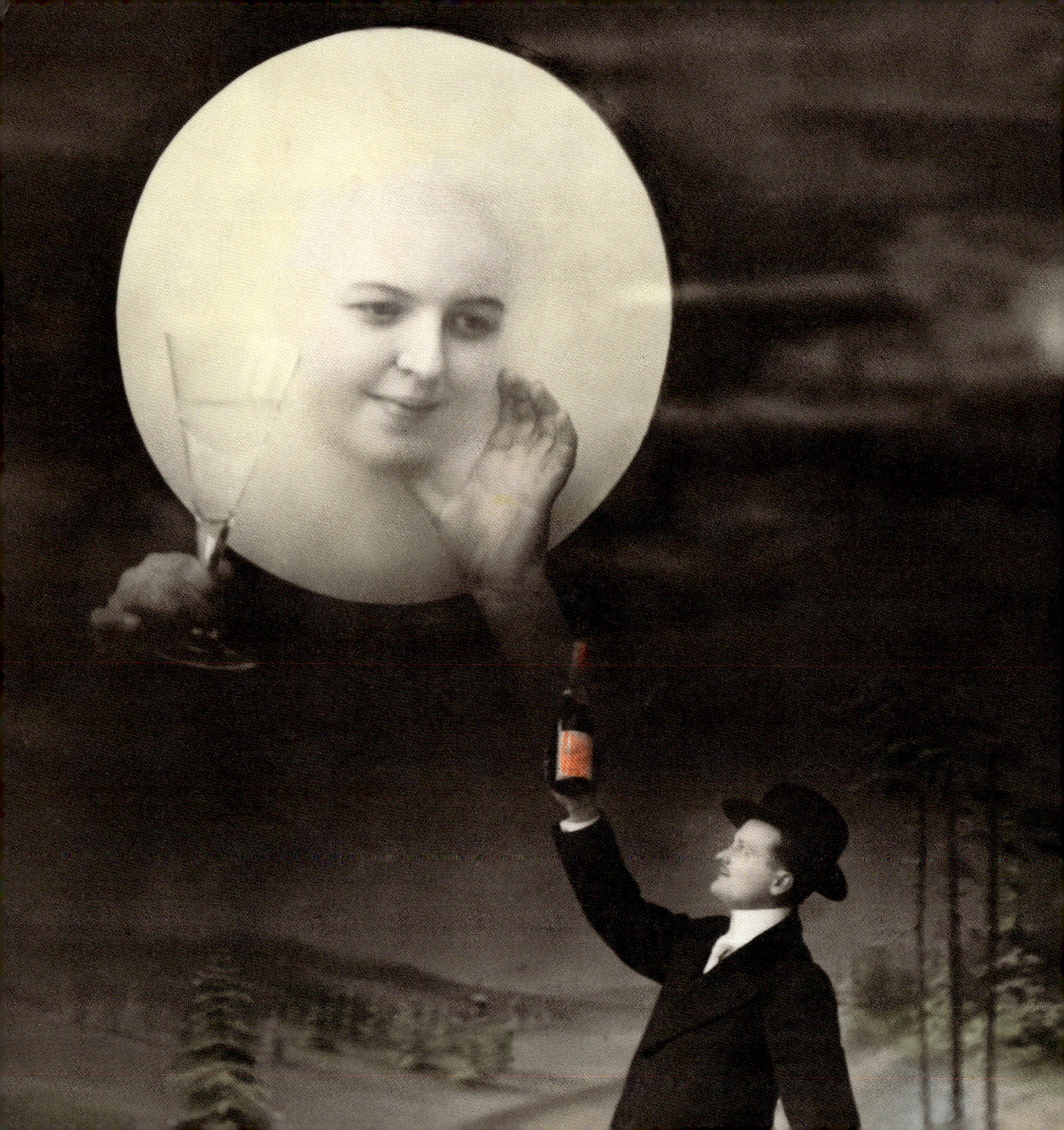

DAYDREAMS BY MOONLIGHT

MIA FINEMAN

In the summer of 1902, the astronomers Maurice Loewy and Pierre Puiseux were hard at work at the Paris Observatory, assembling a photographic atlas of the moon that would be comprehensive and "as truthful and precise as possible."[1] At the same time, in a studio just outside the city, the magician and filmmaker Georges Méliès was producing a whimsical motion picture in which a group of learned astronomers board a cannon-propelled space capsule to the moon, explore its craters and caverns, and flee from an army of insect-like lunar inhabitants. Created in the same cultural moment, Loewy and Puiseux's atlas (see pls. 37, 38) and Méliès's pioneering film *A Trip to the Moon* (*Le Voyage dans la lune*, 1902, pl. 53) neatly represent photography's dual role in the history of lunar imagery. While astronomers harnessed the camera's descriptive accuracy to map the visible surface of the moon in meticulous detail, artists exploited photography's optical realism to create convincing illusions—whether of space travel and life on the moon or of the otherworldly effects of moonlight closer to home.

Since Galileo's time, the study of the moon has mingled observation and imagination, fluctuating between scientific fact and speculative fantasy. To seventeenth-century artists, writers, and religious scholars, the earth's satellite, viewed through a telescope and charted in newly published maps, appeared tantalizingly close, yet achingly unreachable. Discoveries of the early 1600s inspired a profusion of literary and artistic lunar voyages undertaken on fantastical conveyances: demon escorts, ascending vials of dew, fireworks, flying chariots, giant birds, human-powered wings, and the astral projection of

FIG. 7. Frontispiece and title page of Francis Godwin's *The Man in the Moone*, 1638. Woodcut. Houghton Library, Harvard University, Cambridge, Mass.

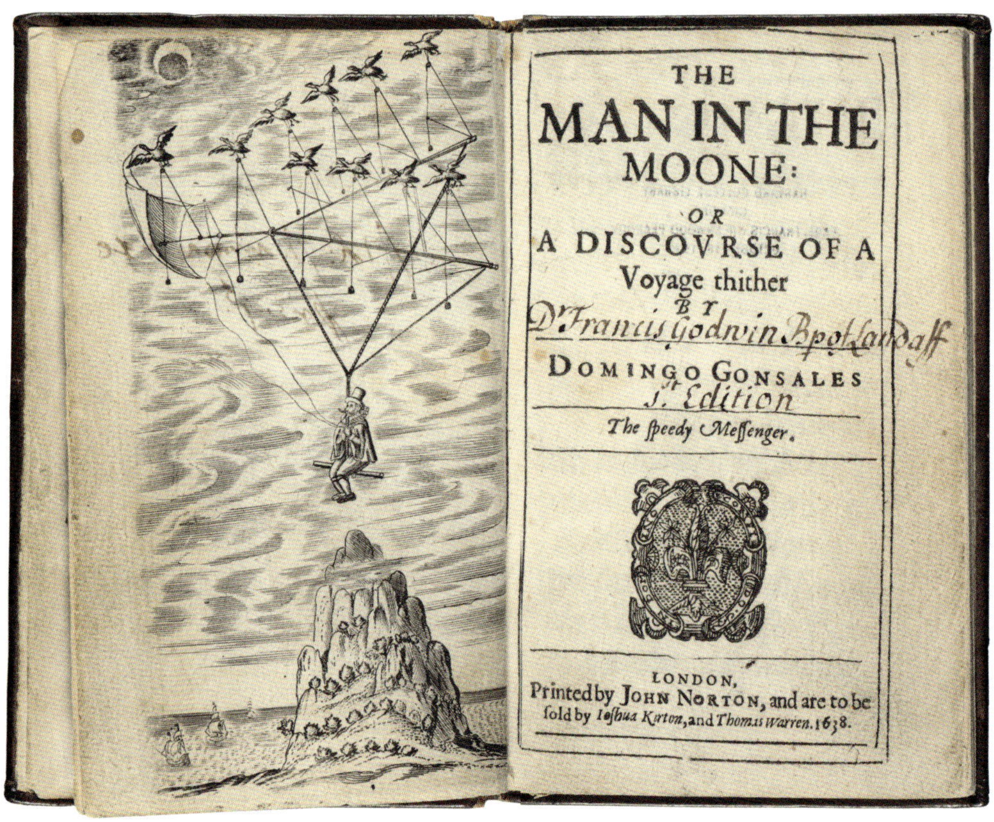

THE MAN IN THE MOONE: OR A DISCOVRSE OF A Voyage thither BY DOMINGO GONSALES The speedy Messenger. LONDON, Printed by JOHN NORTON, and are to be sold by Ioshua Kirton, and Thomas Warren. 1638.

disembodied thoughts. These imaginary narratives were shaped, in part, by European explorers' accounts of strange new worlds in the Americas and Antipodes, as well as by the utopian tradition of literary satire. "The literature of lunar voyaging," notes the historian David Cressy, "was part parodic but mostly sober, combining earnestness and jest."[2]

Among the most influential accounts of lunar travel was *The Man in the Moone* (1638) by Francis Godwin, an English bishop and historian. The narrative follows the fanciful adventures of a Spanish merchant who gets lost in the East Indies and harnesses a flock of "gansas" (Spanish for *geese*) to transport him between islands, as illustrated in the book's frontispiece (fig. 7). The birds fly higher and higher, eventually reaching the moon, where the explorer finds tall Christian people who speak a musical language and enjoy life in a pastoral paradise. Godwin's tale became an international sensation, inspiring a parody by Cyrano de Bergerac and numerous theatrical farces, including Aphra Behn's *The Emperor of the Moon* (1687). Also published in 1638 was *The Discovery of a World in the Moone; or a Discourse Tending to Prove, That 'Tis Probable There May Be Another Habitable World in That Planet*, a philosophical treatise by John Wilkins, an Oxford-educated Puritan bishop. Like many churchmen of his time, Wilkins subscribed to the doctrine of the "plurality of worlds"—the idea that God, in his all-powerful plenitude, would have created an infinite number of populated worlds.[3] Wilkins's treatise proposes that the moon is most likely inhabited by beings he called Selenites, after Selene, ancient Greek goddess of the moon. More than a century later, the Florentine

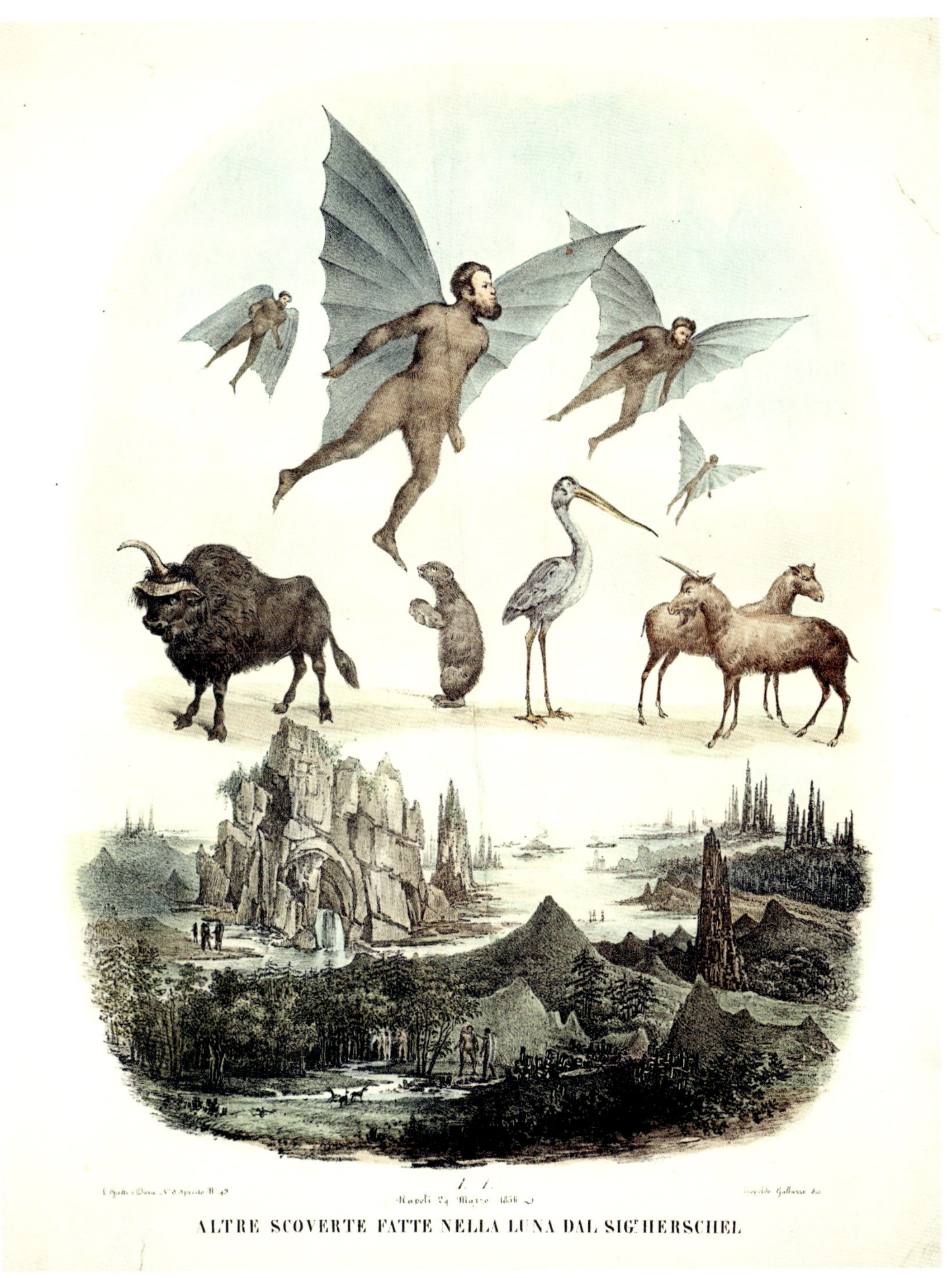

FIG. 8. Leopoldo Galluzzo (Italian, active 1830s), *Other Discoveries Made on the Moon by Sigr. Herschell* [*sic*], 1836. Lithograph, 20 x 24 in. (50.8 x 61 cm). Smithsonian Institution, Washington, D.C.

artist Filippo Morghen used the book as inspiration for a portfolio of nine etchings, *The Collection of the Most Notable Things Seen by John Wilkins, Erudite English Bishop, on His Famous Trip from the Earth to the Moon* (1766–67, pl. 40). Morghen's moon is essentially a New World fantasy embellished with chinoiserie foliage — a tropical paradise in which giant possums cavort with scantily clad natives, who take refuge in dwellings carved out of oversize pumpkins.

The fantasy of life on the moon continued to grip the public imagination at the dawn of photography in the early nineteenth century — an era during which the rapid pace of scientific discovery made it difficult to discern truth from fiction. In August 1835, the *New York Sun* published a series of articles touting "great astronomical discoveries" made by John Herschel through his telescope at the Cape of Good Hope in South Africa. The report, allegedly a reprint from the *Edinburgh Journal of Science*, describes "a gorgeous land of enchantment" populated by diminutive zebras and bison, unicorn-like goats, intelligent bipedal beavers, and, most spectacularly, humanoid creatures with copper-colored hair and bat-like wings.[4] The series was an international media sensation, widely reprinted in publications throughout the United States and Europe, some of which commissioned artists' renderings of the strange new world. Among the most vivid is a set of hand-tinted lithographs by the Neapolitan Leopoldo Galluzzo, who cleverly mimics the idiom of scientific illustration to bolster the reality effect of his images (fig. 8; pl. 41). According to contemporary accounts, the general public greeted the story with "voracious credulity."[5] Edgar Allan Poe, who earlier that summer had published his own moon-travel story, "The Unparalleled Adventure of One Hans Pfaall," reported:

> *Not one person in ten discredited [the report in the Sun], and (strangest point of all!) the doubters were chiefly those who doubted without being able to say why — the ignorant, those uninformed in astronomy, people who would not believe because the thing was so novel, so entirely "out of the usual way." A grave professor of mathematics in a Virginian college told me seriously that he had no doubt of the truth of the whole affair!*[6]

The mastermind behind what became known as the Great Moon Hoax was Richard Adams Locke, an editor at the *Sun*, who later revealed that he intended the story as a satire of the "plurality of worlds" theory. Locke's only misstep in his supremely artful satire was underestimating the gullibility of his readers.[7]

While many artists and writers were daydreaming about the possibility of life on the moon, others were equally captivated by moonlight's aura of mystery and enchantment here on earth. Among the most moon-besotted artists of the early nineteenth century was the German painter Caspar David Friedrich. For him and his late-Romantic cohort, the distant celestial body became an emblem of intense melancholic longing — a popular sentiment William Blake had playfully satirized in his 1793 engraving *I want! I want!* (fig. 9). Between 1819 and the early 1830s, Friedrich painted three masterful variations on the motif of moon gazing; in each, a waxing moon hovers in a cloudless sky, framed by a gnarled oak tree and observed by a pair of figures who stand together, backs to the viewer, on a mountain path (pl. 44).[8] In the version at The Met, the scene is suffused with gentle lavender light, and the full lunar orb is faintly illuminated with what astronomers call "earthshine" (sunlight reflected from the earth onto the shadowed part of the moon). Like many moon-inspired works, Friedrich's paintings are amalgams of observation and invention. The landscape is an imaginary composite, its individual elements — the uprooted oak, the fir tree, the boulder, and the glowing moon itself — based on the artist's sketches after nature, made at different times and places.

The Romantic's fascination with enchantment by moonlight coincided with a fashion for transparencies — paintings or prints illuminated from behind by daylight or lamplight — that swept across Europe around the turn of the nineteenth century. The London printsellers Rudolph

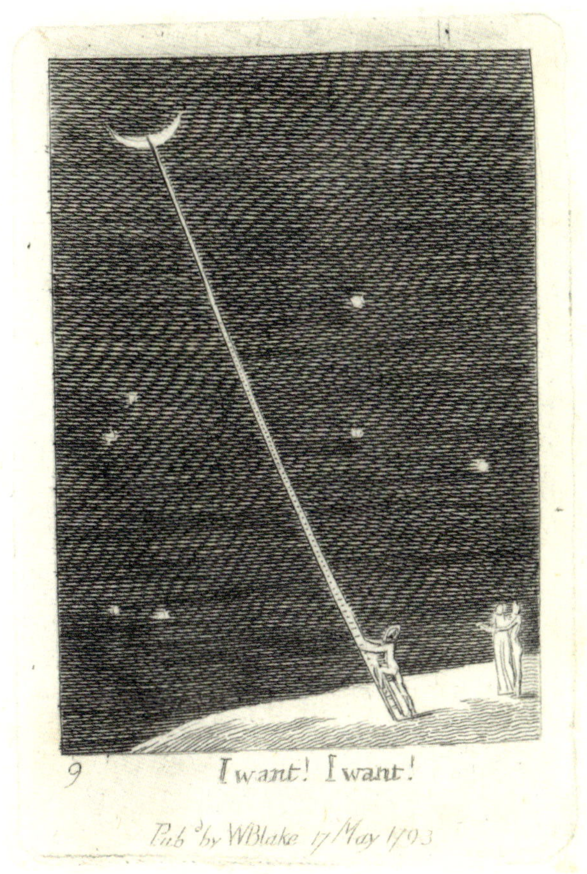

FIG. 9. William Blake (British, 1757–1827), *I want! I want!*, from *For Children: The Gates of Paradise*, 1793. Engraving, 2¾ x 2 in. (7.2 x 5.1 cm). Yale Center for British Art, New Haven, Conn.

Ackermann and Edward Orme were leading producers of transparent prints, and both men published kits and instructional manuals that encouraged amateurs, especially women, to create their own versions at home. The technique involved painting large areas of color on the back of a print, then adding varnish to selected details to make the paper see-through when held up to light. The recommended subjects were picturesque scenes with dramatic atmospheric effects: "abbeys in ruin, Gothic painted windows, illuminated temples and gardens, moonlight, fires, and groups of gypsies sitting round a fire in a thick wood with the moon glimmering behind the trees."[9] Among the artists who capitalized on this fad was the English caricaturist Thomas Rowlandson, who published his own series of transparent etchings in 1799. In *The Assignation,* he pokes fun at conventional ideas about the romance of moonlight, conjuring an illicit rendezvous in which a young woman of modest means sneaks off into the night with a gentleman whose carriage waits by a moonlit pond (pl. 45).

Transparencies count among several optical amusements from this period that directly prefigured photography and cinema, including magic lanterns, phantasmagorias, and Louis-Jacques-Mandé Daguerre's dioramas of the 1820s, which used shifting lights and translucent paintings on thin fabric to create spectacular visual effects.[10] As in photography, light was both the subject and the means of pictorial representation. At the same time, the spread of gas lamps in the early 1800s began to challenge the moon's preeminence as a source of nighttime illumination. The invention of electric light less than a century later further diminished the moon's sovereignty as the silver lamp of the night sky; this situation is playfully evoked in Man Ray's 1931 photogram of a shining moon floating above an electric light switch (pl. 43).

In the mid-nineteenth century, just before the arrival of electric streetlights in major cities, photographers continued to experiment with optical devices to simulate the cool glow of moonlight. These effects were especially sought after by commercial studios in Rome and Venice, which supplied tourists with Romantic *veduti* (views) of architectural landmarks and vistas. In 1862 the Venice-based Carlo Ponti invented the "megalethoscope," a boxlike apparatus through which a photograph could be transformed into a day or a night scene by means of reflected or transmitted light. Because emulsions were not sensitive enough to record images at night, Ponti and his main competitor, Carlo Naya, made photographs during the day, then created the illusion of moonlight through a variety

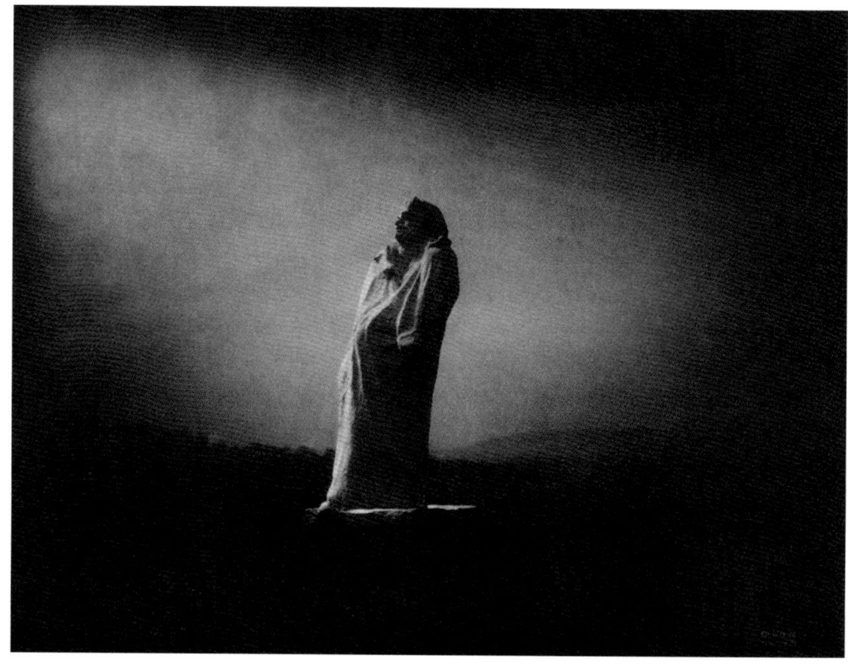

FIG. 10. Edward J. Steichen (American, b. Luxembourg, 1879–1973), *Balzac, Towards the Light, Midnight*, 1908. Direct carbon print, 14⅜ x 19 in. (36.5 x 48.2 cm). The Metropolitan Museum of Art, Alfred Stieglitz Collection, 1933 (33.43.38)

of darkroom tricks, including underexposure, printing on blue paper, tinting, and retouching. In Naya's ghostly view of the Grand Canal, the sun masquerades as a gently glowing moon, and hand-painted highlights glitter on the water and cathedral domes (ca. 1875, pl. 46). Although day-for-night photographs were popular among tourists, serious critics derided their obvious artifice. One commentator in the *British Journal of Photography* wrote scathingly of Naya's "wretched clap-trap moonlights, patched up with painting-in of skies on the negative and tinting the prints blue and green, with a white wafer spot in the sky, till they are just able to catch the uneducated eye with a crude and ghastly travesty of moonlight."[11]

Notwithstanding such critical contempt for darkroom fakery, photographers and filmmakers in the early twentieth century continued to use similar techniques to enhance their moonlit views. For the American Pictorialist Edward Steichen, the woods at dusk was a favorite subject to which he returned time and again, in both paintings and photographs. "The romantic and mysterious quality of moonlight, the lyric aspect of nature made the strongest appeal to me," he wrote in his autobiography.[12] Steichen's attraction to the moon's lyricism found its consummate expression in *The Pond — Moonrise* (1904, pl. 47), a photograph of the celestial body hovering on the horizon line beyond a small body of water in the woods near Mamaroneck, New York. To heighten the mood of mysterious quietude, he brushed layers of Prussian blue and platinum solution onto the print by hand and accentuated the sliver of moon with careful retouching. A few years later, the sculptor Auguste Rodin asked Steichen to produce a series of moonlight studies of Rodin's sculpture of the French author Honoré de Balzac. For two nights, Steichen worked from dusk to dawn, exposing his negatives for up to an hour. The resulting prints, which he toned and retouched to enhance the lambent atmosphere, transform the plaster cast into a spiritual vision, a monument to artistic genius towering over a nocturnal landscape (fig. 10).

FIG. 11. Cover of the official Luna Park viewbook, 1903, featuring the airship *Luna* flying over the park. Halftone, 5½ x 8¼ in. (14 x 21 cm). Coney Island History Project Collection, New York

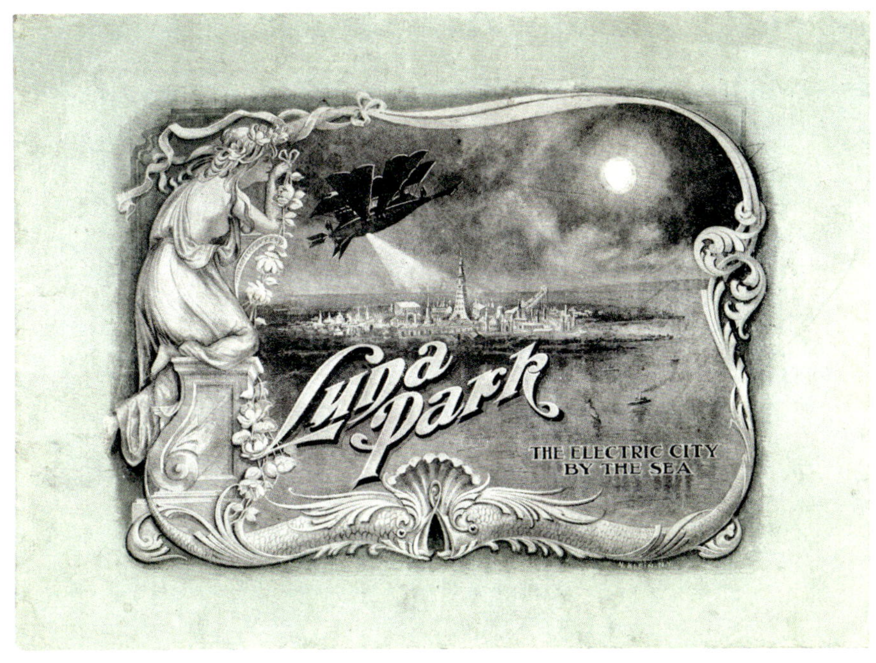

As an aficionado of camera trickery, aviation, and popular novelties, Steichen was almost certainly among the enthusiastic crowds who attended the first screenings of *A Trip to the Moon* when it opened in New York in 1902, shortly after its Paris premiere. Méliès, who wrote, directed, and produced the film, cast himself as Professor Barbenfouillis, a wizard-like astronomer who leads a group of companions on a fantastic journey to the moon and back. As the story unfolds, the audience is treated to a spectacular array of ingenious tableaux and special effects, such as the famous sequence in which the spaceship pierces the eye of a crater-faced man in the moon. Every aspect of the imagery, including the elaborately painted theatrical backgrounds, was based on detailed sketches by Méliès, which he re-created in 1930 at the request of the French film archivist Henri Langlois (pls. 48–51). Although he was surely familiar with contemporary photographs of the lunar surface, Méliès had no interest in scientific verisimilitude. Instead, he envisioned the moon as the dreamlike setting for a dazzling series of artistic effects and imaginative "attractions."

The most immediate precedent for Méliès's film was a fairground attraction, also called A Trip to the Moon, which had opened in 1901 at the Pan-American Exposition in Buffalo, New York.[13] Designed by architect Frederic Thompson and promoter Elmer "Skip" Dundy, A Trip to the Moon was a cyclorama, an immersive experience composed of theatrical sets, painted panoramas, projected images, sound effects, and live action. Upon entering, visitors boarded a thirty-seat winged spaceship called the *Luna*, which simulated blast-off while a series of rotating canvas backdrops displayed images of the earth rapidly disappearing into the distance. Upon landing on the moon, passengers were greeted by sixty little people costumed as Selenites, who guided them through a maze of stalactites to the City of the Moon, where they could sample green cheese and purchase souvenirs in the gift shop (pl. 52). The journey concluded with an extravagant stage show in the palace of the Man in the Moon. Over the course of six months, the spectacle welcomed four hundred thousand visitors, including Thomas Edison and President William McKinley.

Thompson and Dundy reconstructed their moon attraction at the Coney Island fairgrounds in New York, where it served as the centerpiece of Luna Park from 1903 until 1914 (fig. 11), when the ride was destroyed by fire. After riding in the spacecraft, visitors might have continued their lunar journey at one of the amusement park's photography studios, where they could pose for a portrait with a cardboard crescent moon — usually enhanced with a smiling face — set against a starry painted backdrop (pl. 54). The photographs were printed quickly on postcard stock and could be sent through the mail or treasured as souvenirs. Although the motif of a figure seated on a crescent moon extends back to Renaissance images of the Virgin Mary, it seems likely that "Man in the Moon" portraits, as they were known at the time, were directly inspired by Méliès's film — specifically, a scene in which the astronomers fall asleep and dream of a lunar deity suspended on a crescent moon, flanked by Saturn on one side and a golden star on the other (pl. 53). At the height of the picture-postcard craze, which lasted from about 1905 to 1915, studios at carnivals, state fairs, beach resorts, and downtown arcades throughout Europe and North America created countless variations on the "Man in the Moon" theme.[14] Some backdrops featured a small airplane (the Wright brothers had made their first powered flight in 1903); others included planets or Halley's Comet, which crossed the earth's atmosphere in 1910. By the time Harold Arlen wrote the song "It's Only a Paper Moon" in 1932 for *The Great Magoo*, a Broadway play set in Coney Island, the waning fad had become fodder for nostalgia.

In a 1930 interview, Méliès cited Jules Verne's seminal science-fiction novel *From the Earth to the Moon* (*De la terre à la lune*, 1865) and its sequel *Around the Moon* (*Autour de la lune*, 1870, pl. 56) as primary inspirations for his film.[15] Like the novels, Méliès's movie begins with a spaceship launched from a giant cannon and ends with a splashdown at sea, but the similarities end there. While *A Trip to the Moon* belongs to the long tradition of satirical lunar fantasy, Verne's books mark the beginning of a new genre of space fiction rooted in technical and scientific verisimilitude. "With the publication of Verne's novel, the possibility of spaceflight was instantly transformed from the realm of the fantastic to a mere exercise in Victorian engineering," notes science-fiction illustrator Ron Miller. "For the first time, the problem of space travel had been put on a firm mathematical and technological basis."[16]

Verne was himself an aviation enthusiast and a founding member of the Society for the Encouragement of Aerial Locomotion by Means of Machines Heavier Than Air, a group of visionaries dedicated to raising funds for the construction of helicopter-like flying machines. The society met at the Paris studio of the charismatic portrait photographer Nadar, who in 1858 captured the first overhead views of Paris from the gondola of a hot-air balloon, and on whom Verne modeled the hero of his novels, the anagrammatic Michel Ardan (pl. 57). Translated and published worldwide, Verne's books intensified the interplay between scientific fact and speculative fiction as never before. The pioneers of modern rocketry — Robert Goddard in the United States, Hermann Oberth in Germany, and Konstantin Tsiolkovsky in Russia — were all avid readers of Verne, as were the American astronauts who flew the first missions to the moon. Neil Armstrong began his final television broadcast from Apollo 11 with this tribute to the French author:

> *A hundred years ago, Jules Verne wrote a book about a voyage to the Moon. His spaceship, Columbia [sic], took off from Florida and landed in the Pacific Ocean after completing a trip to the Moon. It seems appropriate to us to share with you some of the reflections of the crew as modern-day Columbia completes its rendezvous with the planet Earth and the same Pacific Ocean tomorrow.*[17]

In the twentieth century, the movement toward technical accuracy in science fiction accelerated, particularly in cinema. Artists and scientists collaborated on motion pictures that were intended not only to entertain but also to convince the public of the plausibility of space travel. After the success of his science-fiction masterpiece *Metropolis* (1926), Fritz Lang conceived of his 1929 movie *Woman in the Moon* (*Frau im Mond*, pl. 59) as "the first 'scientifically accurate' space flight film."[18] Written by his wife, Thea von Harbou, the screenplay imagines the first experimental rocket expedition to the moon, spiced up by a melodramatic love triangle and a villainous scheme to mine gold discovered in lunar caverns. In contrast to the era's popular space-travel fantasies featuring "rocket men," such as Buck Rogers and Flash Gordon, Harbou's script includes among the crew members a female astronomer (played by Gerda Maurus, Lang's love interest at the time).

Lang brought in as scientific advisers Oberth and the space-flight advocate Willy Ley, whose expertise informed the design and construction of the film's massive multistage upright rocket, as seen in a sketch by production designer Otto Hunte (pl. 58).[19] Scientific veracity had its limits, however. When Oberth insisted that the moon had no atmosphere, Lang invoked artistic license: "How could one present a love story taking place on the Moon and have the lead characters talk to each other and hold hands through space suits?"[20] In exchange for Oberth's services, Lang and the UFA studio helped finance his research and the construction of a real liquid-fueled rocket to be launched from a site on the Baltic Sea in celebration of the film's premiere. Although UFA released publicity shots of the rocket under construction (pl. 60), Oberth ran out of time and money, and the project fizzled before the premiere.

The first major Hollywood space-travel movie was *Destination Moon* (1950, pl. 61), a Technicolor feature produced by George Pal and directed by Irving Pichel, with a screenplay cowritten by science-fiction author Robert Heinlein. Like *Woman in the Moon*, the film envisions the first trip to the moon, this time on a nuclear-powered rocket ship built and financed by a group of American industrialists. As technical adviser, Heinlein ensured the film's scientific authenticity: the thrust and trajectory of the rocket, the design of the space suits, the representation of g-force and weightlessness inside the spacecraft. To create the visual effects, Pal recruited Chesley Bonestell, an astronomical illustrator renowned for the photographic realism of his paintings. Bonestell produced a detailed fourteen-foot-long painting of the interior of the crater Harpalus that served as a background in the film. A few years later, he painted an even larger mural of the lunar landscape for the Boston Museum of Science (pl. 62). Bonestell's process was meticulous. For his preliminary studies, he would often sculpt a Plasticine model of the landscape, photograph it, then paint over the photographs. The topography appears convincingly realistic, yet Bonestell could not resist the urge to embellish; the craggy peaks in his lunar landscapes are considerably more dramatic than our satellite's rounded, windswept hills.[21]

The marketing of *Destination Moon* emphasized its quasi-documentary realism. "There is no hokum, no comic-book sensationalism, no pulp magazine fantasy about destination moon. Here is fact," proclaimed a promotional brochure distributed to theaters.[22] Even more impressive than the film's scientific verisimilitude, however, is its prescient portrayal of the political and economic realities of the Cold War. Although Communism and the Soviet Union are never named, the movie's trip to the moon is framed in explicitly political terms. "The race is on and we'd better win it," declares a military general as he exhorts the American businessmen to support the project. "Because there is absolutely no way to stop an attack from outer space. The first country that can use the moon for the launching of missiles will control the earth." After the American space travelers return from their fictional lunar journey, the film's final title card heralds the dawn of the coming space race: "THIS IS THE END . . . OF THE BEGINNING."

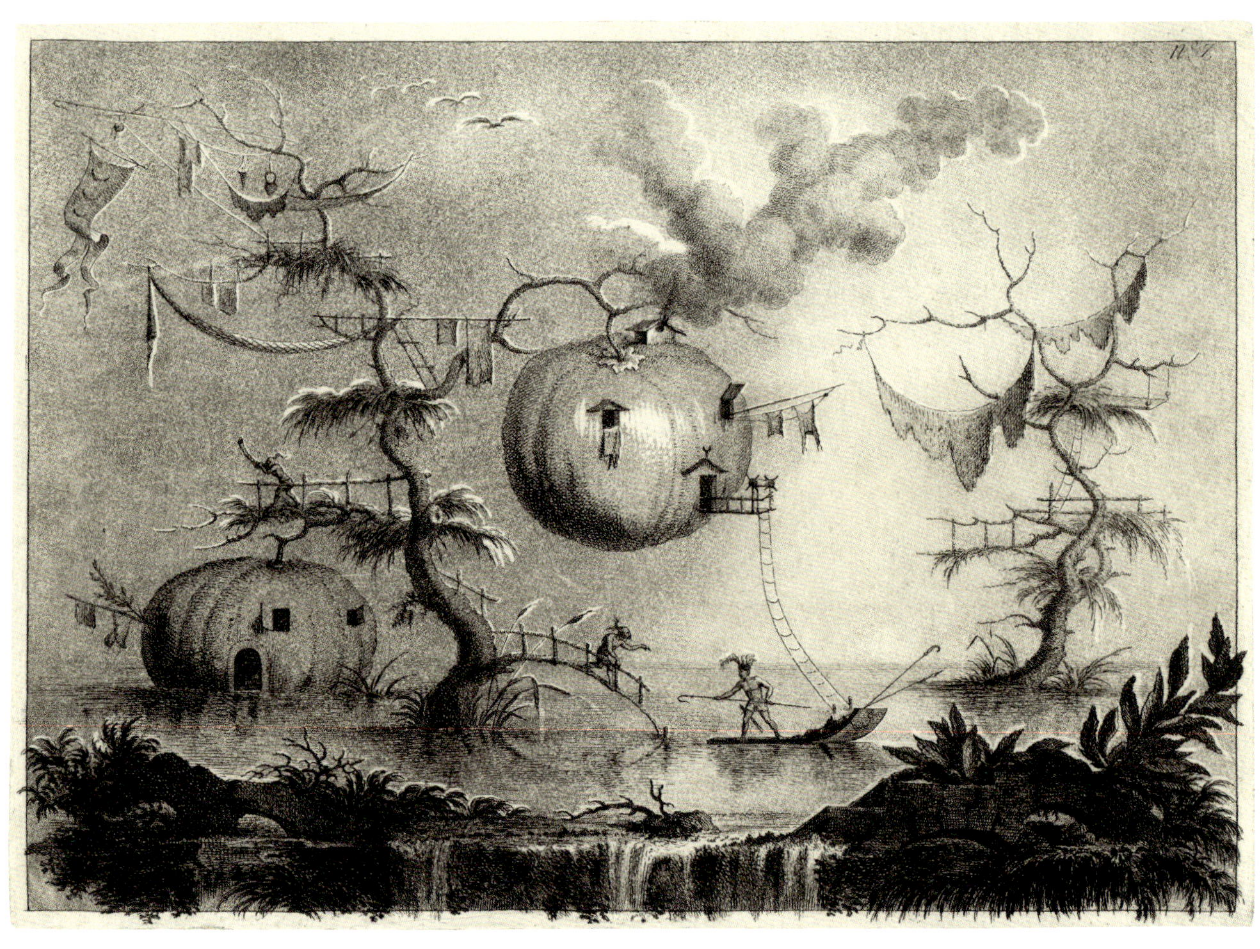

40 | FILIPPO MORGHEN | *Pumpkins Used as Dwellings to Secure against Wild Beasts*, from *The Collection of the Most Notable Things Seen by John Wilkins, Erudite English Bishop, on His Famous Trip from the Earth to the Moon*, 1766–67

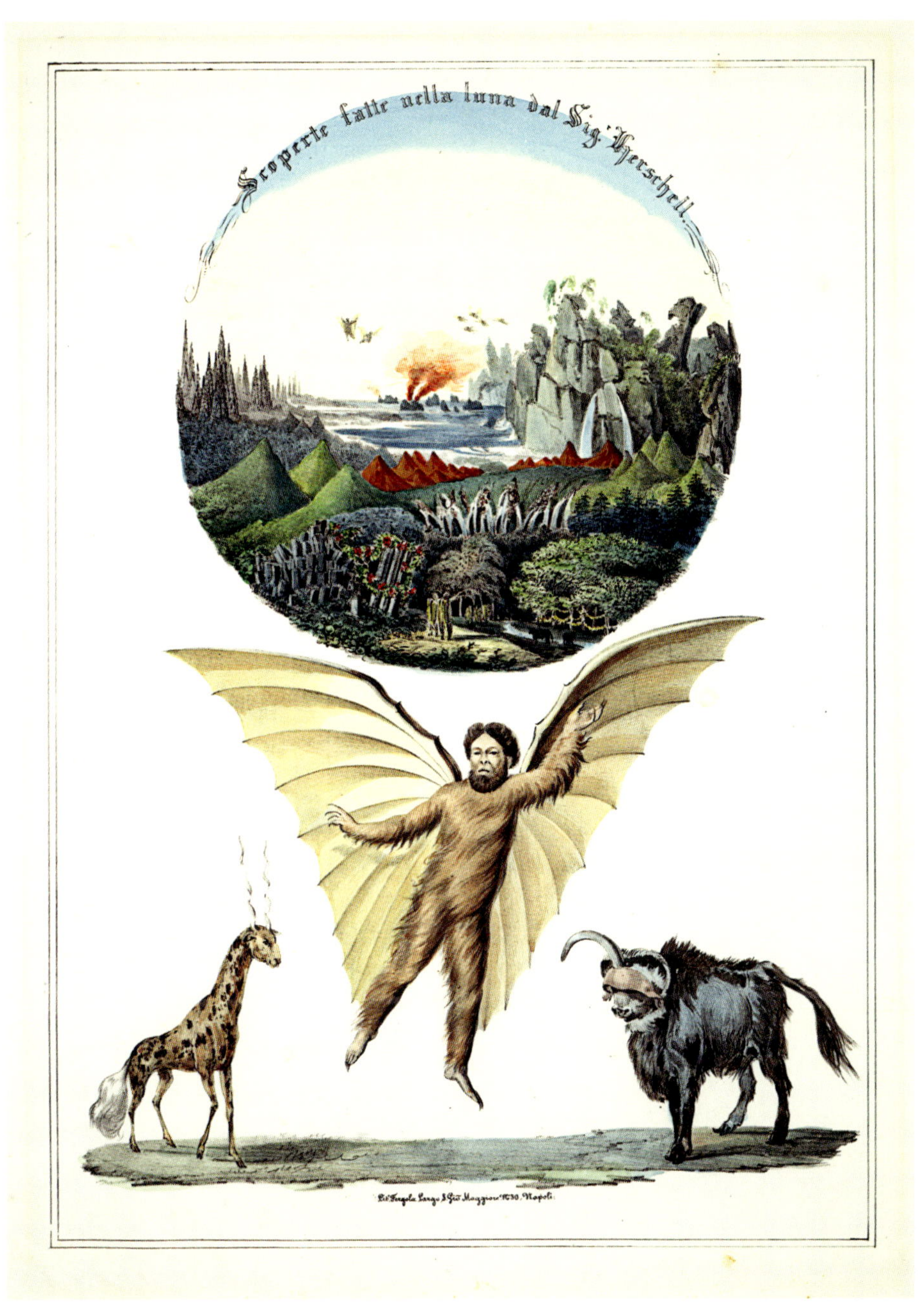

41 | LEOPOLDO GALLUZZO | Plate from *Other Discoveries Made on the Moon by Sigr. Herschell* [*sic*], ca. 1836

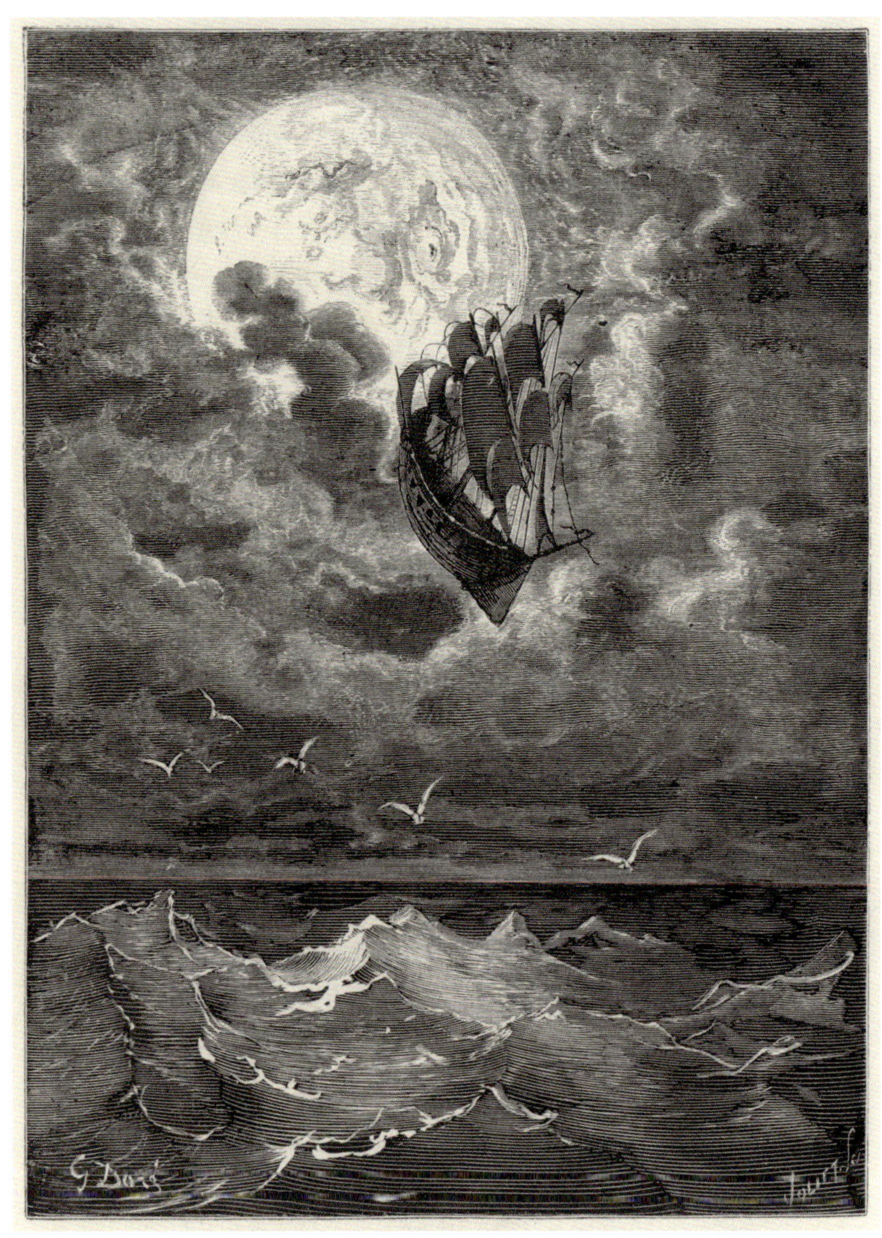

42 | GUSTAVE DORÉ | *It Looked Round and Shining Like a Glittering Island,* in *Aventures du Baron de Munchhausen*, by Rudolf Erich Raspe, 1862

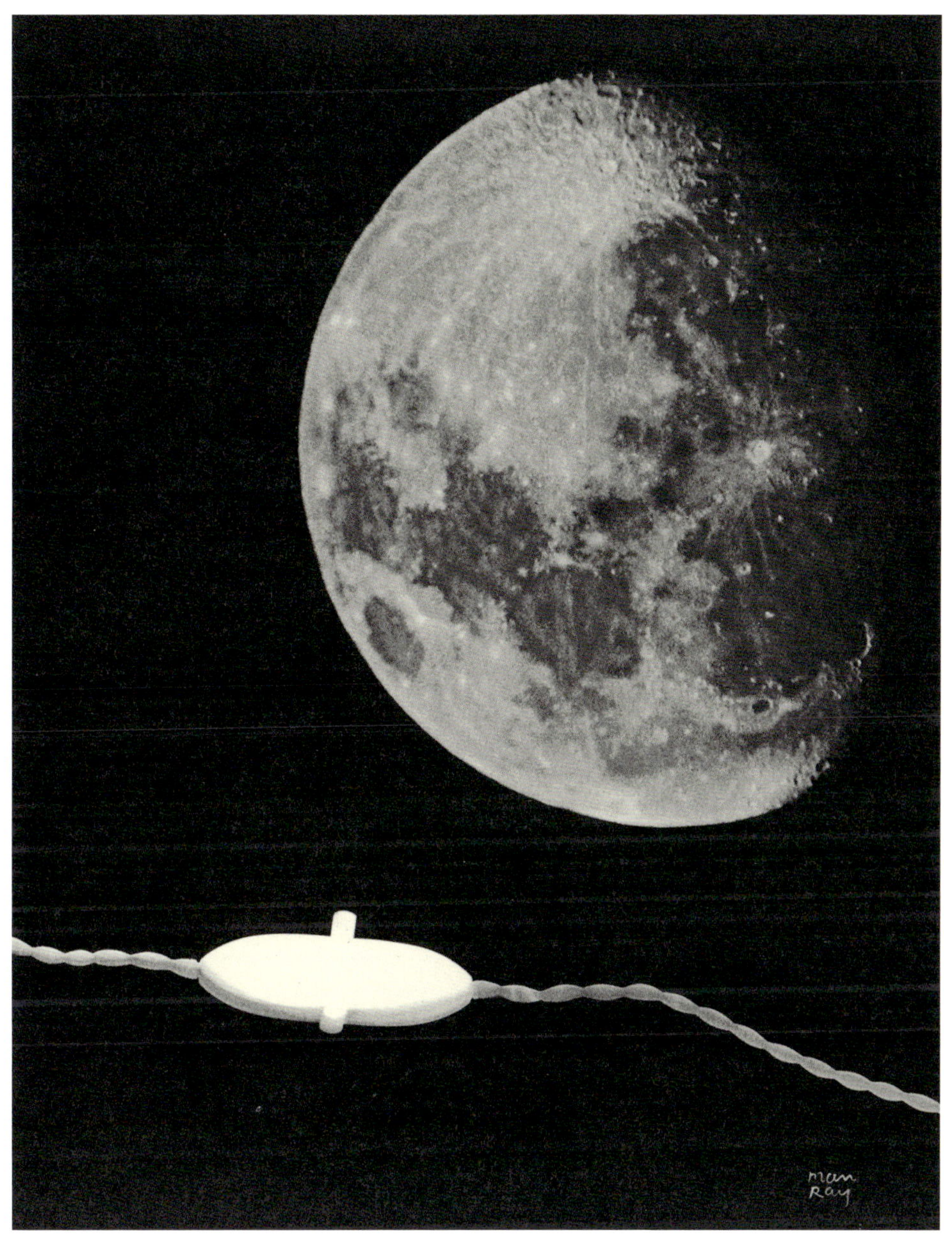

43 | MAN RAY | *The World* (*Le Monde*), in *Electricité*, 1931

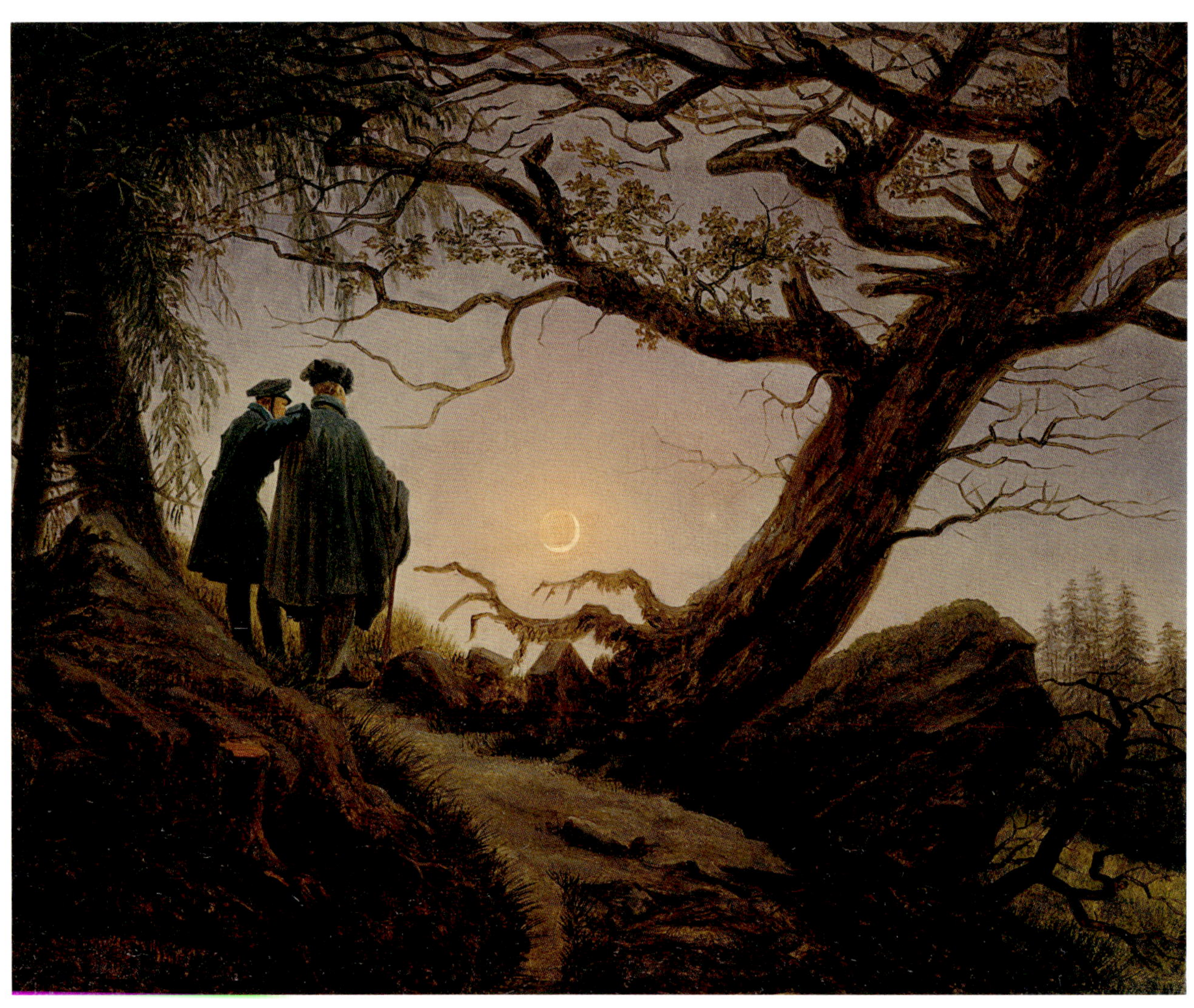

44 | CASPAR DAVID FRIEDRICH | *Two Men Contemplating the Moon*, ca. 1825–30

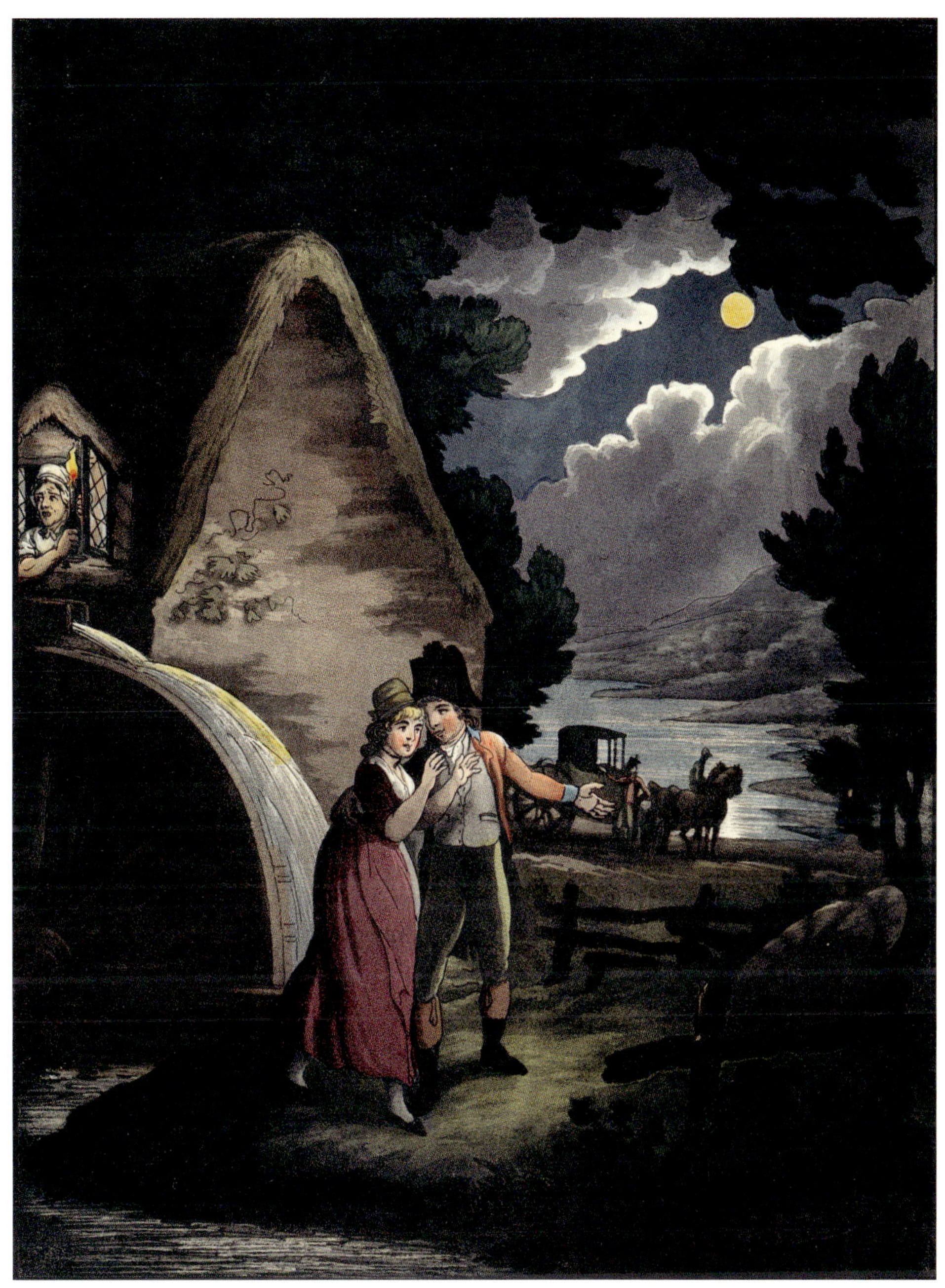

45 | THOMAS ROWLANDSON | *The Assignation*, 1799

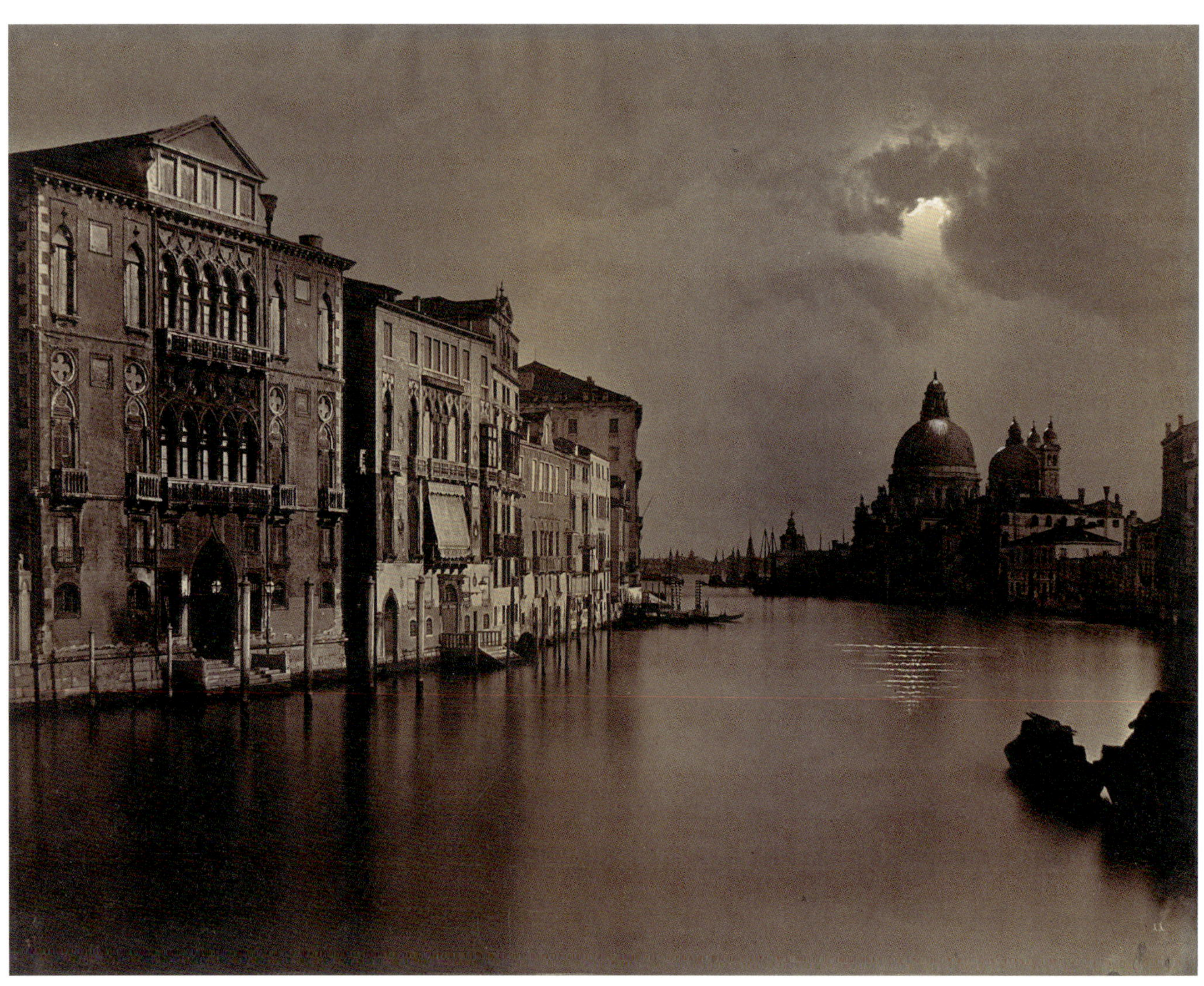

46 | CARLO NAYA | *Night View of the Grand Canal, Venice*, ca. 1875

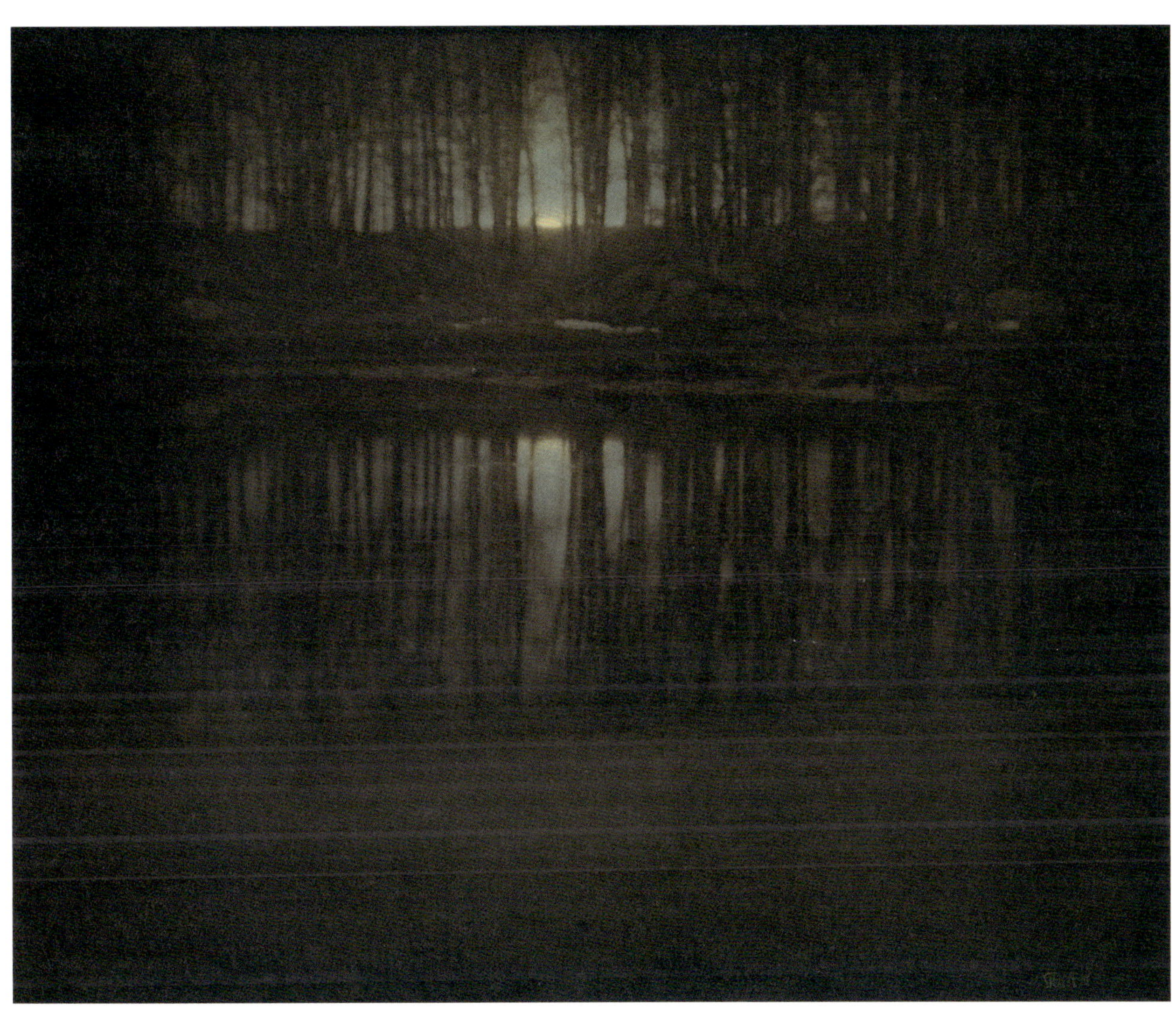

47 | EDWARD J. STEICHEN | *The Pond—Moonrise*, 1904

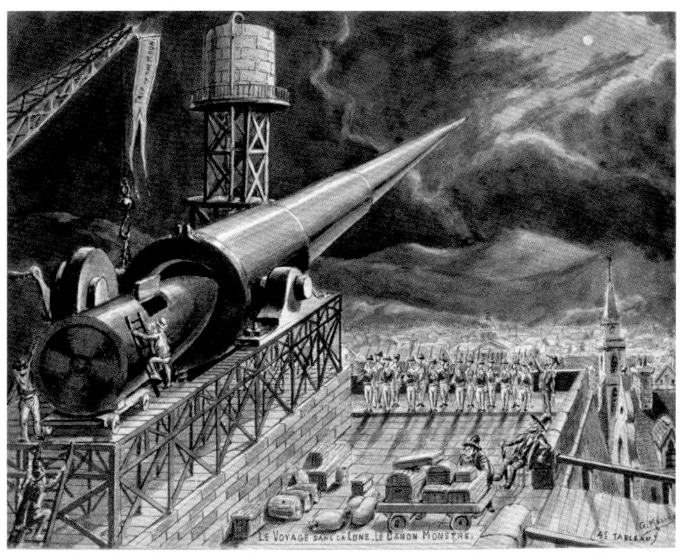

The Giant Cannon

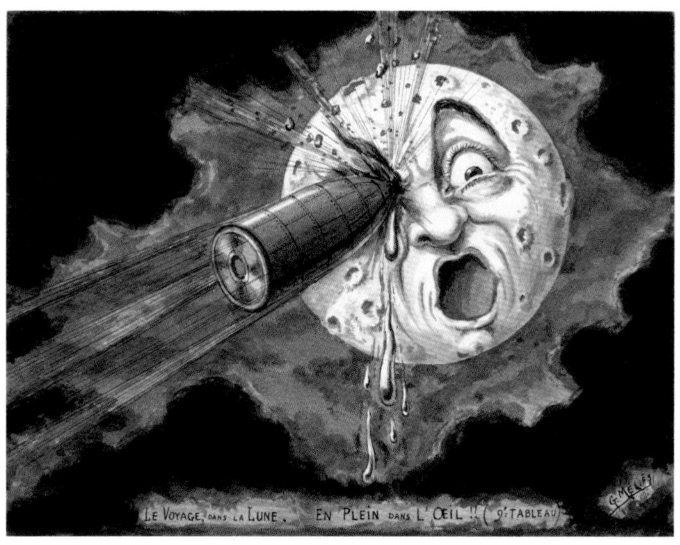

Square in the Eye

48–51 | GEORGES MÉLIÈS | Four preparatory drawings for the film *A Trip to the Moon* (*Le Voyage dans la lune*, 1902), re-created 1930

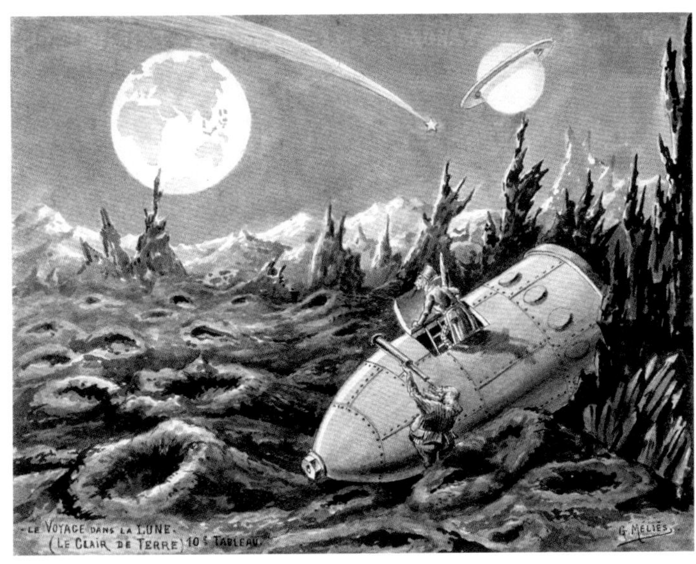

Earthlight

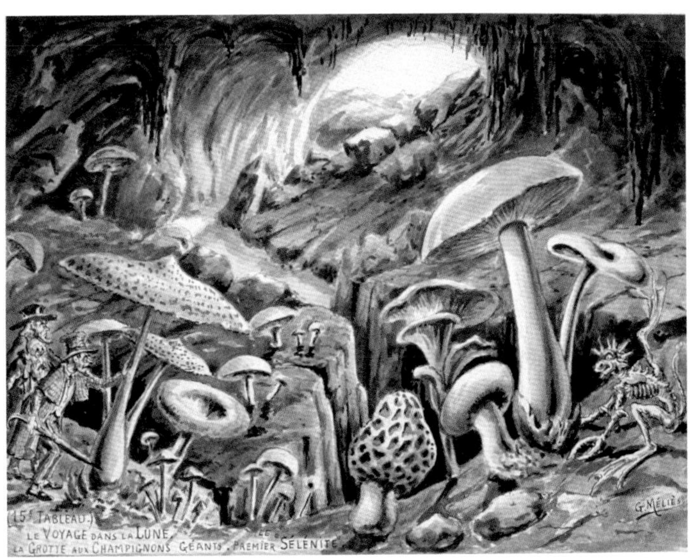

The Grotto of Giant Mushrooms

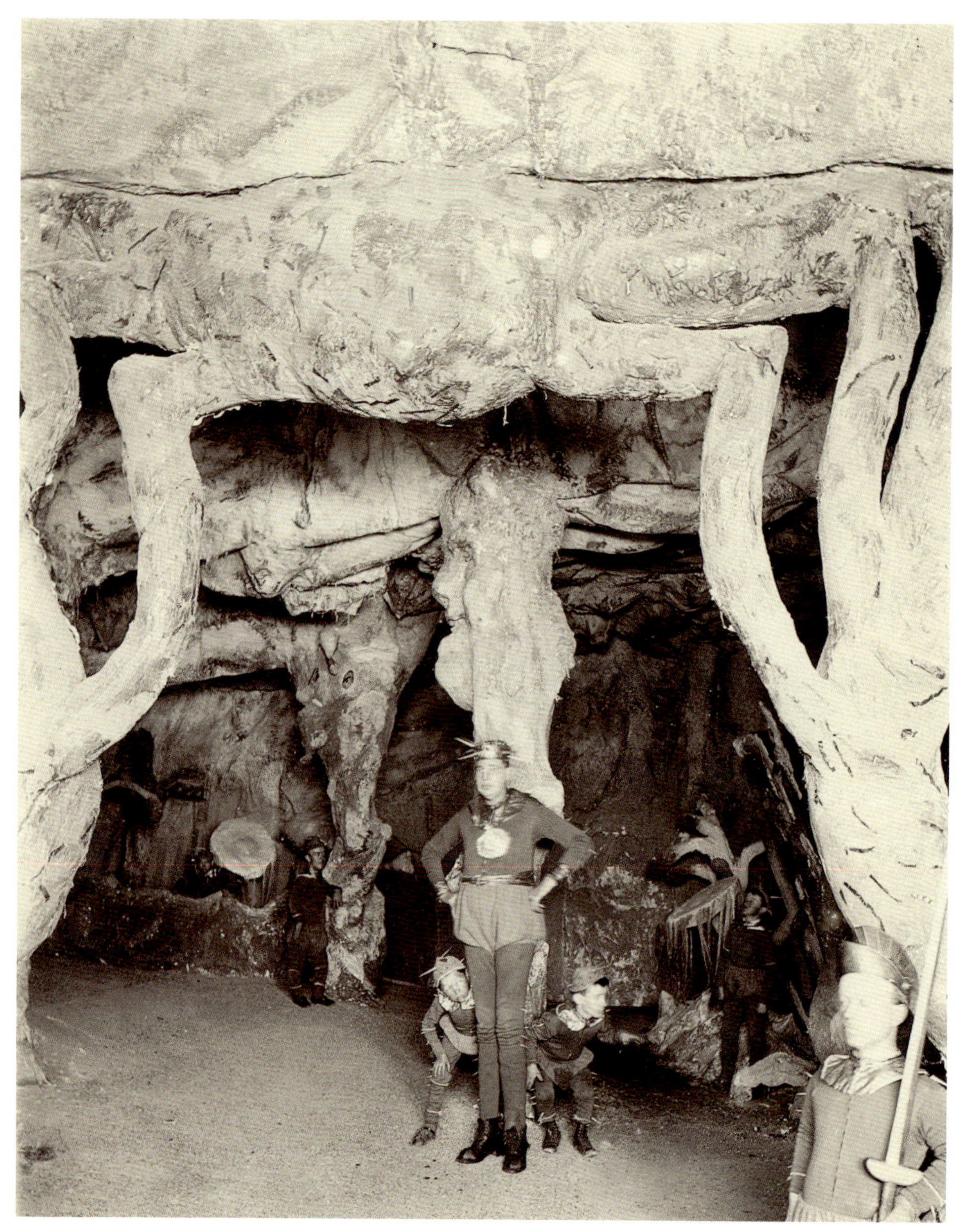

 | BLISS BROTHERS STUDIO | *Gorge, A Trip to the Moon, Pan-American Exposition, Buffalo, New York*, 1901

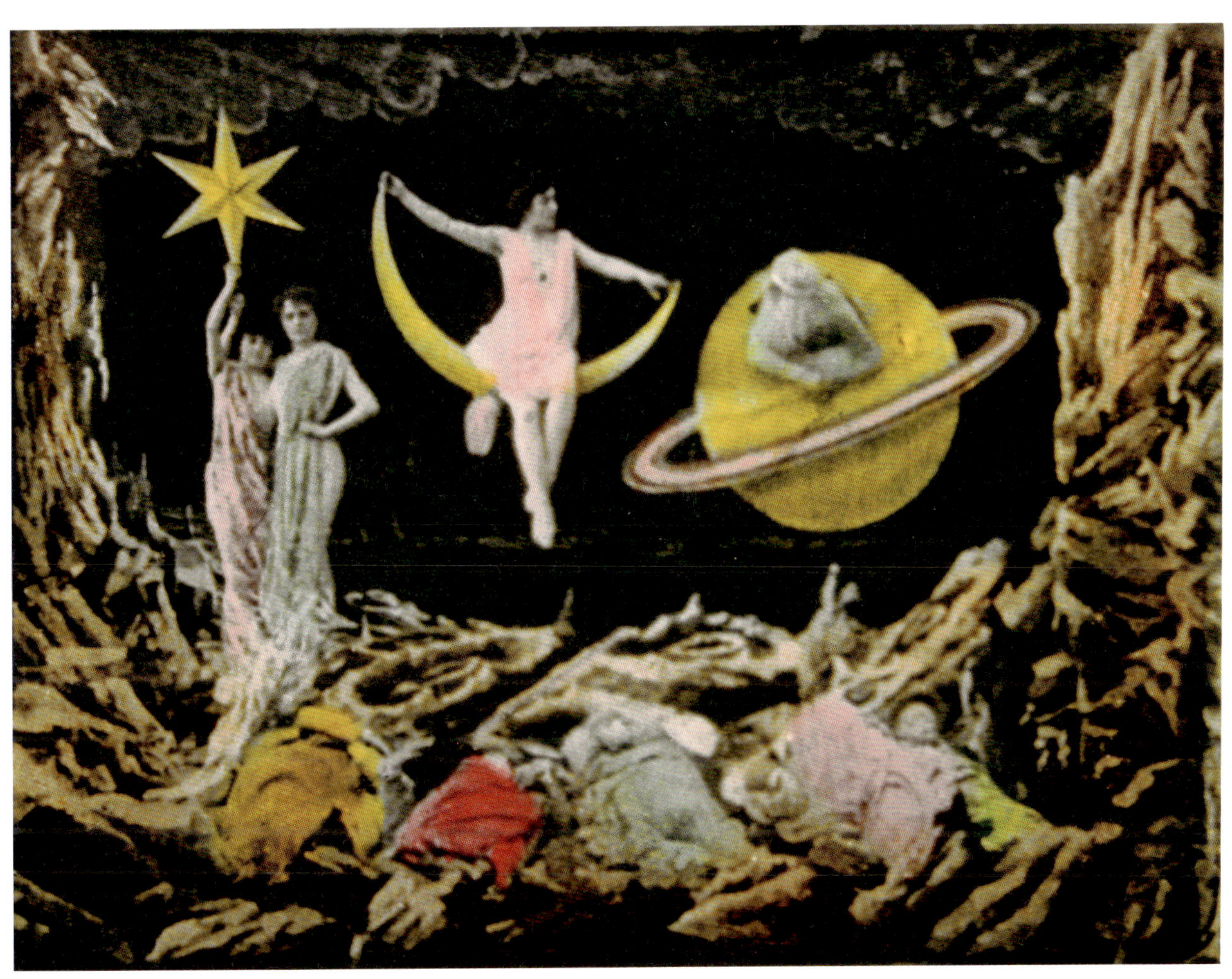

53 | GEORGES MÉLIÈS | Still from *A Trip to the Moon* (*Le Voyage dans la lune*), 1902

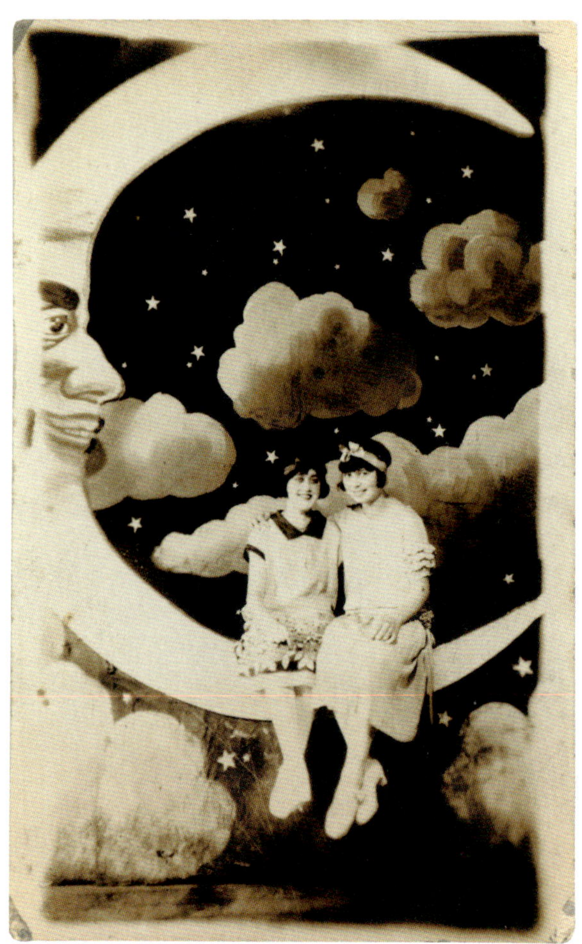

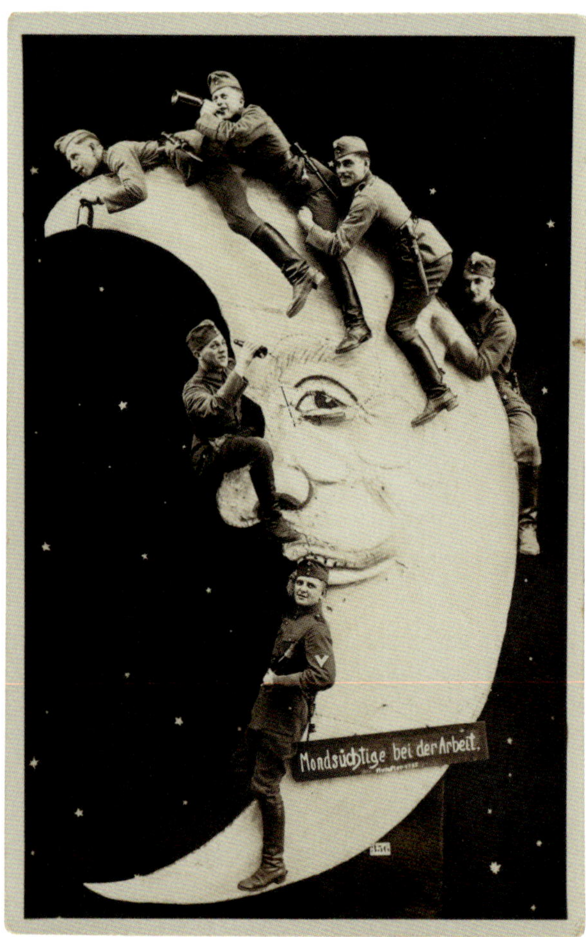

 | *“Man in the Moon” Postcards*, 1900s–1940s (this spread and next)

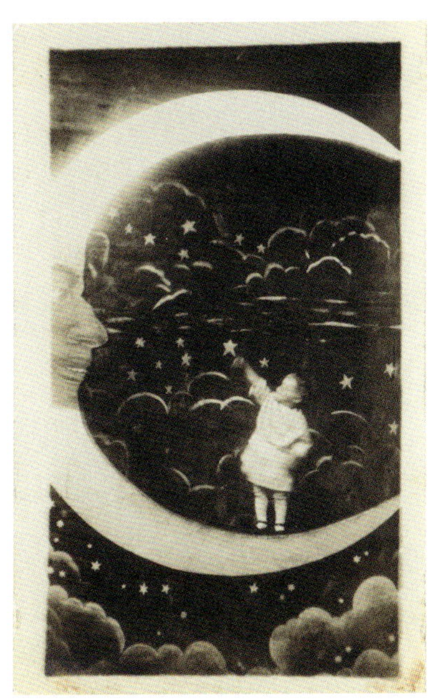
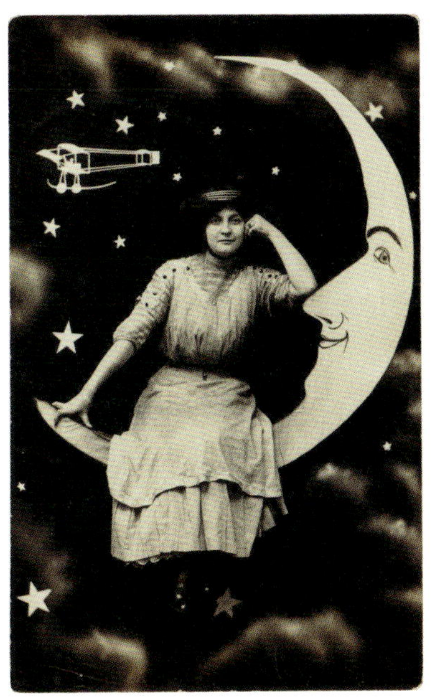
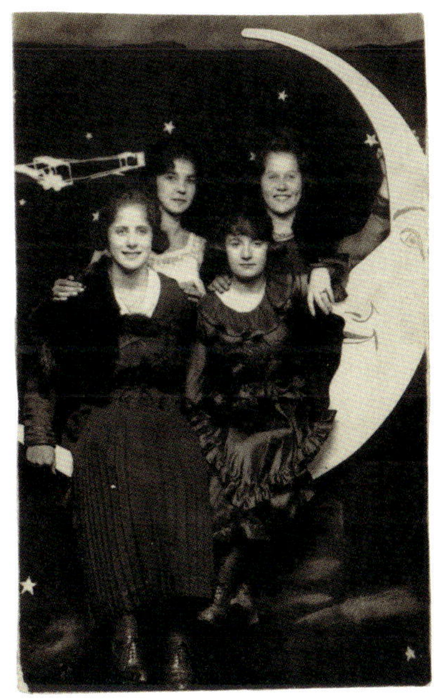
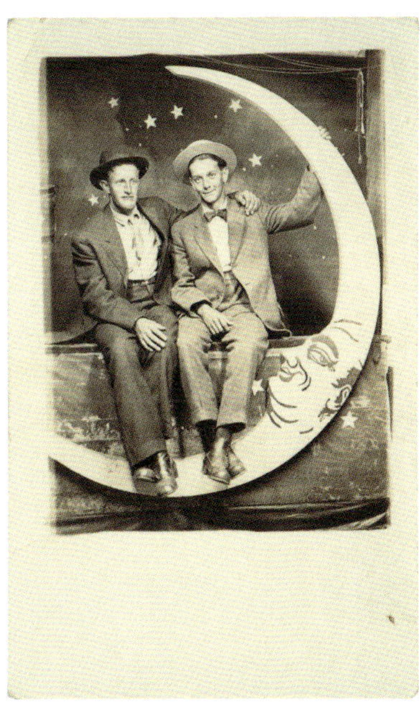

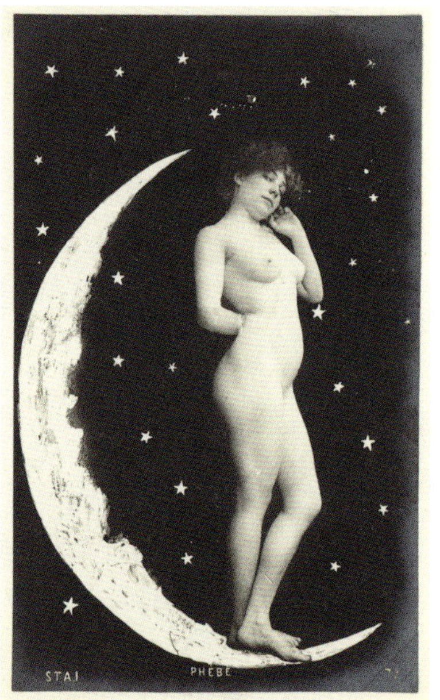

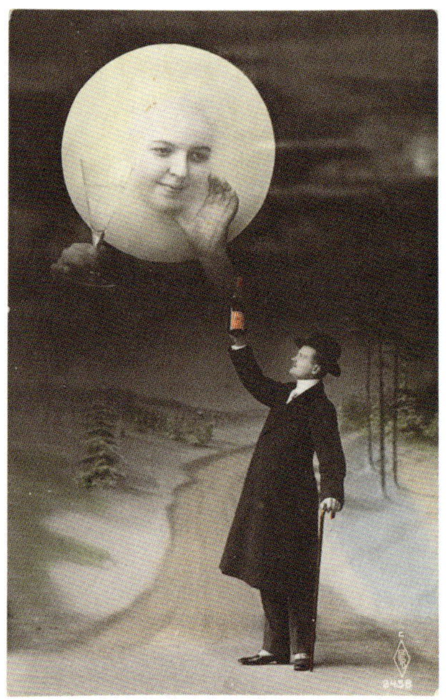

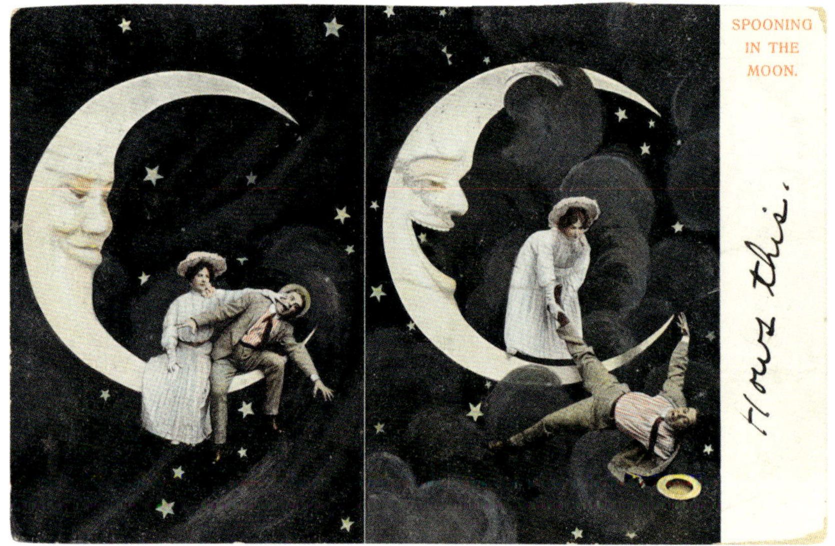

55 (upper right) | *Drinking with the Moon*, 1910s

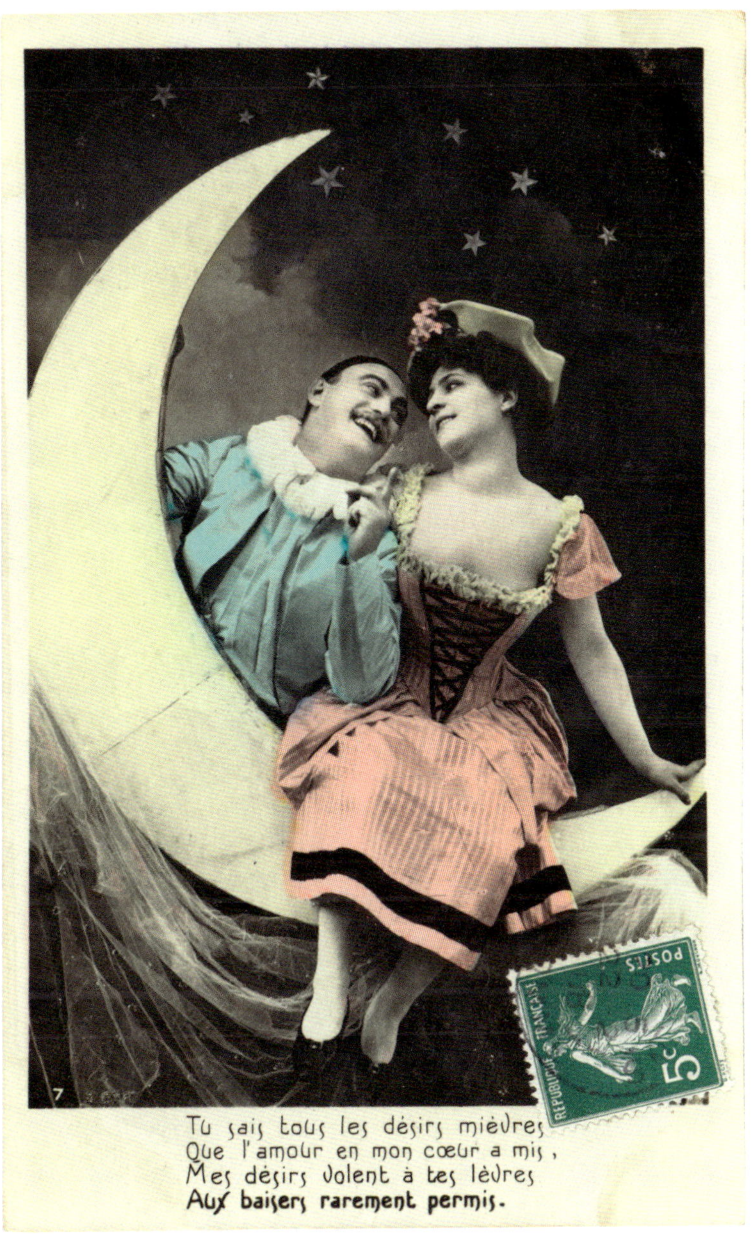
Tu sais tous les désirs mièvres
Que l'amour en mon cœur a mis,
Mes désirs volent à tes lèvres
Aux baisers rarement permis.

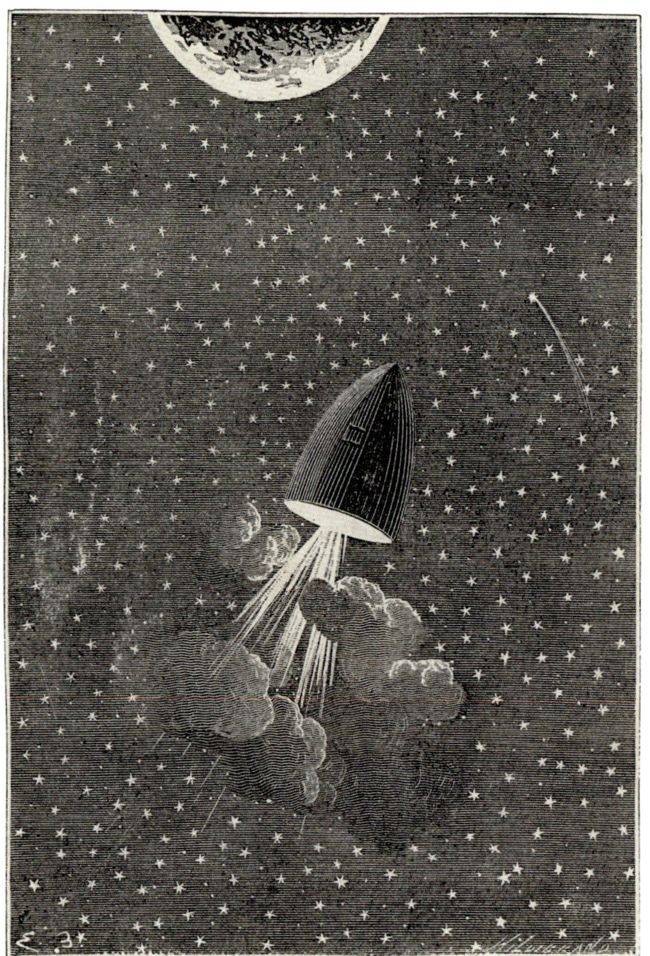

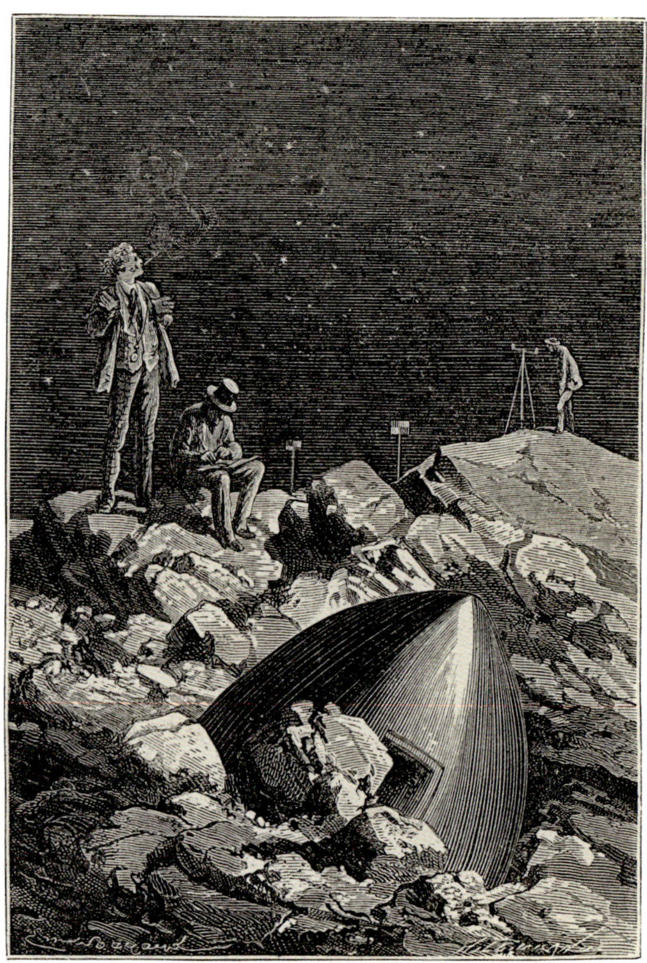

56 | EMILE-ANTOINE BAYARD AND ALPHONSE-MARIE-ADOLPHE DE NEUVILLE |

Illustrations in *From the Earth to the Moon followed by Around the Moon*

(*De la terre à la lune suivi de Autour de la lune*), by Jules Verne, 1872

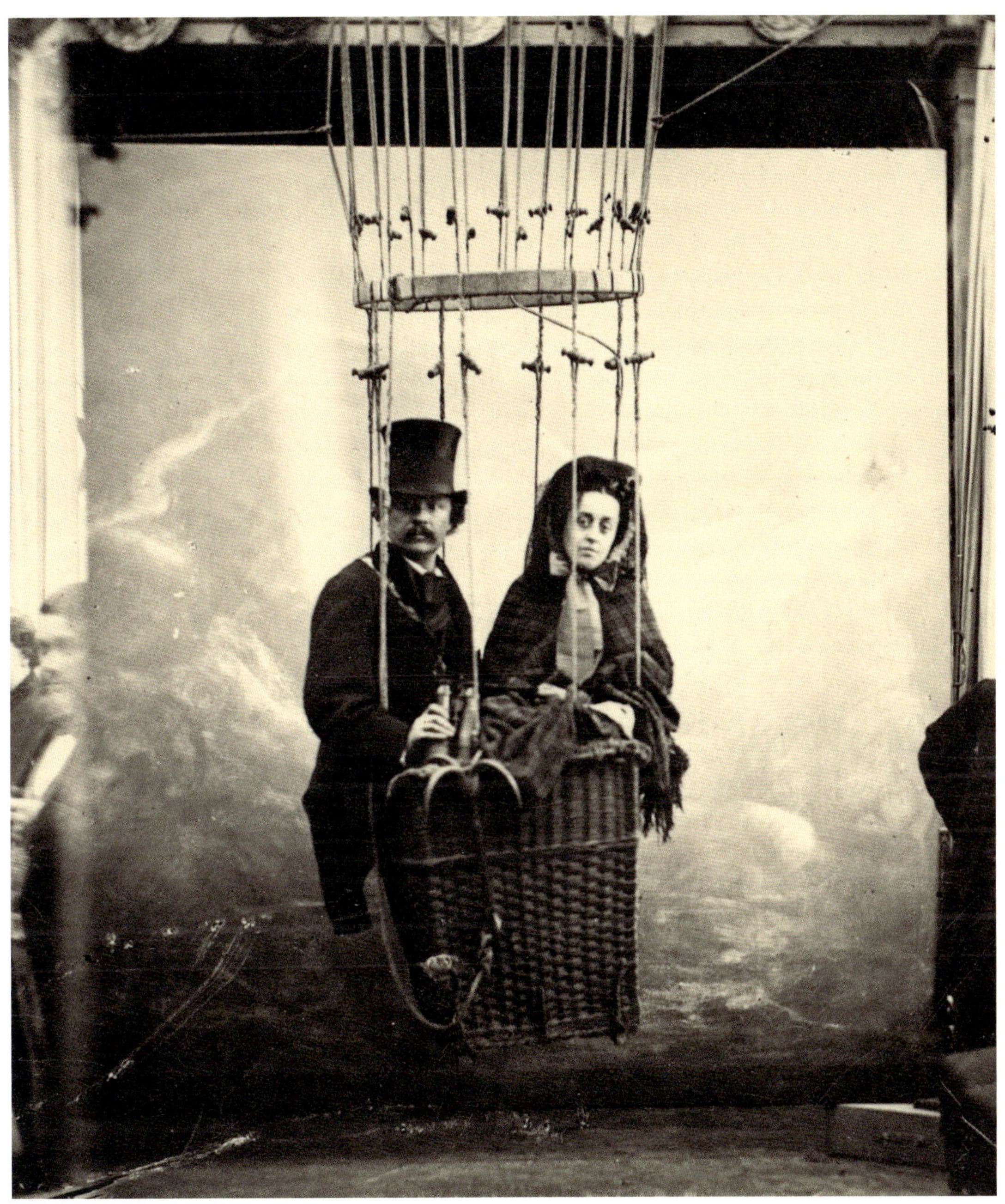

57 | NADAR | *Nadar with His Wife, Ernestine, in a Balloon*, ca. 1865

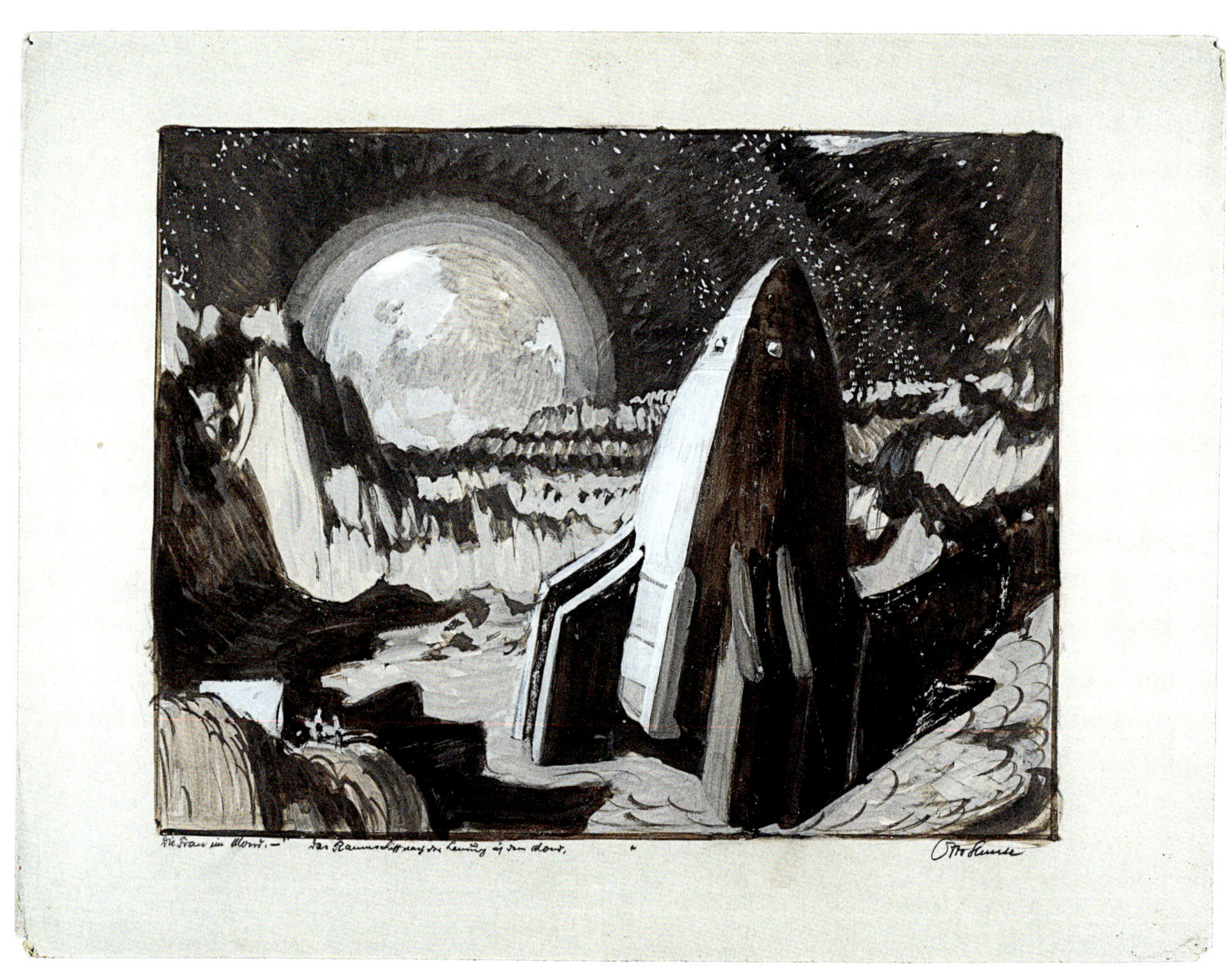

 | OTTO HUNTE | Production sketch for the film *Woman in the Moon* (*Frau im Mond*), 1929

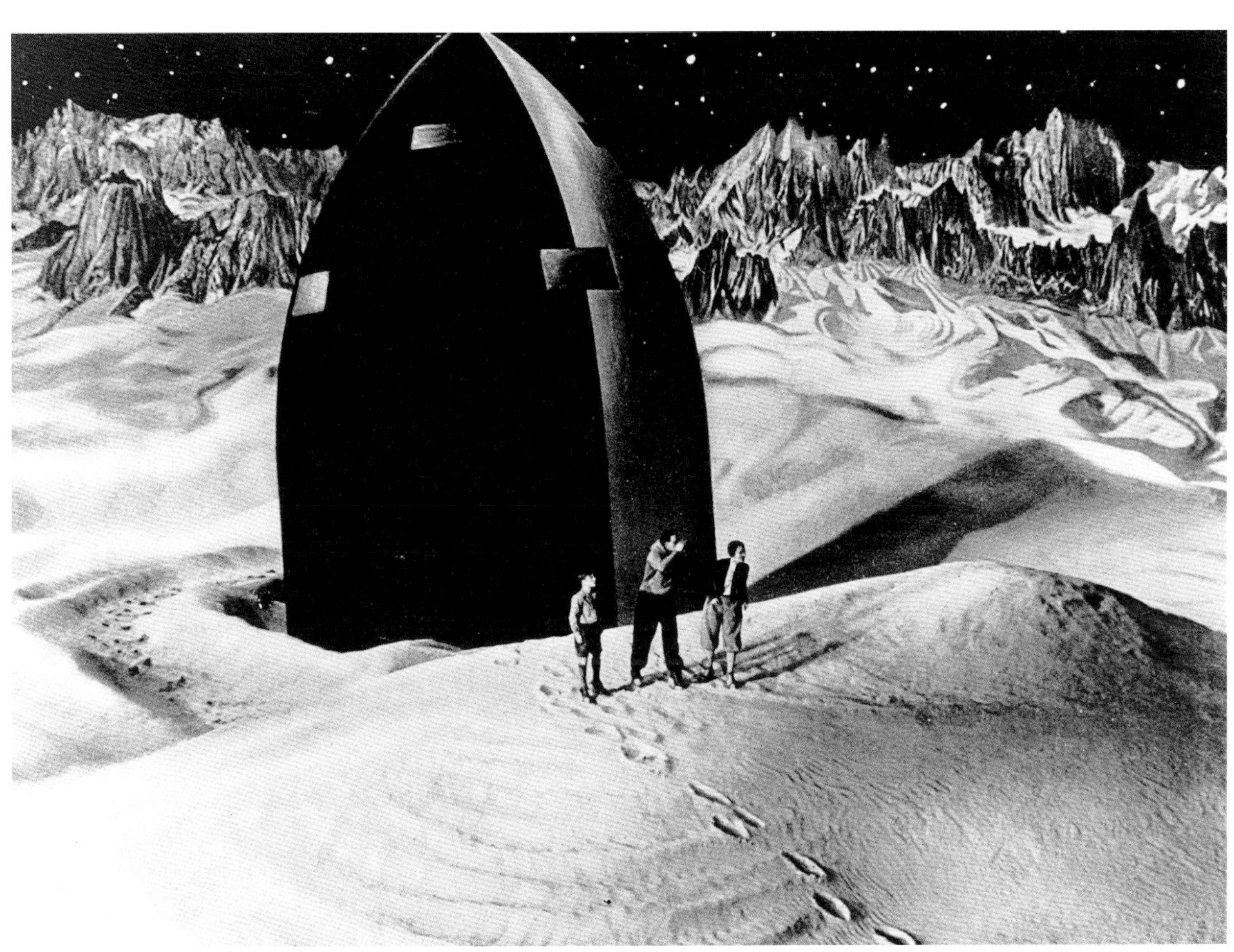

59 | Still from *Woman in the Moon* (*Frau im Mond*), 1929

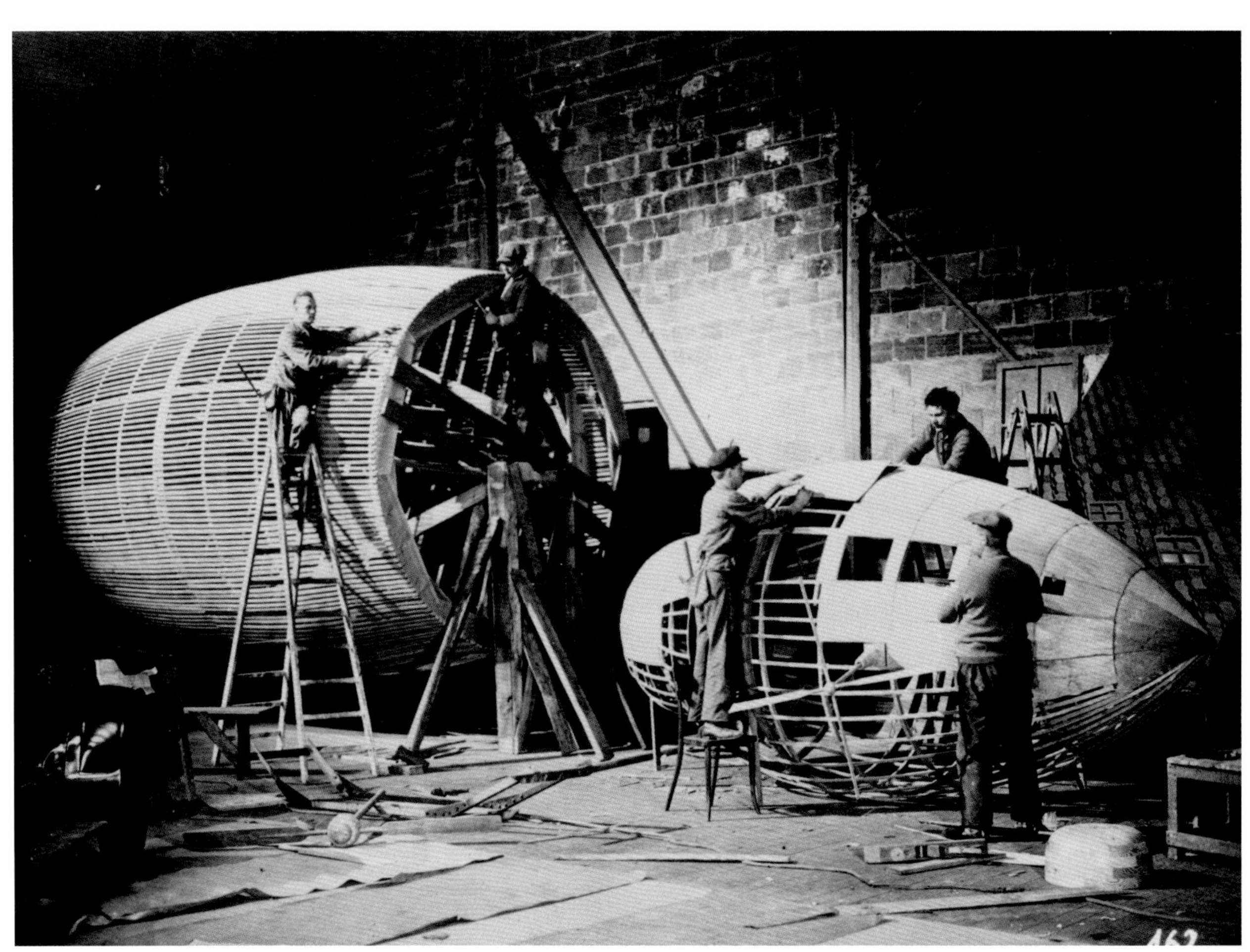

60 | *Constructing the Model for Germany's "Moon Rocket,"* 1929

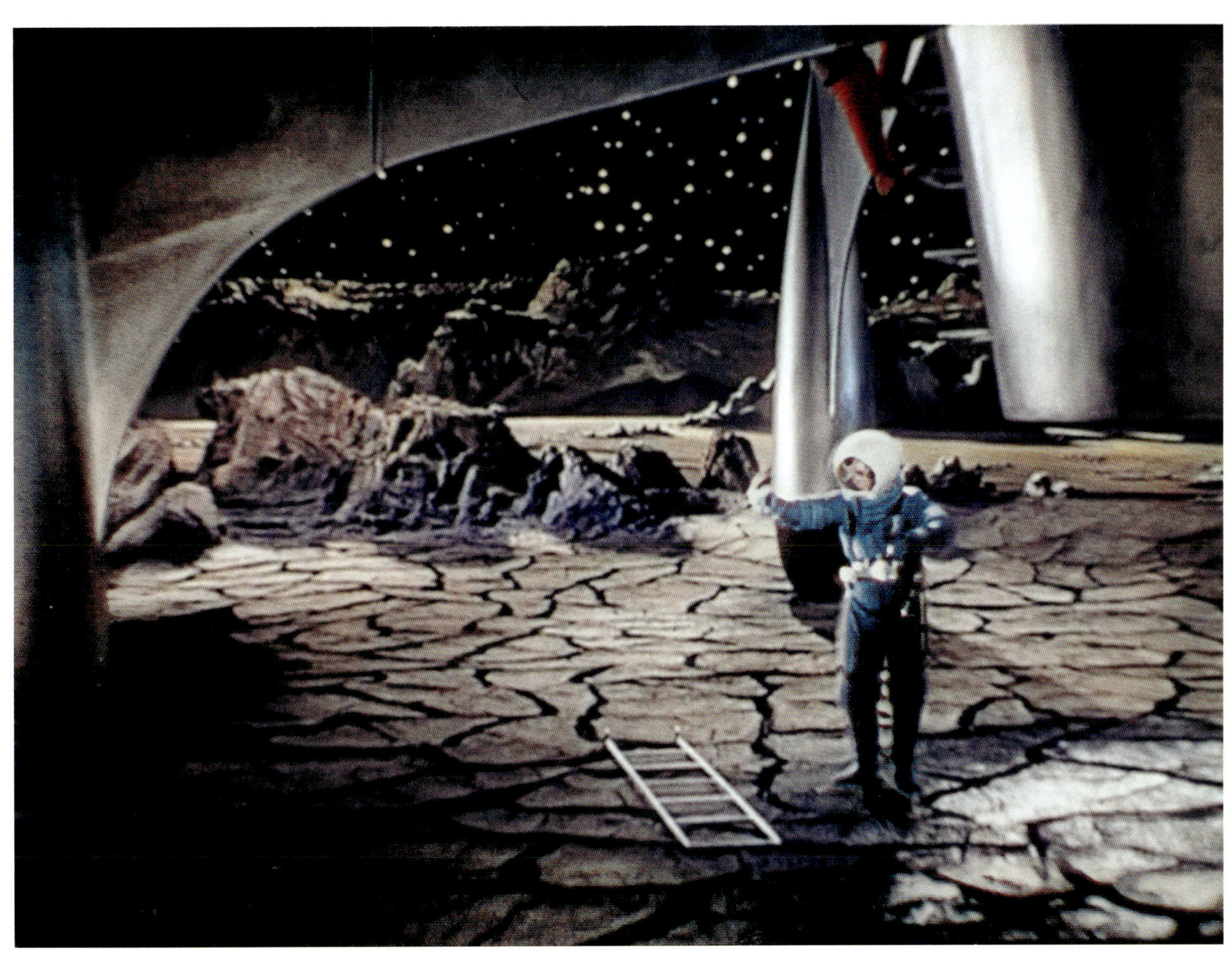

61 | Still from *Destination Moon*, 1950

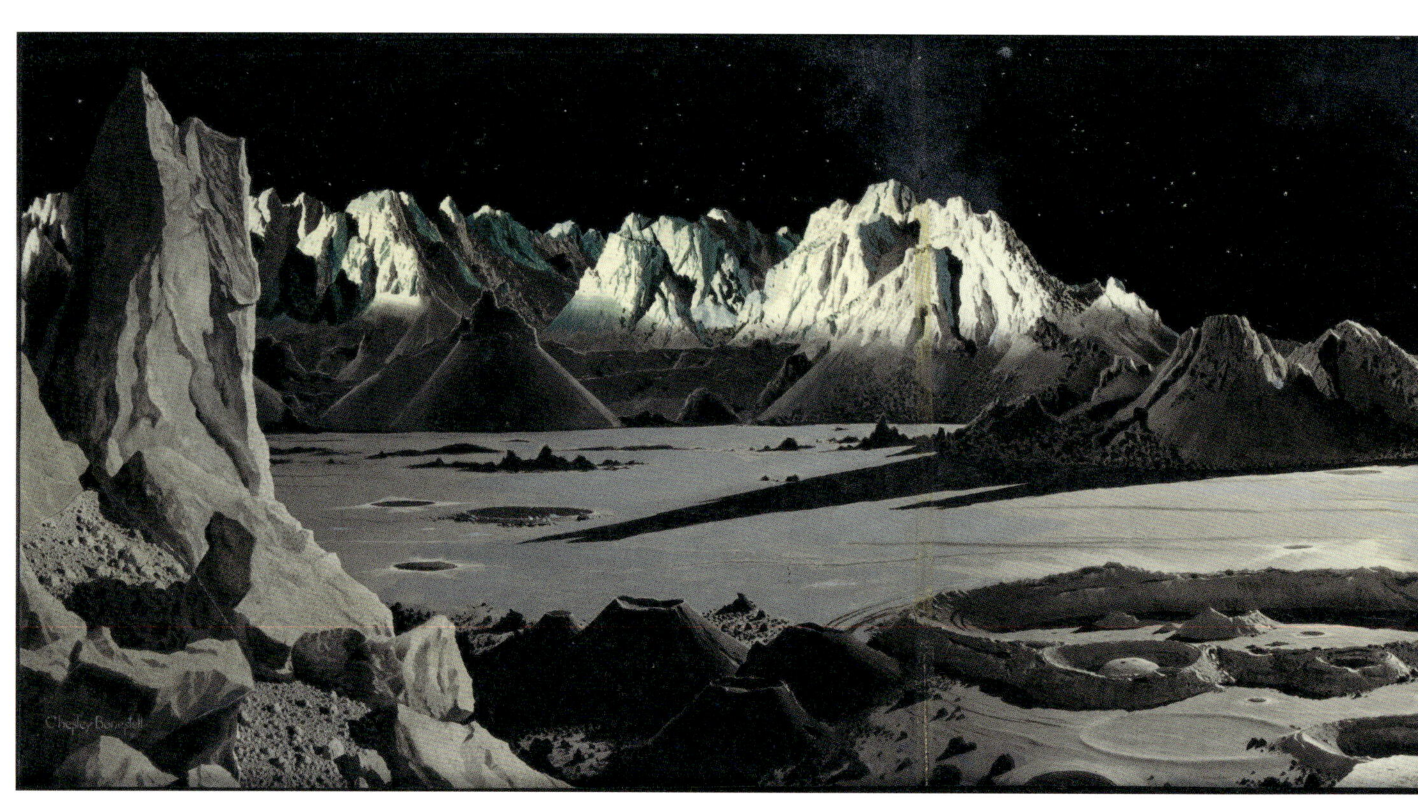

 Study for *A Lunar Landscape*, 1957

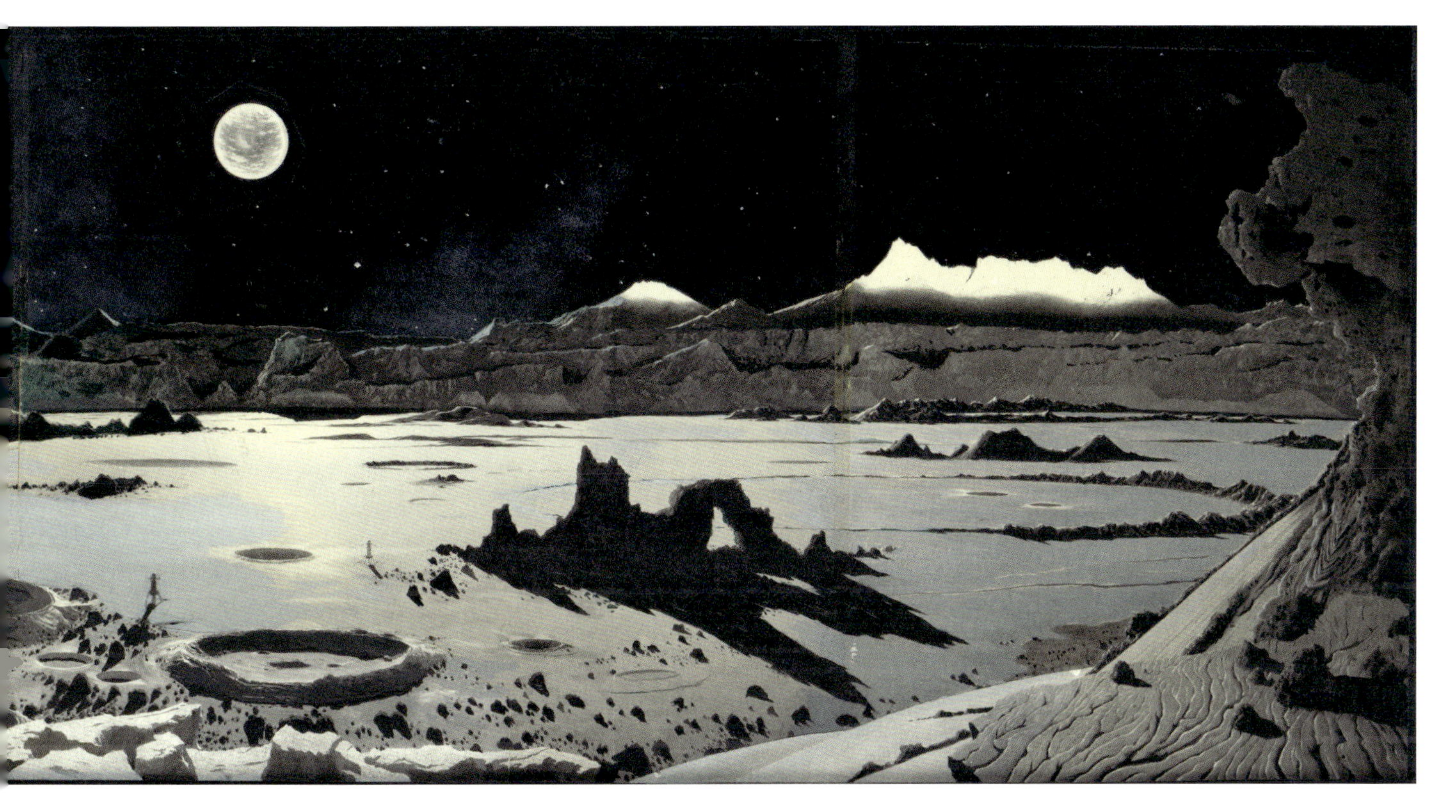

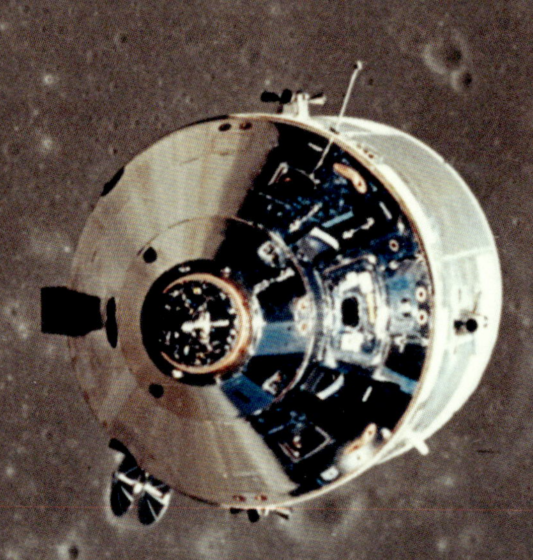

MOONSHOT

MIA FINEMAN

On May 25, 1961, President John F. Kennedy stood before a special joint session of Congress and issued the challenge that set the United States on a course to the moon: "I believe that this nation should commit itself to achieving the goal, before this decade is out, of landing a man on the moon and returning him safely to the earth. No single space project . . . will be more exciting, or more impressive to mankind . . . and none will be so difficult or expensive to accomplish."[1] The decision to send an American astronaut to the moon was a strategic one — a response to the Soviet Union's early triumphs in the Cold War competition for military dominance. Beginning in 1957 with the Soviets' surprise launch of the earth's first artificial satellite, Sputnik 1, space exploration had become a public arena in which rival world powers battled for political prestige and technological supremacy.

Until the mid-1960s, the Soviet Union appeared to be winning the space race. Under the direction of an official identified only as the Chief Designer — later revealed to be the pioneering rocket engineer Sergei Korolev — the Soviet space program achieved a string of headline-making "firsts." They were first to send into orbit an earthborn creature (the canine cosmonaut Laika in 1957), a man (Yuri Gagarin in 1961), and a woman (Valentina Tereshkova in 1963; pls. 63–65). In September 1959, nearly two years before Kennedy's proclamation, the Soviets' Luna 2 probe had become the first human-made object to hit the moon. One month later, the USSR launched Luna 3, a small, cylindrical spacecraft equipped with a dual-lens 35 mm camera and an onboard image-processing system and scanner. Its mission objective was to capture and transmit the first photographs of the far side of the moon.

Because the moon is tidally locked to our planet—meaning it rotates on its axis at the same rate that it orbits around the earth—the same side of the moon is always turned toward us. As the Soviet Luna 3 probe flew past the moon in a long elliptical orbit, its camera made twenty photographs, capturing roughly 70 percent of the uncharted lunar far side. Although the images it beamed back were noisy and indistinct, with a resolution close to that of Galileo's early telescopes, they revealed a terrain starkly different from the near side: rugged and densely cratered, with only two visible maria (respectively dubbed the Sea of Moscow and the Sea of Dreams). The single photograph released by the Soviet wire service TASS (Telegraph Agency of the Soviet Union) in late October 1959 and published around the globe represents a groundbreaking moment in the history of visual culture (pl. 67). "Once again, one of the secrets of the universe has been revealed," trumpeted *Ogonyok*, the Soviet Union's leading weekly picture magazine, in a front-page headline.[2] Blurry, heavily retouched, and scrawled with annotations, the photograph resembles nothing so much as an early fetal sonogram—a fitting image for humankind's very first glimpse of the moon's long-hidden face.

Aside from that trailblazing picture, however, most of the widely published photographic imagery associated with space-age lunar exploration was captured by U.S. missions. The reasons for this are rooted in the ideological divide of the Cold War itself. Unlike NASA (National Aeronautics and Space Administration), which was created in 1958 as a civilian agency with a mandated policy of openness, the Soviet space program was part of a highly secretive military-industrial complex. The latter's achievements were couched within a deeply ingrained culture of censorship intended to protect military secrets while projecting a heroic image of technological supremacy. Soviet space missions were rarely announced in advance. If a rocket launch failed, it was quietly forgotten; only successes were publicized. "Secrecy was necessary so that no one would overtake us," notes Russian science journalist Yaroslav Golovanov. "But later when they did overtake us, we maintained secrecy so that no one knew that we had been overtaken."[3]

Both the Soviets and NASA recognized early on that photography was an indispensable tool for rallying public support for their respective space programs. For this purpose, images of Soviet cosmonauts and American astronauts proved to be far more effective than photographs of the moon itself. Throughout the 1960s, *Ogonyok* ran hundreds of human-interest stories on cosmonauts, featuring photographs—often colorized or heavily retouched—from training programs, public appearances, and the space explorers' own family albums.[4] A personal album assembled by Vasily Baturin, a cinematographer hired to document the cosmonauts' training, offers an intimate view of the self-congratulatory visual culture surrounding the Soviet space program (pls. 63–66). NASA put forward a similarly heroic image of the "all-American flyboys" who piloted its crewed missions.[5] In 1959 the agency struck a controversial deal with *Life* magazine, granting exclusive access to the personal stories of the astronauts and their families in exchange for approval prior to publication and generous payments to the subjects. The coverage began in August 1960 with an eighteen-page spread featuring color photographs by Ralph Morse of the original seven Project Mercury crew members in training (fig. 12); the next week's issue ran a cover story on the astronauts' wives.

About a year after President Kennedy's "moonshot" speech, *Life* continued its coverage of the space race with a lavishly illustrated feature on the preparations for Project Apollo, heralded with the rousing cover line, "MAN'S JOURNEY TO THE MOON: Preview of the Greatest Adventure of All Time." The articles offer a heady mix of science and speculation about space travel, from the hazards of weightlessness to the need for "portable distilling plants" to recycle urine into drinking water.[6] In contrast

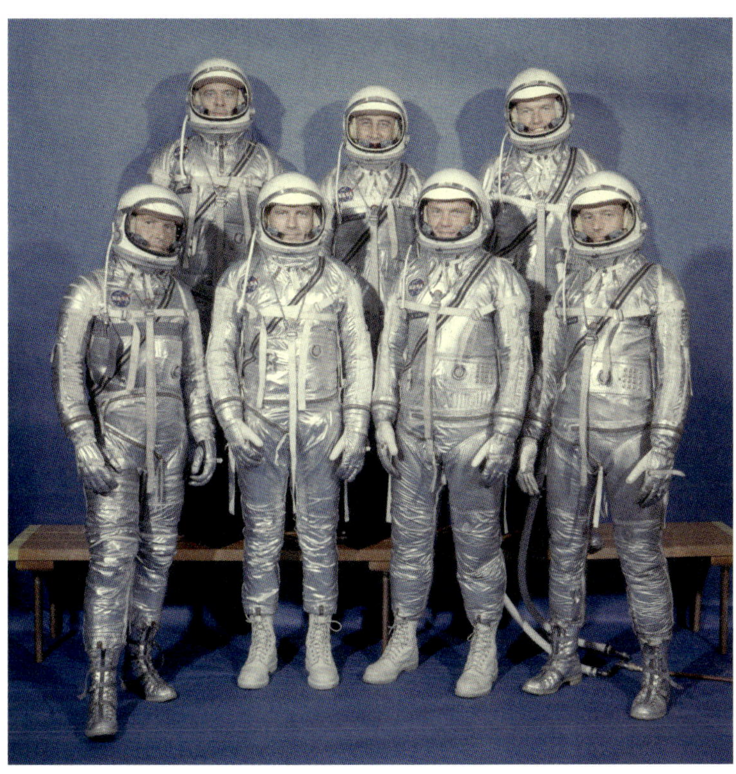

FIG. 12. Ralph Morse (American, 1917–2014), *Project Mercury Astronauts at Langley Air Force Base, Virginia*, 1960. Published in *Life* (August 1, 1960), p. 37

to the Soviets, who suppressed nearly all images of their aerospace hardware, NASA and its corporate partners allowed the magazine to illustrate prototypes of rockets, lunar modules, moon mobiles, and other space technology. The color photograph on the cover shows a man encased in a barrel-like space suit, an image straight out of a 1950s science-fiction adventure film. Designed by Allyn B. Hazard, a development engineer at Space-General, a subsidiary of Aerojet-General Corporation, the two-hundred-pound prototype carried within it food, water, a radio, a tiny stove, and an air-conditioning system to protect the wearer from extreme temperature fluctuations. According to the 1962 edition of the Aerojet newsletter, Hazard designed the space suit to integrate with a tractor-like moon mobile that would transport future astronauts on a ten-day, 500-mile expedition across the lunar surface (pls. 68, 69).[7] While it never made it to the moon, the Lunar Exploration Space Suit Mark 1 became a popular sensation and, in 1966, the costume of Major Matt Mason, an astronaut action figure introduced by Mattel.

Aside from garnering support for space exploration through public relations, photography played an important practical role in NASA's primary objective of sending a man to the moon. In the mid-1960s NASA organized three crewless lunar reconnaissance programs equipped with different video and photographic systems to survey the terrain and collect information about potential landing sites. Together the Ranger, Surveyor, and Lunar Orbiter programs captured and relayed back to earth nearly one hundred thousand photographs in preparation for the Apollo missions. The Ranger program, nicknamed "Shoot and Hope," consisted of a series of hard-landing probes carrying television cameras programmed to record and beam back still images of the lunar surface during the final moments of the flight before crashing into the moon.[8] Of the nine Rangers launched between 1961 and 1965, only the last three succeeded, but the images they returned were one thousand times more detailed than those obtained with the best earthbound telescopes. Their sharpness was due not simply to the close shooting distance but also to the fact that there was no earthly atmosphere to interfere with the camera's sights.

To lead the scientific analysis of the Ranger photographs, NASA appointed Gerard Kuiper, a distinguished astronomer who had recently founded the Lunar and Planetary Laboratory at the University of Arizona in Tucson. In 1964 Kuiper invited Ralph Turner, a young artist teaching sculpture at the university, to translate the NASA images into three-dimensional scale models, similar to those created by James Nasmyth in the 1870s (see pp. 24–25). Working with a group of about seventy photographs taken by Ranger 9 in March 1965, Turner sculpted a high relief

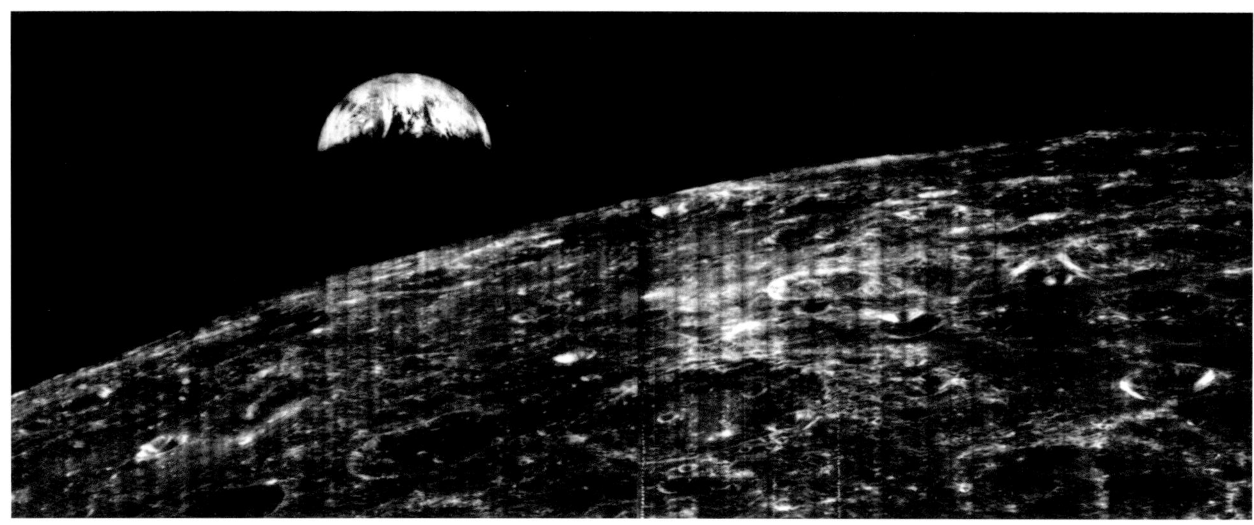

FIG. 13. NASA (Lunar Orbiter 1), *Original Earthrise*, 1966

of the central peak of Alphonsus, a large crater near the center of the moon's earth-facing side (pl. 70). To determine the elevations, he used a lighting system that simulated the angle of the sun on the lunar surface, building up the Plasticine model until it matched the photographs; the finished models were cast in epoxy. Like Nasmyth, Kuiper and Turner forged a productive partnership between science and art, translating visual into tactile understanding. "My eyes have searched long over the soft rilles, broken plains and steep rims of craters to recognize the real shapes on this satellite," Turner wrote in 1972. "My fingers have roamed over them in an attempt to reconstruct countless features. In the piles of photographs . . . the silent Moon waits to be appreciated."[9]

More technically advanced than the hard-landing Rangers were the soft-landing spacecraft of the Surveyor program, launched from 1966 to 1968. These probes were tasked with investigating the structure and topography of the lunar surface to understand the risks and challenges of the terrain for crewed landings. Four months after the Soviets' Luna 9 made the first semisoft landing on the moon, Surveyor 1 touched down gently on the lunar surface and began transmitting still images, which were broadcast live on some television networks. Each Surveyor was equipped with a slow-scan video camera topped with a mirror that could rotate 360 degrees to capture sweeping panoramas. Back on earth, individual video frames were copied onto black-and-white Polaroid film; the image chips were then assembled into mosaics by technicians from the U.S. Geological Survey at Jet Propulsion Laboratory in Pasadena, California. These fish-scale-like composites, with their distinctive elliptical curves—the result of the swiveling camera—present a machine's-eye view of the lunar surface, flattened, fragmented, and perspectivally distorted (pl. 71).

The Lunar Orbiter program of 1966–67 was the final and most successful photographic operation before the crewed Apollo landings. Over the course of one year, five probes systematically photographed 99 percent of the lunar surface. The Orbiters investigated potential landing sites, revealing features as small as one foot in diameter. Rather than video, this program relied on a film-based robotic imaging system designed by the Eastman Kodak Company, similar to that used on the Soviets' Luna and Zond (1964–70) spacecraft. The photographs were captured

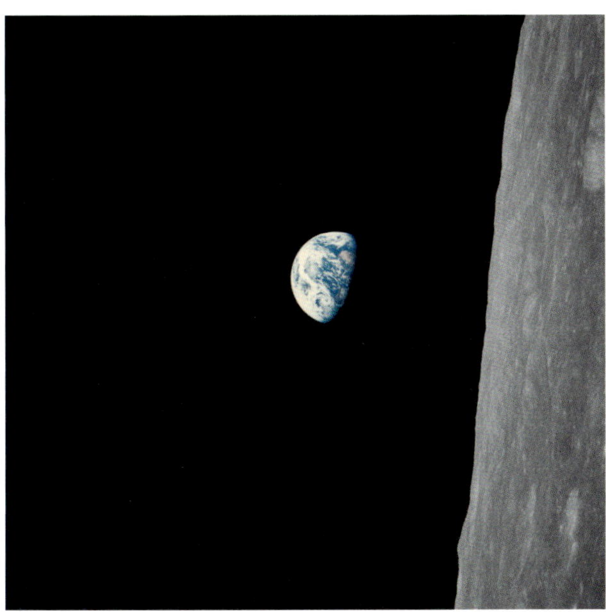

FIG. 14. William Anders (American, b. Hong Kong, 1933; NASA Apollo 8), *Earthrise* (uncropped), 1968

on 70 mm film with a dual-lens camera, chemically processed on board, scanned in strips, and transmitted via radio back to earth. At Eastman Kodak headquarters in Rochester, New York, technicians transferred the scanned strips onto sheets of large-format film from which they printed the final photographs.

The Orbiter probes returned about two thousand images of unprecedented clarity and precision, which together constituted the most detailed and comprehensive photographic map of the moon produced since Maurice Loewy and Pierre Puiseux's 1910 atlas (see pls. 37, 38).[10] Orbiting from 25 to 4,900 miles above the lunar surface, the cameras captured stunning vistas and vertiginous overhead views of craters and valleys, with shadows and details sharply defined by the unfiltered sunlight (pls. 72–77). Among the most celebrated images is an oblique view inside the Copernicus crater, captured by Lunar Orbiter 2 on November 23, 1966 (pl. 72). Billed by NASA as "the picture of the century," the photograph depicts a barren mountainous landscape seen at close range, as in Chesley Bonestell's paintings of the lunar surface from the late 1950s (see pl. 62). When the Copernicus picture was published in *Time* magazine, the caption made explicit the resemblance to the landscape of the American West: "Except for the black sky in the background, the photograph might have been mistaken for a composite of the scenic grandeur of the Grand Canyon and the barren desolation of the Badlands of South Dakota."[11]

Just a few months earlier, on August 23, 1966, Lunar Orbiter 1 had beamed back the first photograph of our planet rising beyond the moon's horizon, an image that required an unscheduled and somewhat risky maneuver executed by mission controllers in Langley, Virginia (fig. 13). Although the black-and-white photograph was widely published at the time—NASA also reproduced it as a poster—it never achieved the cultural impact of a color shot of the same phenomenon made by astronaut William Anders two years later (pl. 78). In monochrome, the earth looked like any other planet. But in color, the lunar earthrise, seen through the eyes of a homesick explorer a quarter of a million miles away, was nothing short of a revelation.

On Christmas Eve 1968, Apollo 8 became the first crewed spacecraft to enter lunar orbit. Mission commander Frank Borman later recalled:

> *I happened to glance out of one of the still-clear windows just at the moment the earth appeared over the lunar horizon. It was the most beautiful, heart-catching sight of my life, one that sent a torrent of nostalgia, of sheer homesickness, surging through me. It was the only thing in space that had any color to it. Everything else was either black or white, but not the earth.*[12]

Anders, who was positioned closest to the window, quickly grabbed the Hasselblad camera and photographed the view, first in black-and-white and then in color. What the astronauts actually saw that day was the planet floating just to the left of the moon's surface, with the north pole at the top (fig. 14). When the image was published, however, it was rotated to a landscape format more familiar to earthbound viewers used to watching the sun rise and set along a horizon line. Since there is no absolute up or down in space, neither orientation is more correct. The final image of our vibrant blue planet, glowing against the barren surface of the moon and surrounded by the black void of space, struck many with the force of an epiphany. The photograph known as *Earthrise* became a touchstone for environmentalism and global justice, a powerful visual symbol of human beings as "riders on the earth together," in the words of poet Archibald MacLeish.[13] The final scene of Stanley Kubrick's film *2001: A Space Odyssey*, released a few months earlier, had featured a dramatic shot of the earth viewed from afar by an embryonic "star child," which enhanced the new photograph's power — science fiction had become fact.[14]

Exactly four years later, on Christmas 1972, NASA released what would become an even more famous photograph of the earth from space, *Blue Marble* (pl. 79). Shot by astronaut Harrison Schmitt from Apollo 17, the last crewed mission to the moon, the crystal-clear image of the planet is believed to be the most widely reproduced photograph in human history. In one of the great ironies of the Apollo era, the photographs that struck public and astronauts alike as the most beautiful, poignant, and awe-inspiring were not of the moon but of our home planet — cosmic selfies shot from a quarter million miles away. "Our Earth was quite colorful, pretty, and delicate compared to the very rough, rugged, beat-up, even boring lunar surface," Anders recalled. "I think it struck everybody that here we'd come 240,000 miles to see the Moon and it was the Earth that was really worth looking at."[15]

From today's perspective, it is difficult to imagine traveling anywhere without a camera. Yet in the early days of the space race, neither the Soviet nor the American program included plans for astronauts to carry photographic equipment. Against engineers' objections that cameras would add extra weight and distract from the primary mission objectives, however, early space travelers insisted on bringing them along as part of their personal baggage. The first to do so was cosmonaut Gherman Titov, who used a 35 mm Konvas motion-picture camera to shoot footage of the earth through the window of the Vostok 2 space capsule in 1961. The following year, when John Glenn piloted the first American orbital flight, he brought with him a modified point-and-shoot that he had picked up at a drugstore in Cocoa Beach, Florida.[16] After consulting with photojournalists at *Life* and *National Geographic*, Walter Schirra, the next astronaut to fly, purchased a Hasselblad 70 mm camera, which he used to record the view of our planet through the window of the Mercury 8 space capsule in October 1962.

Thanks to its superb mechanics and sharp Zeiss lens, the square-format 70 mm Hasselblad, outfitted with an external viewfinder and oversize controls, became NASA's camera of choice for still photography in space (fig. 15a). The interchangeable film magazines were loaded with extra-thin film manufactured for NASA by Eastman Kodak, which increased the capacity from twelve to nearly two hundred frames on each roll (fig. 15b). From Apollo 11 onward, cameras used by astronauts on the lunar surface were painted silver to prevent them from overheating in the unfiltered sunlight, and each was fitted with a reseau plate, a register glass that superimposed a grid on the focal plane to aid in calibration of the images. In the early 1960s NASA brought in Richard Underwood, an aerial photography specialist, to train astronauts in camera technique and image interpretation. With cameras firmly bracketed to the chests of their space suits to prevent them from floating away in low gravity, crew members practiced operating the controls with gloved hands and framing their pictures without looking through a viewfinder. Astronaut Michael Collins recalls, "They would loan us a Hasselblad and we would take it home weekends, take pictures of the kids, fly around in our airplanes, shoot pictures out the window, and become familiar enough with it so it was second nature."[17]

When Edward White became the first American to perform a space walk in 1965, NASA viewed the event as a major

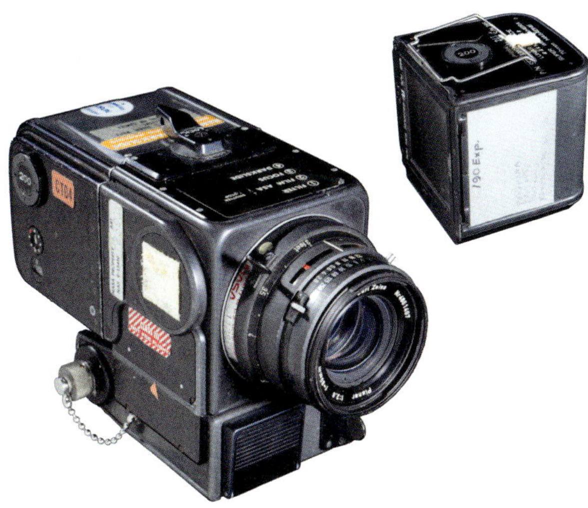

FIG. 15A, B. Hasselblad 70 mm camera used for astronaut training and film magazine used on Apollo 11, 1969

photo opportunity. With a Hasselblad aimed through the window of the Gemini 4 spacecraft, James McDivitt captured his copilot floating in the vacuum of space, with darkness above and our vibrant blue planet below him (pl. 80). "We were joking with some of the NASA management before the flight," McDivitt later recalled, "and the consensus was that if I didn't take some good pictures I would get killed."[18] Fortunately for McDivitt, his stunning images ran as a cover story in the June 18, 1965, issue of *Life*, and they helped cement NASA's commitment to astronaut photography as a powerful tool for promoting the human spaceflight program.

By the time of the first crewed Apollo flights, in 1968, handheld photography had become an established part of mission protocol. When Apollo 11 lifted off at Cape Kennedy on the morning of July 16, 1969, its cargo included two 16 mm movie cameras, two television cameras (one color, one black-and-white), a Kodak stereo camera for recording close-ups of lunar rocks and soil, and two 70 mm Hasselblad EL (electric-drive) cameras. The astronauts' use of this equipment on the moon was tightly choreographed, guided by a timeline that prescribed exactly when and where photographs were to be made. During their two-hour-and-forty-minute extravehicular activity (EVA), or time spent outside the spacecraft, Neil Armstrong and Edwin "Buzz" Aldrin set up a television camera, which transmitted live video of their moonwalk to a global audience of about half a billion people, many of whom made snapshots of their home screens to commemorate the event (pl. 82).[19] With the still cameras, the astronauts documented the condition of the lunar module, captured panoramas of the lunar terrain, and methodically recorded their scientific experiments and the sites of the rock samples they collected (pls. 83–87). Although the emphasis was on operational and scientific reporting, even the most straightforward images conveyed the magnitude of the event and evoked a sense of wonder in the general public. Aldrin shot his now-iconic photograph of his boot print as part of the mission's investigation of the properties of lunar soil, as is evident in his careful before-and-after documentation (fig. 16a–c). Widely reproduced in the popular press, the image came to symbolize humankind's first triumphant steps into the cosmos. In addition to providing scientific records, the men also took time to make a few "standard home pictures for the folks back on Earth," as Aldrin put it, including Armstrong's photograph of his colleague facing the American flag (pl. 86).[20]

Within NASA there was still considerable ambivalence about the importance of "public affairs" photography, especially on crewed missions. "Part of the 'right stuff' image was not to care about that sort of thing," recalled Brian Duff, who served as NASA's director of public affairs in the late 1960s.

> *It may surprise you to know that it was necessary to demand that color film be carried to the Moon on* Apollo 11. *It was not considered "scientifically accurate." . . . Only when it was asked whether NASA could accept a black and white photograph of the first man on the Moon on the cover of* LIFE *was it agreed to include a quantity of color film.*[21]

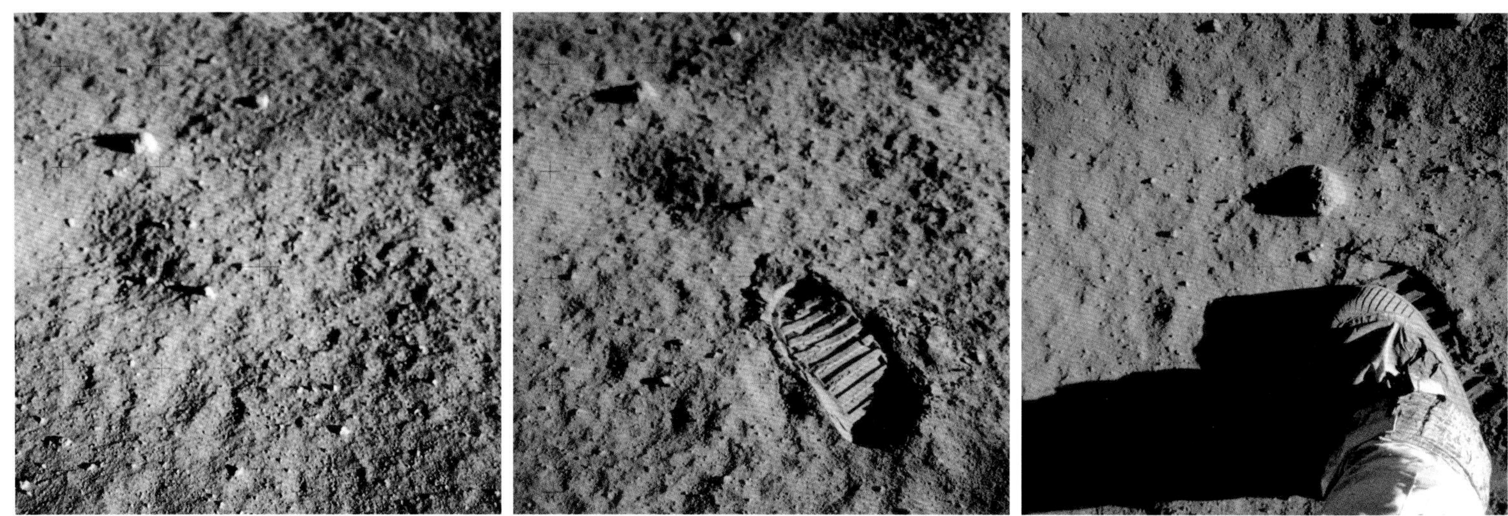

FIG. 16A–C. Edwin "Buzz" Aldrin (American, b. 1930; NASA Apollo 11), sequence of photographs documenting Aldrin's footprints on the surface of the moon, 1969

As it turned out, when the film was processed and analyzed upon Apollo 11's return, NASA made an embarrassing discovery: there was no high-quality color photograph of Armstrong on the moon. Duff attributed the failure to "a series of simple human oversights."[22] During most of the EVA, the Hasselblad was bracketed to Armstrong's chest; swapping the camera while wearing bulky space suits in low gravity was a difficult maneuver. More importantly, Duff notes, "Those kind of 'touristy' snap shots were not programmed into the timeline. They were supposed to be taken as a matter of course in among the 'scientific' or 'documentary' photography."[23] Underwood, the NASA photography specialist, recalls the Public Affairs office asking, "Why don't we say this picture by the flag is Armstrong? How do you know? You can't see his face or anything." Underwood advised against it: "Some nine-year-old . . . space groupie" was bound to spot it and identify the figure as Aldrin.[24]

Although Armstrong never had the opportunity to pose for a heroic portrait on the lunar surface, his legacy as the first photographer on the moon is borne out by the hundreds of remarkable images he captured and brought back to earth. Among them is his iconic shot of Aldrin standing next to the gold-foil-covered strut of the *Eagle*—a visual emblem of the Apollo 11 mission that has since become one of the most famous photographs of the twentieth century (pl. 85). In a Twitter post marking the forty-eighth anniversary of the moon landing, Aldrin recalled that the making of the image was spontaneous: "He said 'stop right there' & I turned. You can see the motion of the strap."[25] Looking closely, one can also see reflected in the mirrored visor of Aldrin's helmet a tiny spectral image of Armstrong in his gleaming white space suit, Hasselblad strapped to his chest—the first human being to set foot on the moon, seen through a glass, darkly.

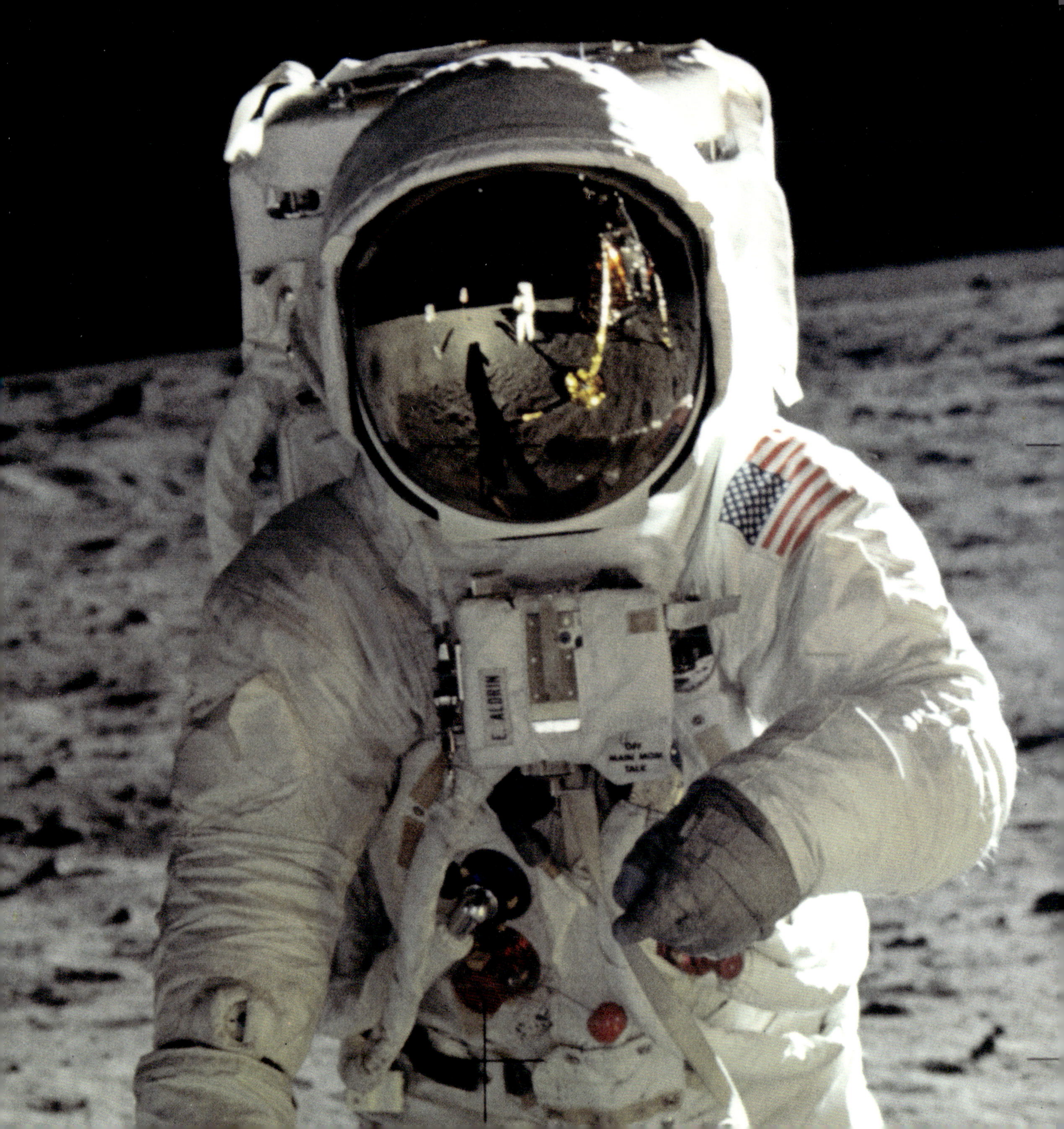
E. ALDRIN

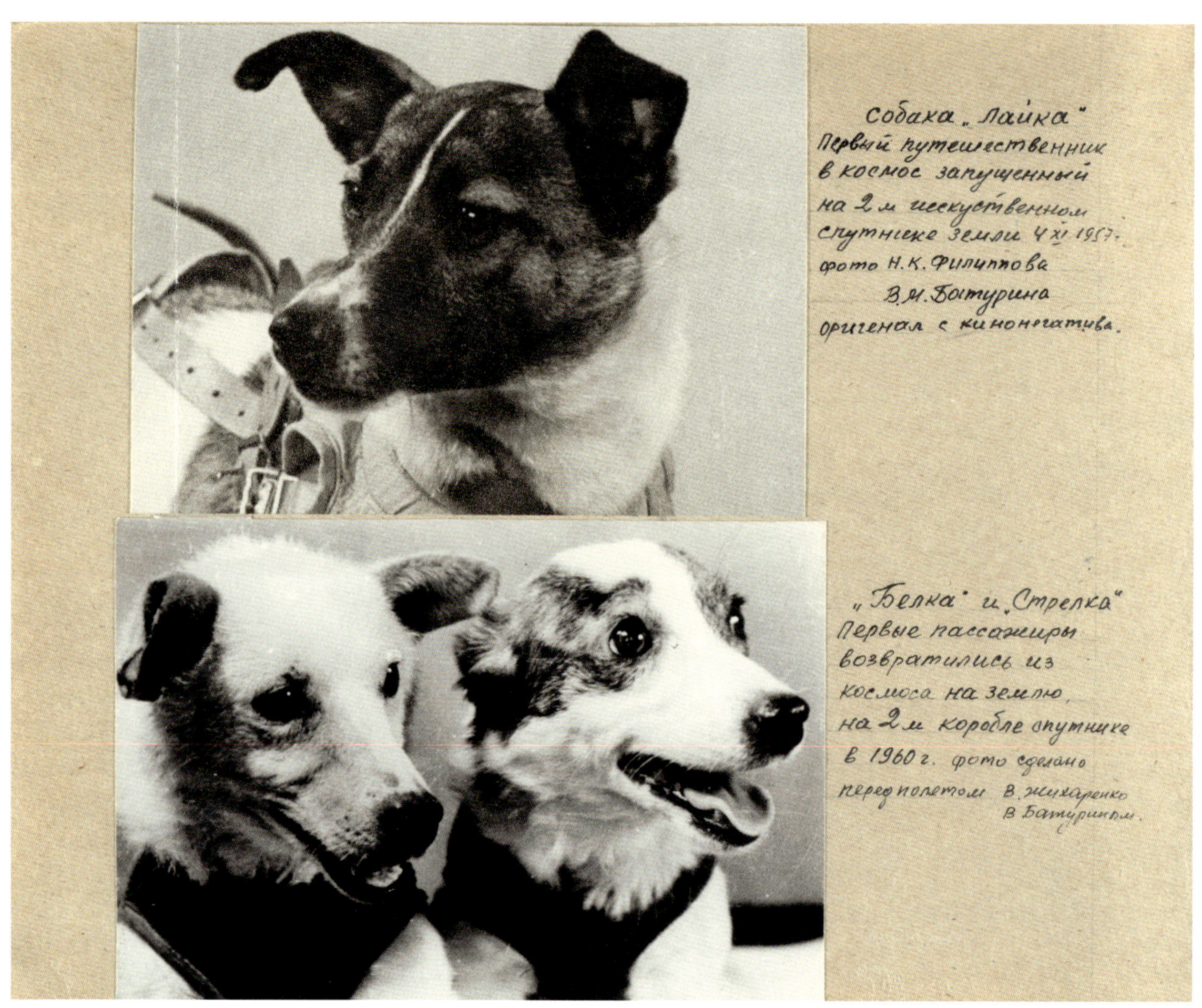

63 | VASILY M. BATURIN | Page from *Brothers of the Cosmos*, 1969, featuring images of Laika, 1957 (top), by Baturin and N. K. Filippov; and Belka and Strelka, 1960 (bottom), by Baturin and V. Zhiharekno

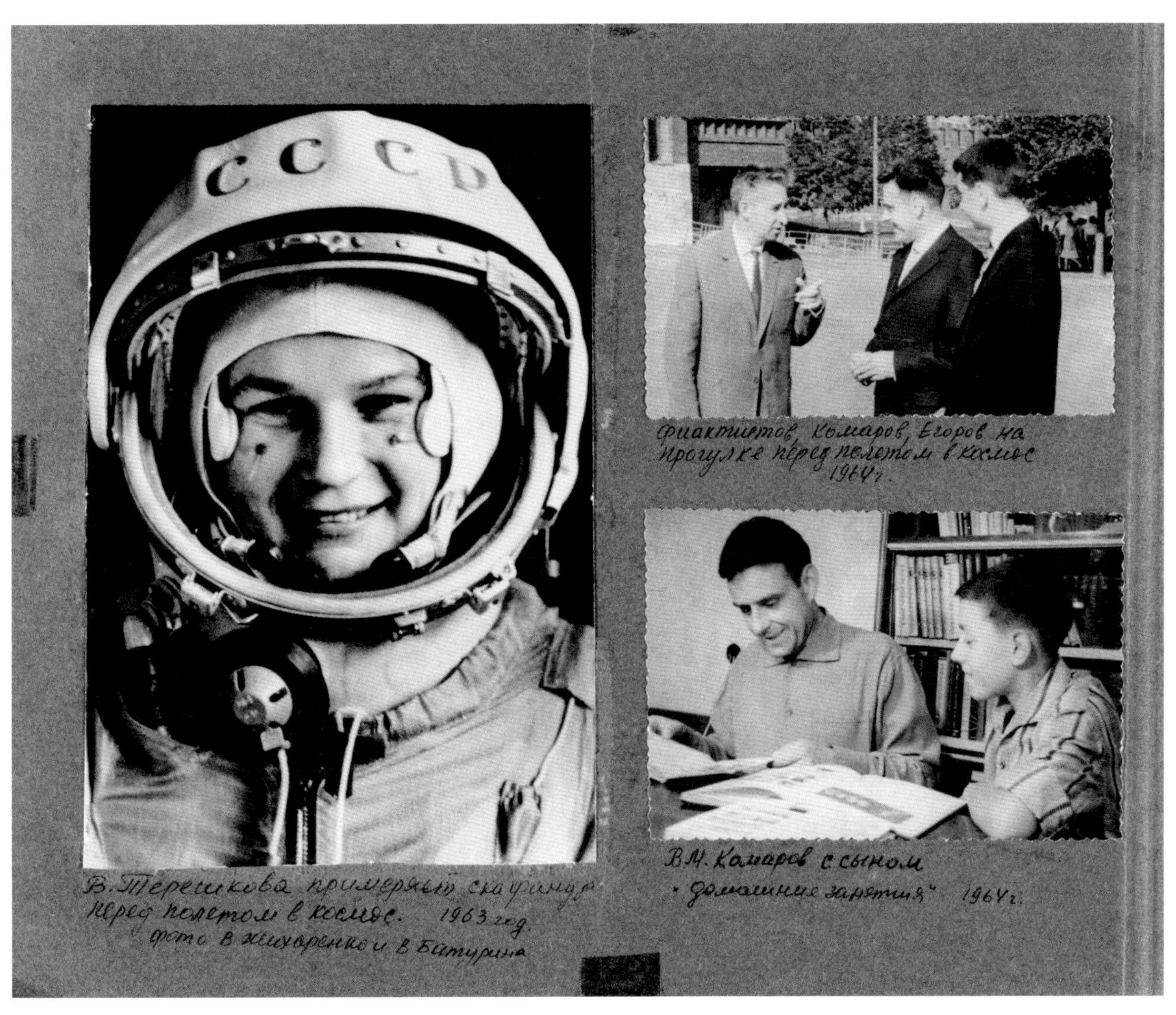

 | VASILY M. BATURIN | Page from *Brothers of the Cosmos*, 1969, featuring images of Valentina Tereshkova, 1963 (left); Konstantin Feoktistov, Vladimir Komarov, and Boris Yegerov, 1964 (top right); and Komarov with his son, 1964 (bottom right), all by Baturin and V. Zhiharekno

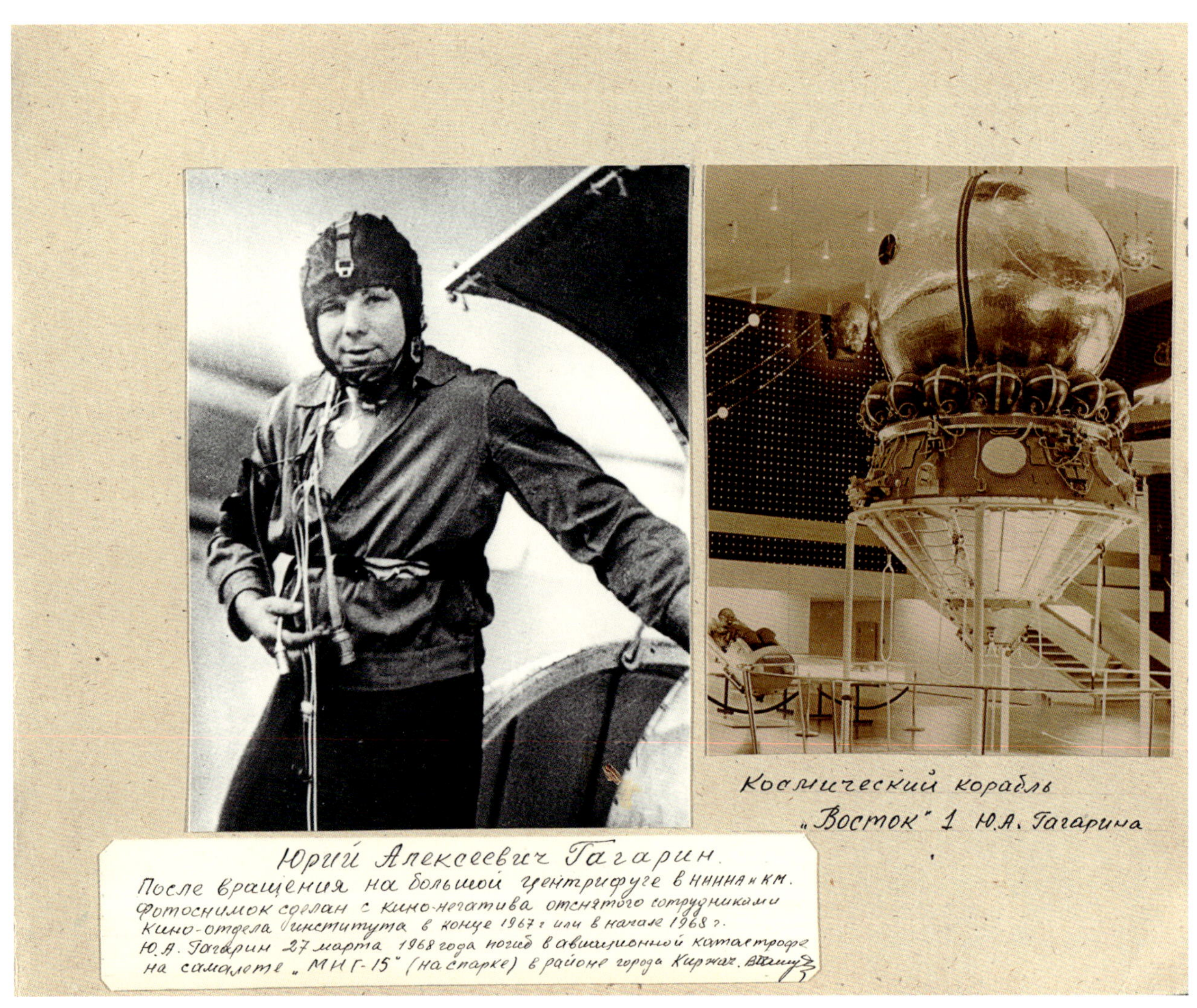

 | VASILY M. BATURIN | Page from *Brothers of the Cosmos*, 1969, featuring images of Yuri Gagarin and Space Shuttle Vostok 1 at the Yuri Gagarin Cosmonaut Training Center, 1967–68

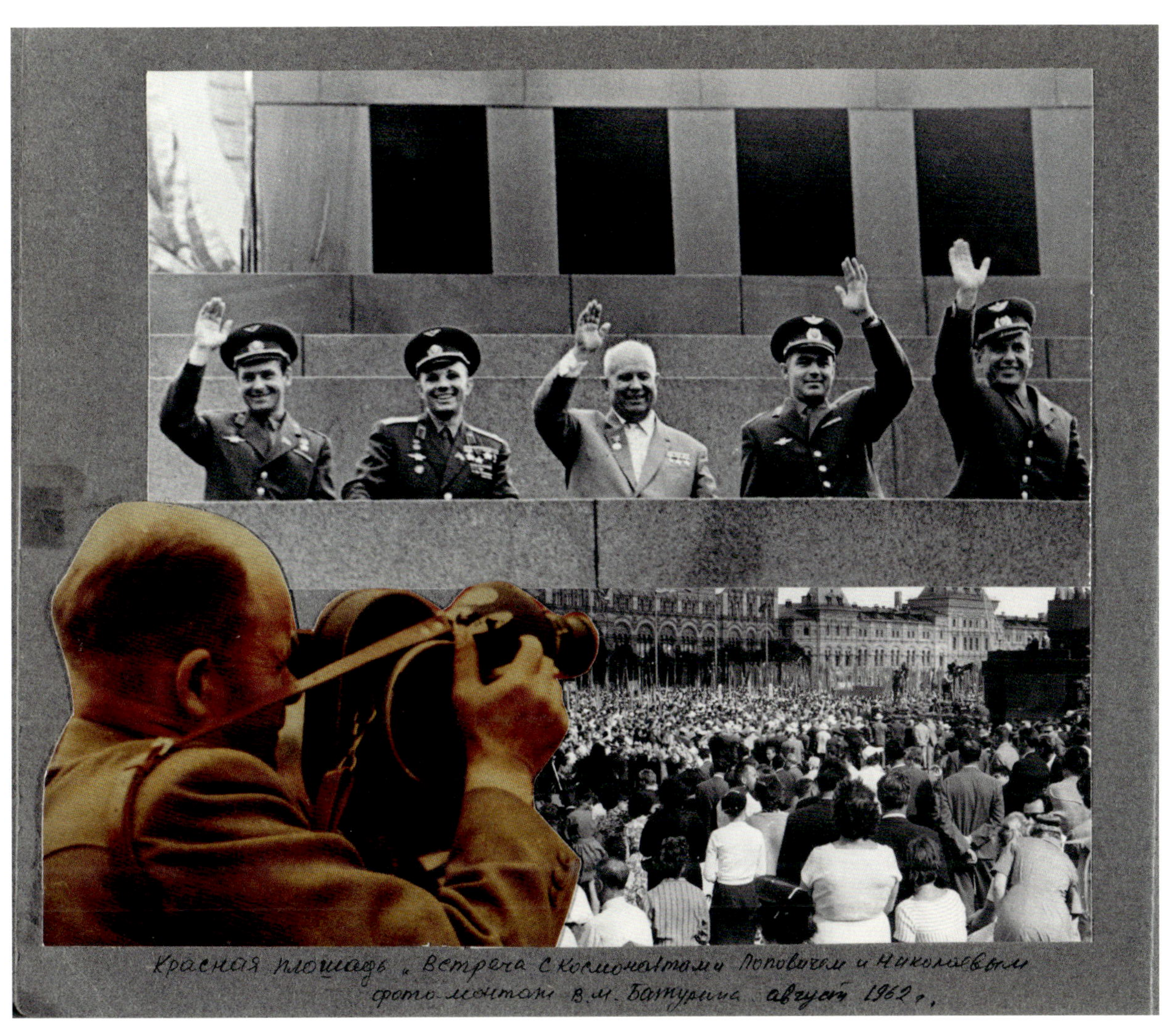

66 | VASILY M. BATURIN | Page from *Brothers of the Cosmos*, 1969, featuring a photomontage of Baturin with a camera and cosmonauts Pavel Popovich and Andriyan Nikolayev meeting Soviet Premier Nikita Khrushchev in Moscow, 1962

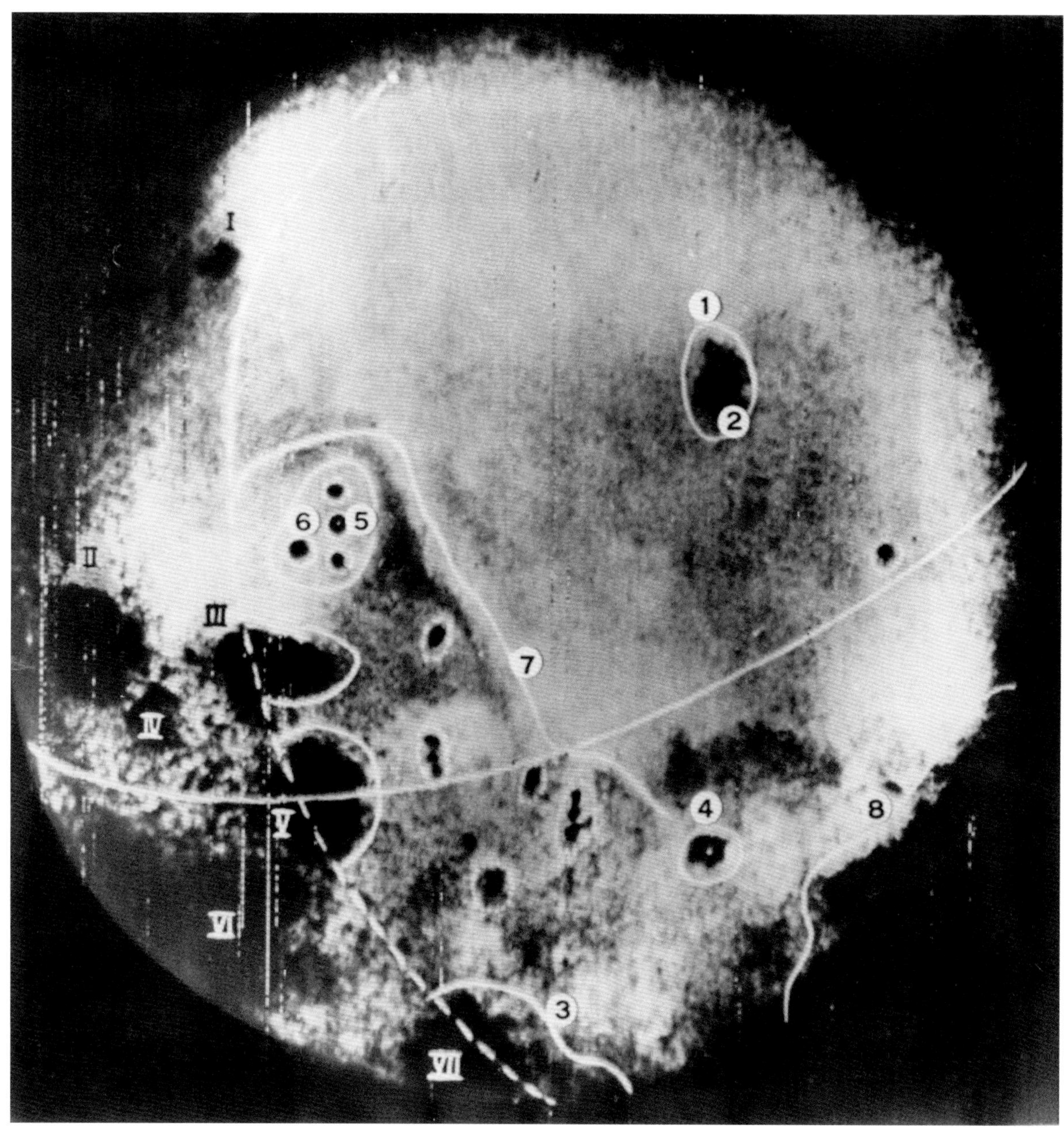

67 | TASS (USSR LUNA 3) | *The Far Side of the Moon*, 1959

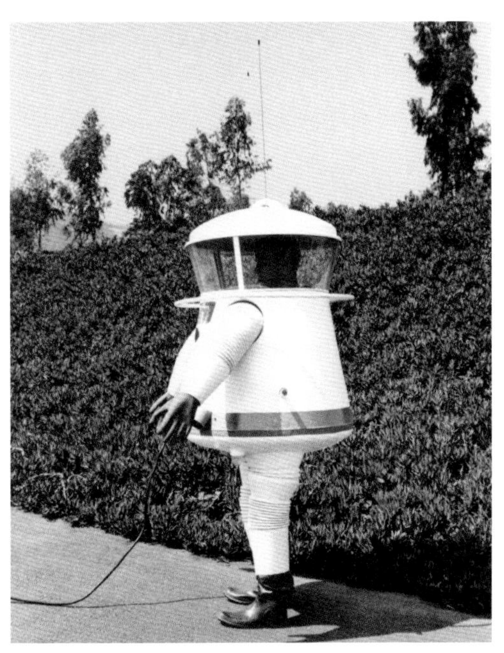

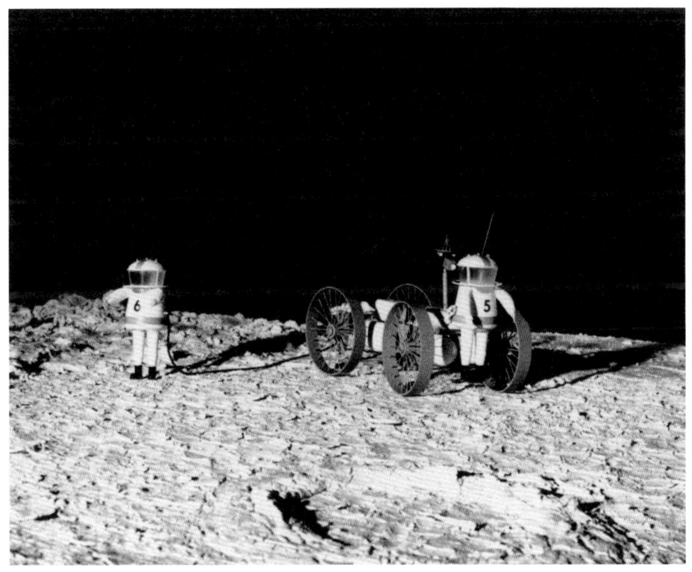

68, 69 | AEROJET-GENERAL CORPORATION | *Mock-Up of Moon Suit in Action* and *Moon Mobile Carries Two Explorers, Life-Support Systems*, 1962

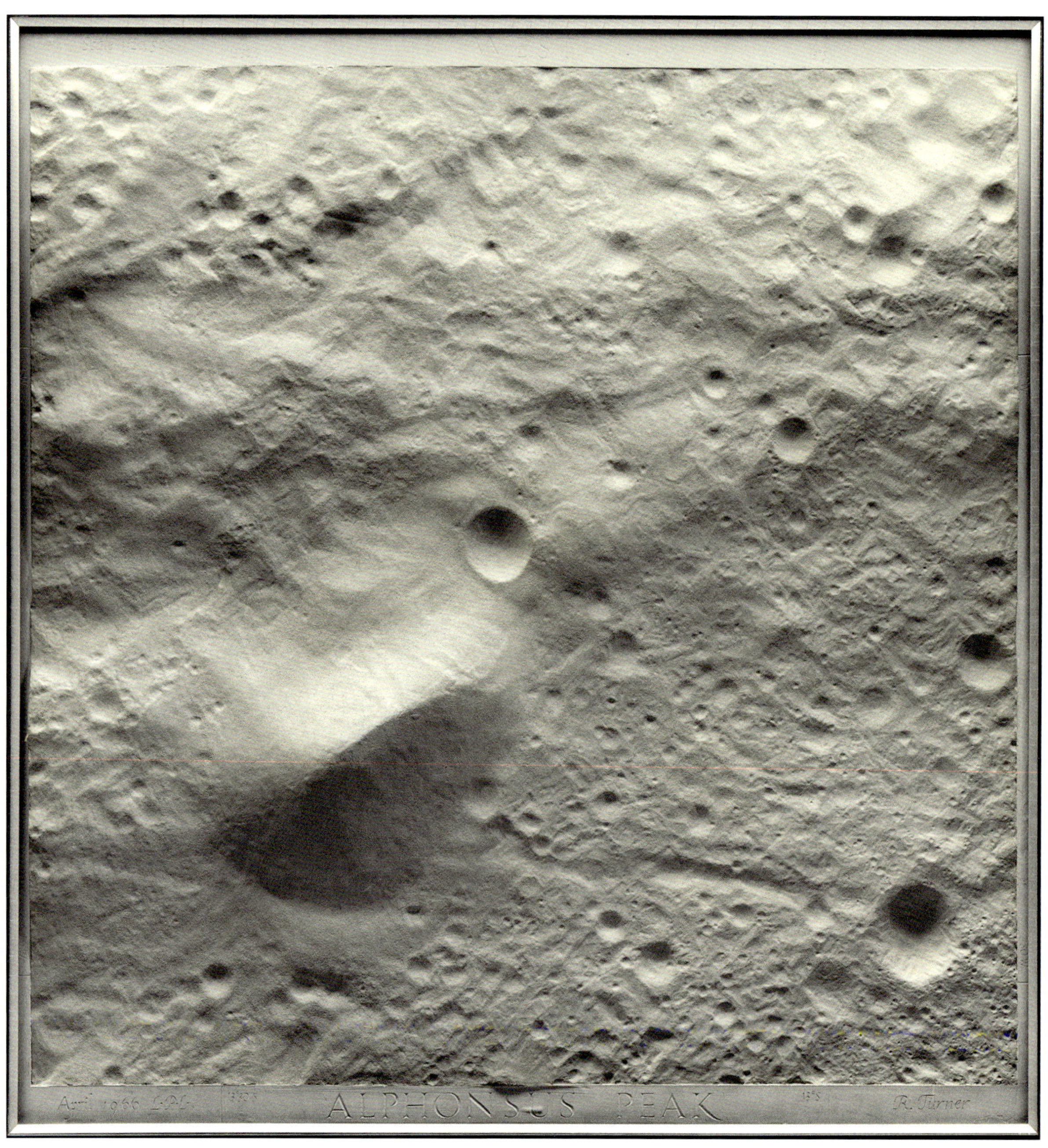

 High Relief of Alphonsus Peak, Made at the Lunar and Planetary Laboratory, Tucson, 1966

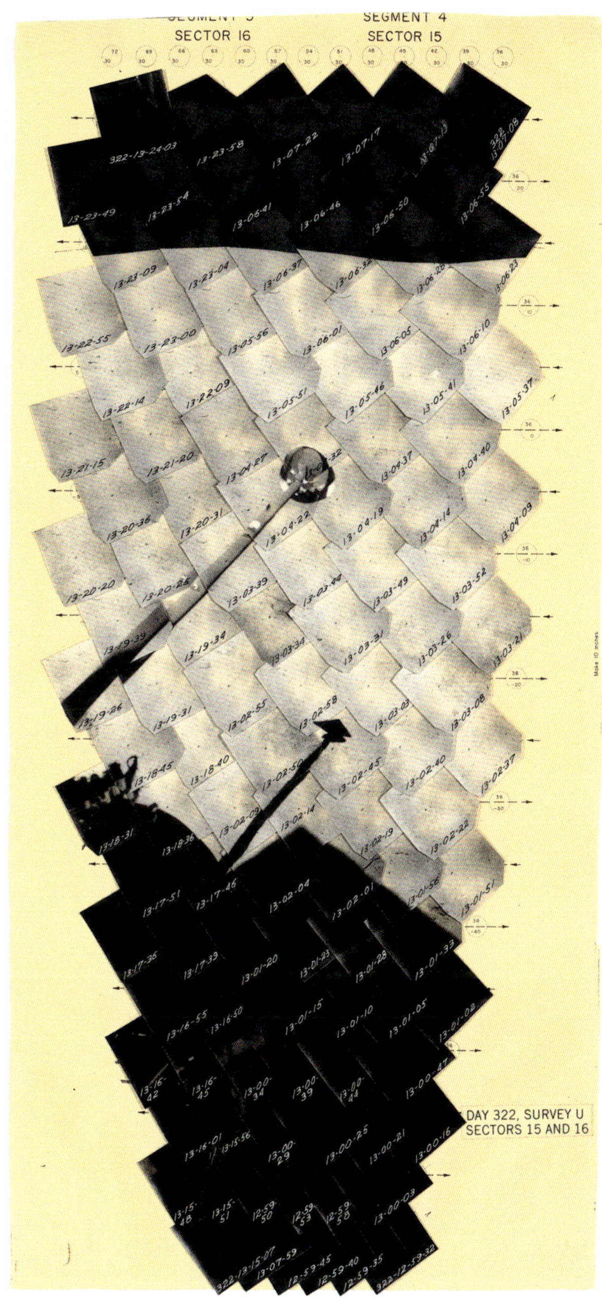

71 | NASA SURVEYOR 6 | *Day 322, Survey U, Sectors 15 and 16*, 1967

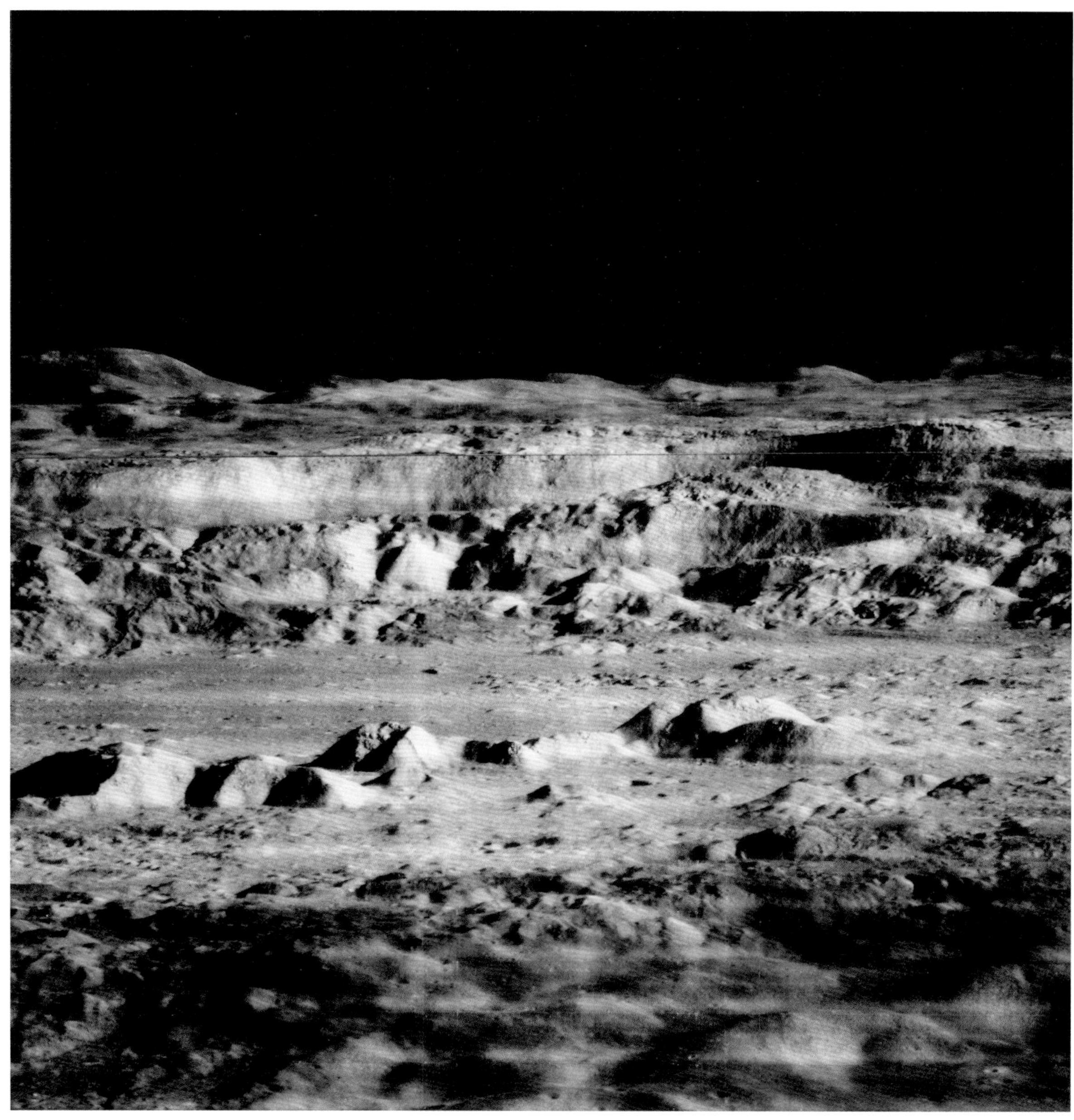

 | NASA LUNAR ORBITER 2 | *Close-Up of Crater Copernicus*, 1966

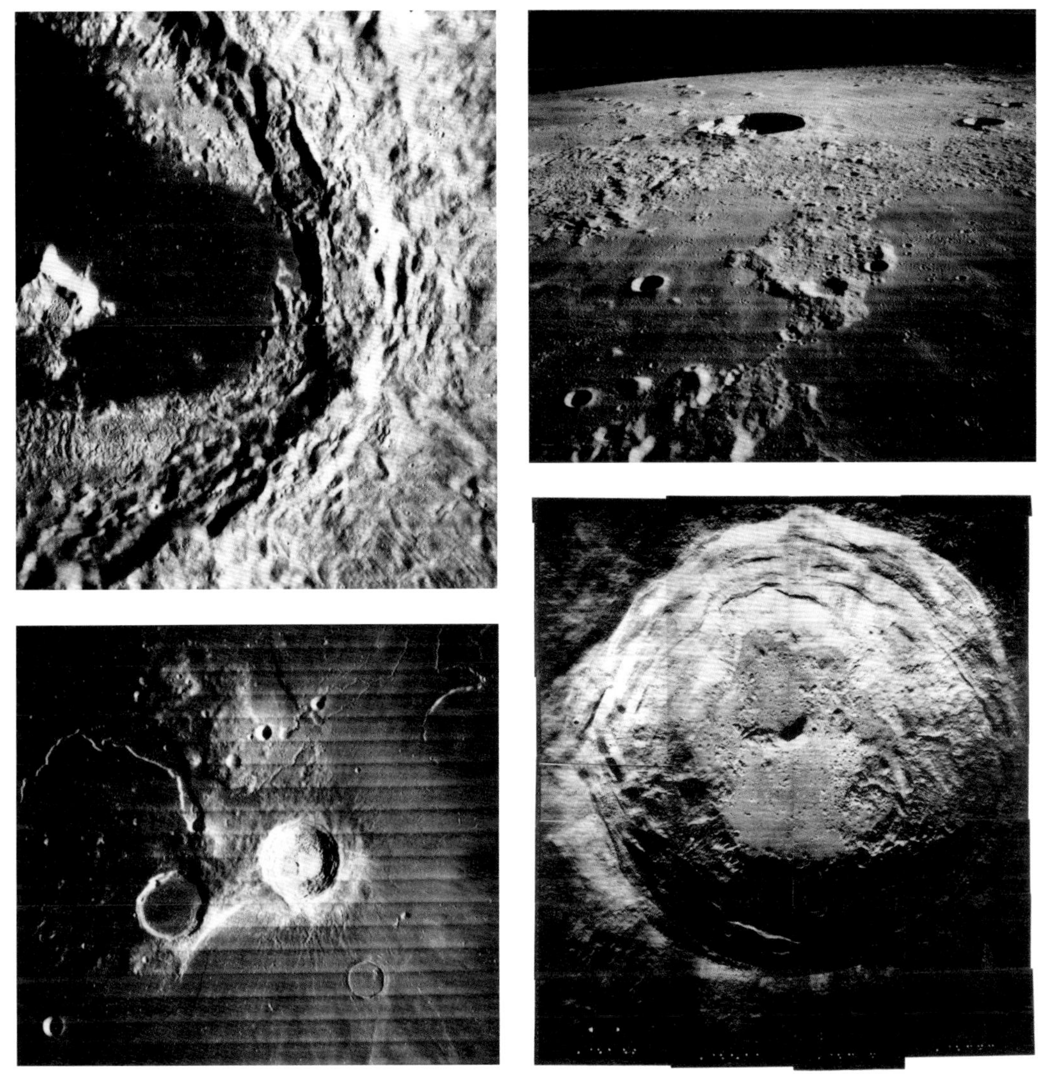

73, 74 | NASA LUNAR ORBITER 3 | *Far Side of the Moon at Apolune* and *Crater Kepler and Vicinity*, 1967

75 | NASA LUNAR ORBITER 4 | *Crater Aristarchus, Schroter's Valley, and Vicinity*, 1967

76 | NASA LUNAR ORBITER 5 | *Crater Aristarchus*, 1967

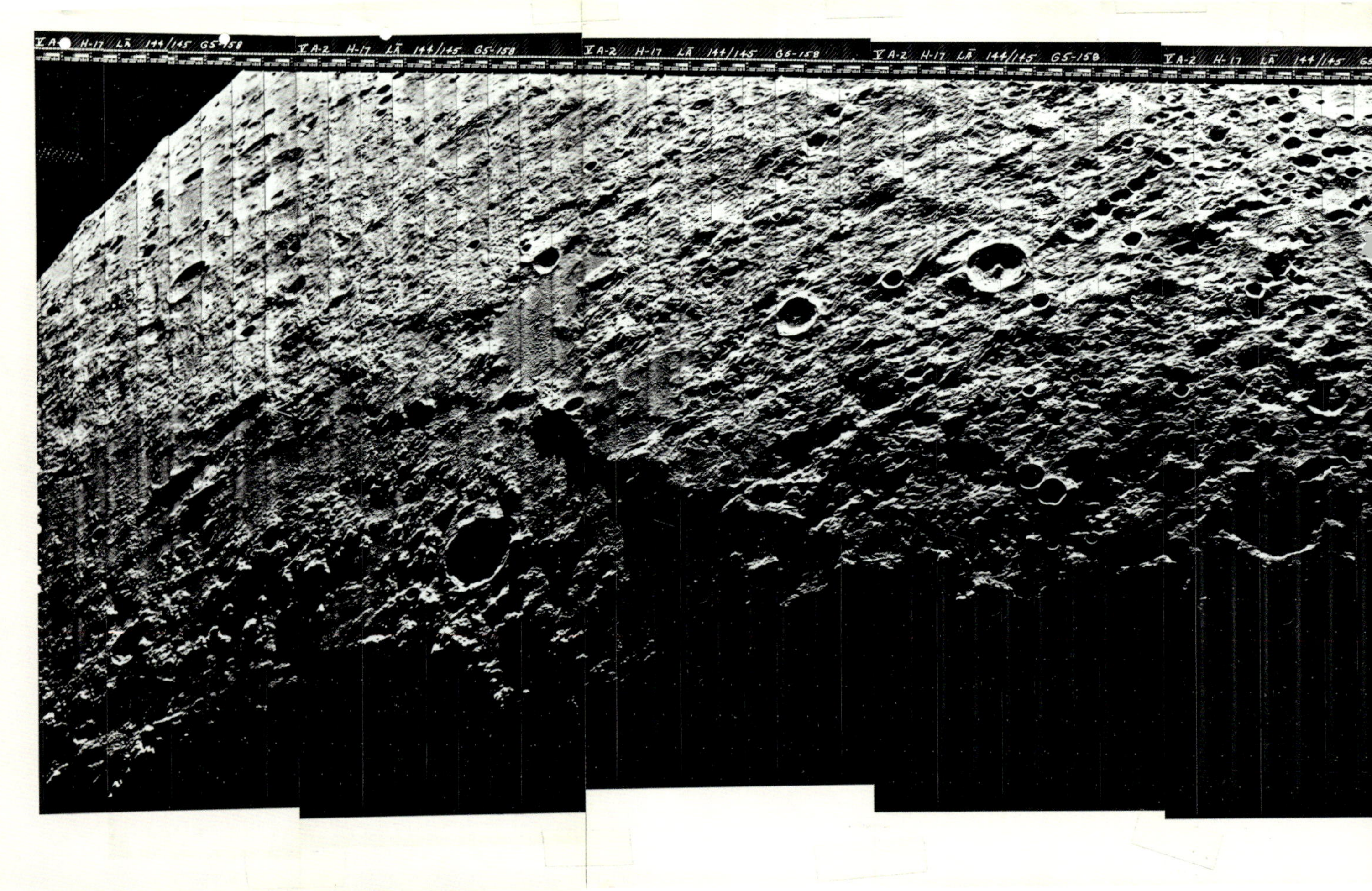

77 | NASA LUNAR ORBITER 5 | *Lunar Panorama #158*, 1967

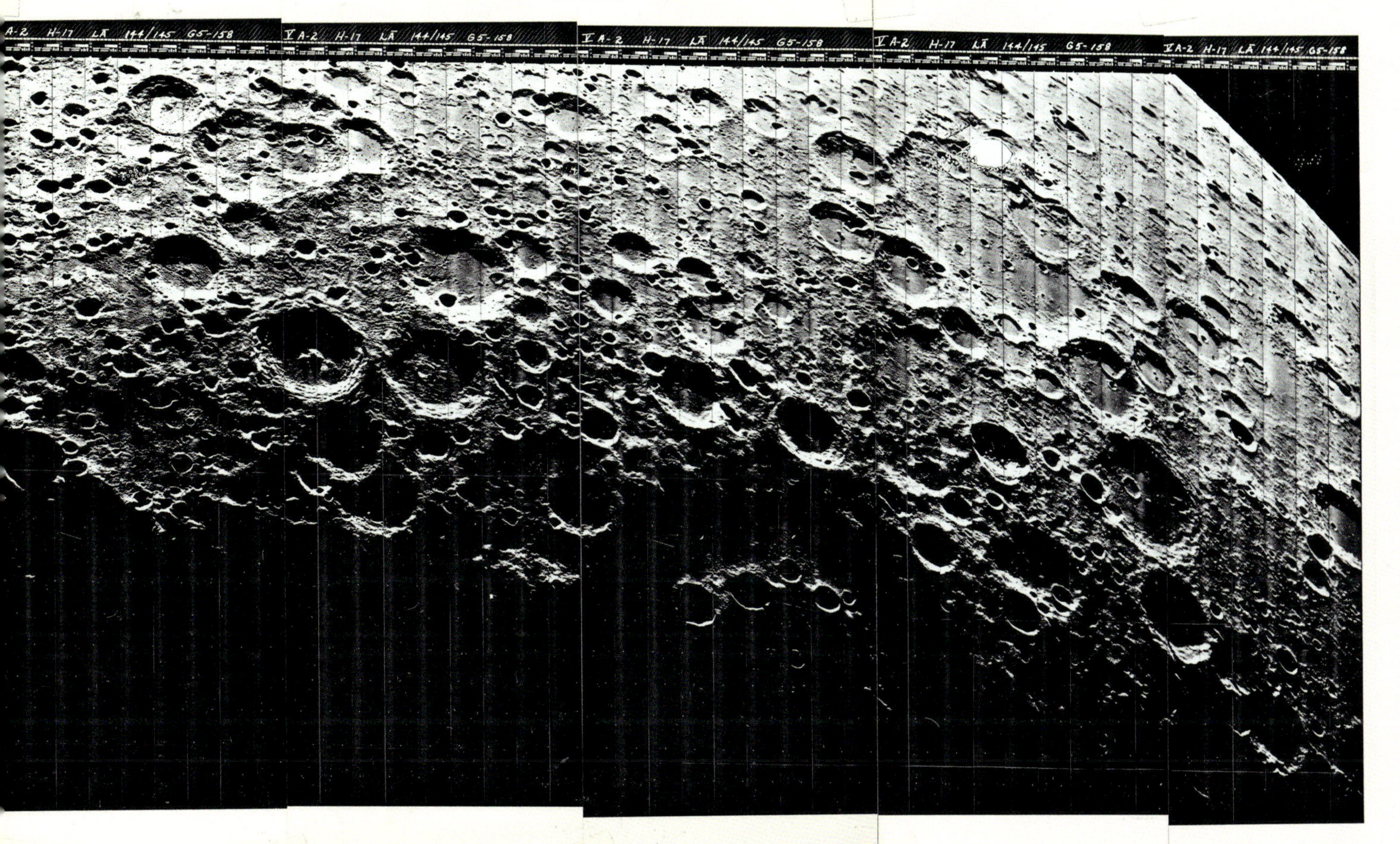
A-2 H-17 LĀ 144/145 G5-158
V A-2 H-17 LĀ 144/145 G5-158
V A-2 H-17 LĀ 144/145 G5-158
V A-2 H-17 LĀ 144/145 G5-158
V A-2 H-17 LĀ 144/145 G5-158

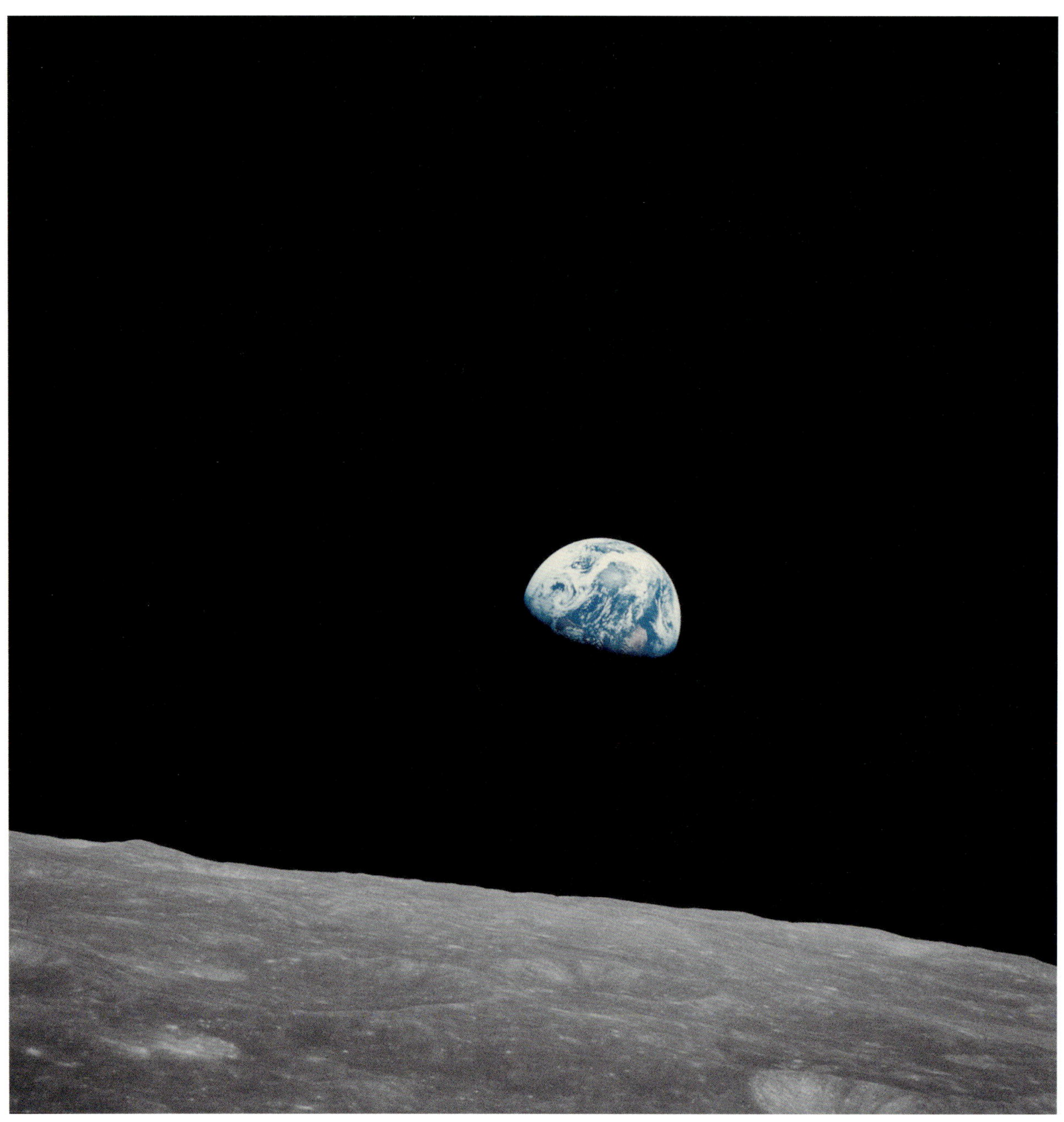

78 | WILLIAM ANDERS (NASA APOLLO 8) | *Earthrise*, 1968

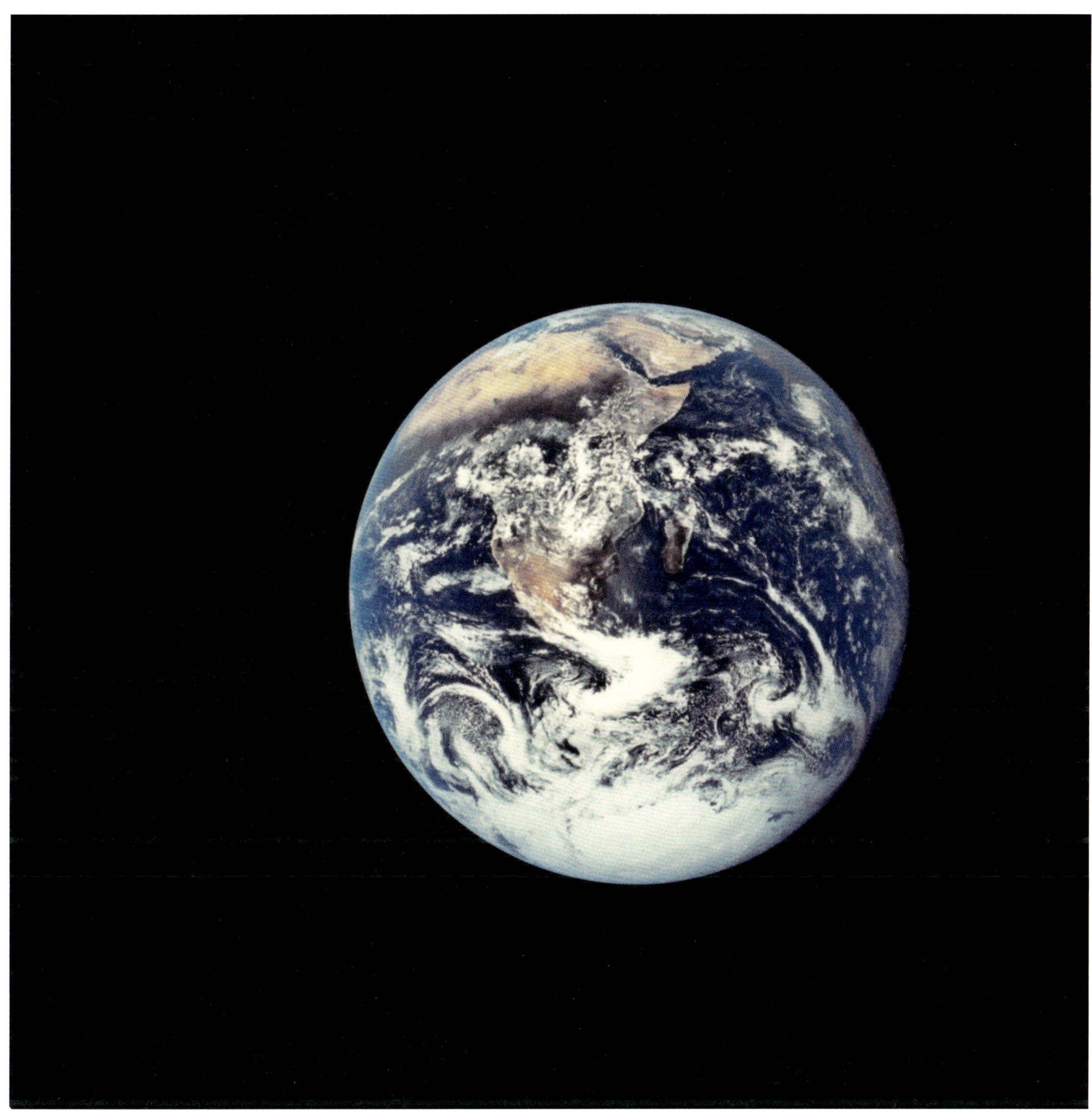

79 | HARRISON SCHMITT (NASA APOLLO 17) | *Blue Marble*, 1972

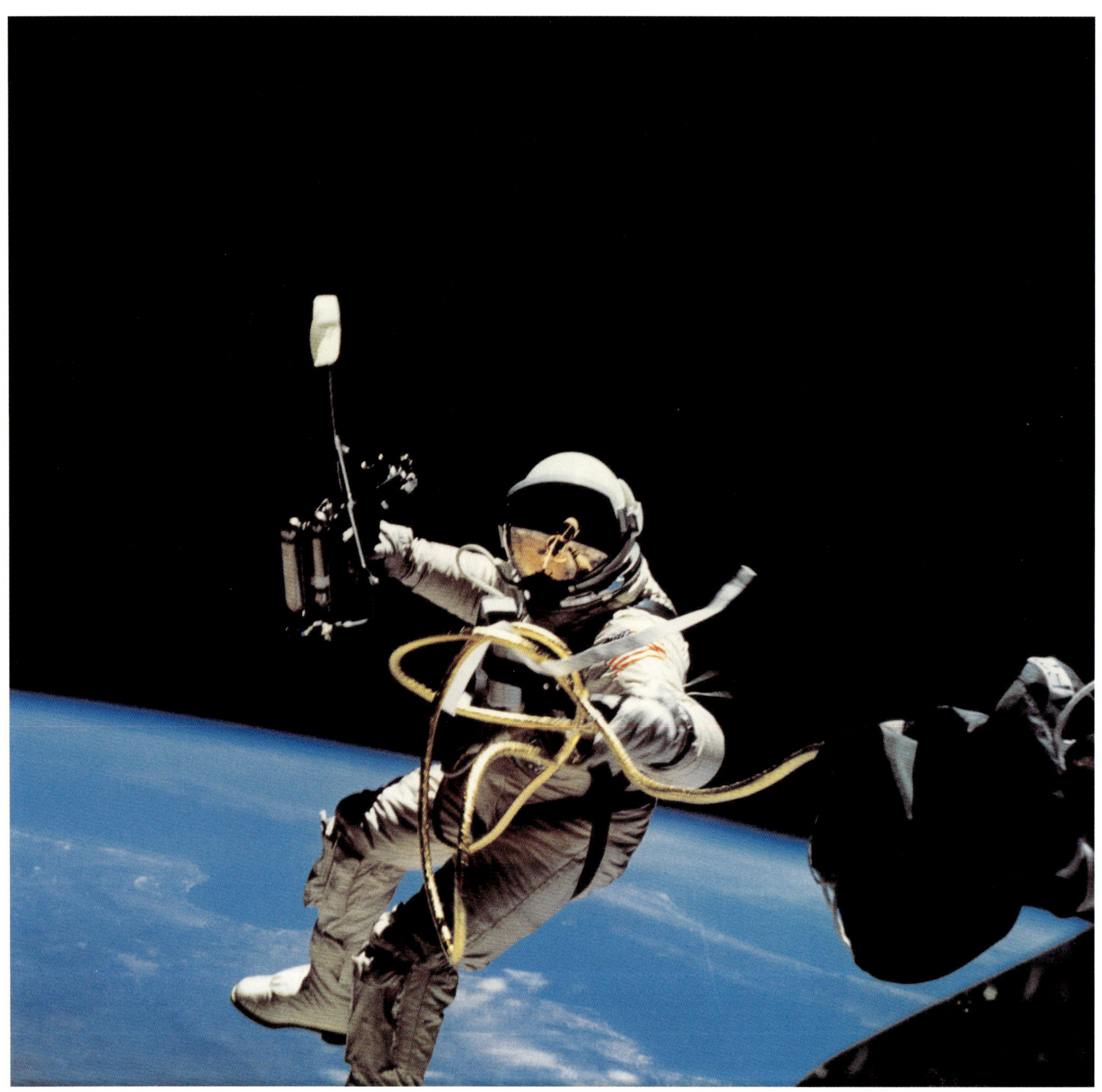

80 | JAMES MCDIVITT (NASA GEMINI 4) | *Ed White Extravehicular Activity (EVA)*, 1965

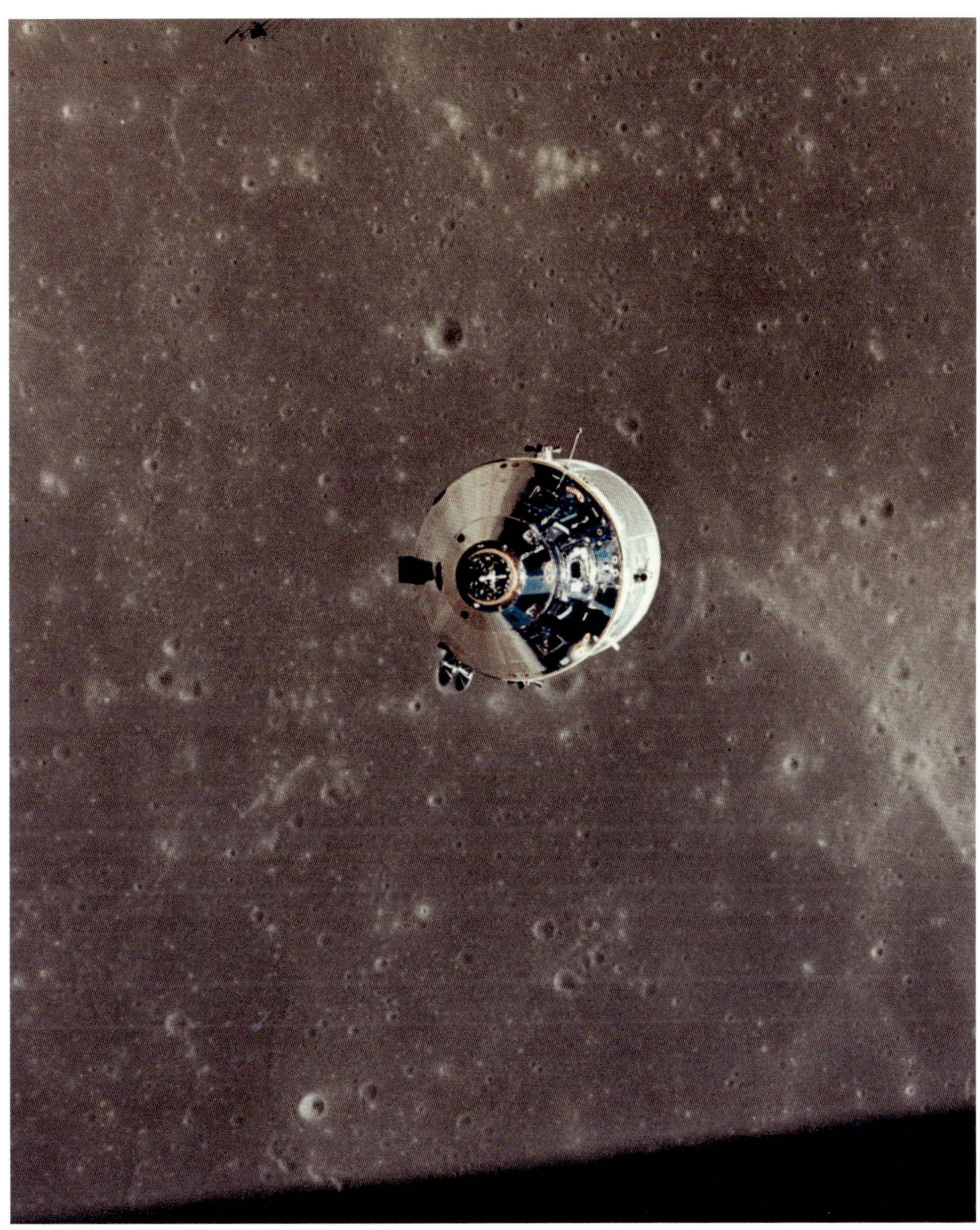

 | NASA APOLLO 11 | *Apollo 11 Command and Service Modules Photographed from the Lunar Module in Orbit*, 1969

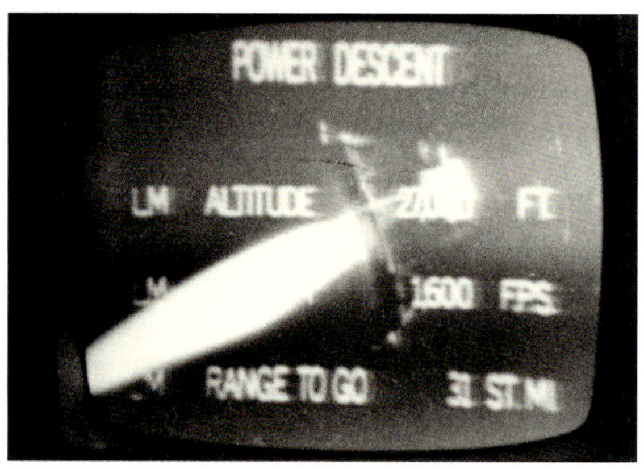

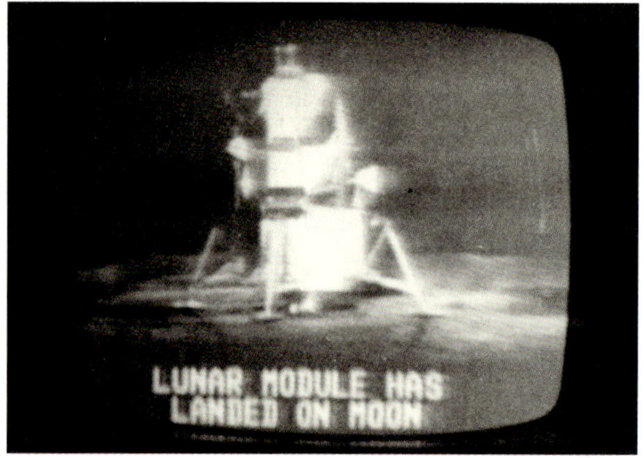

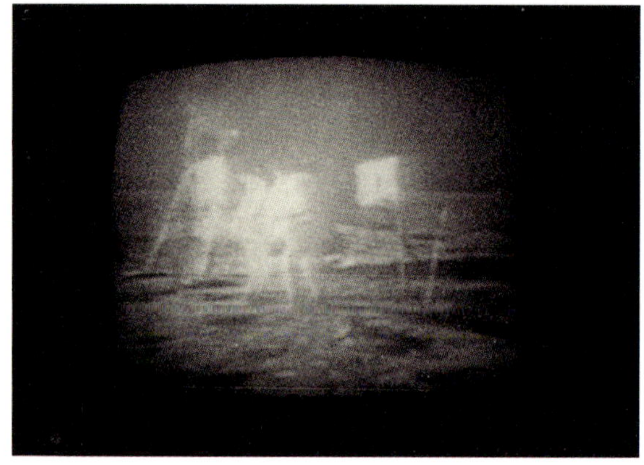

82 | *Apollo 11 Moon Landing on Television Screens*, 1969

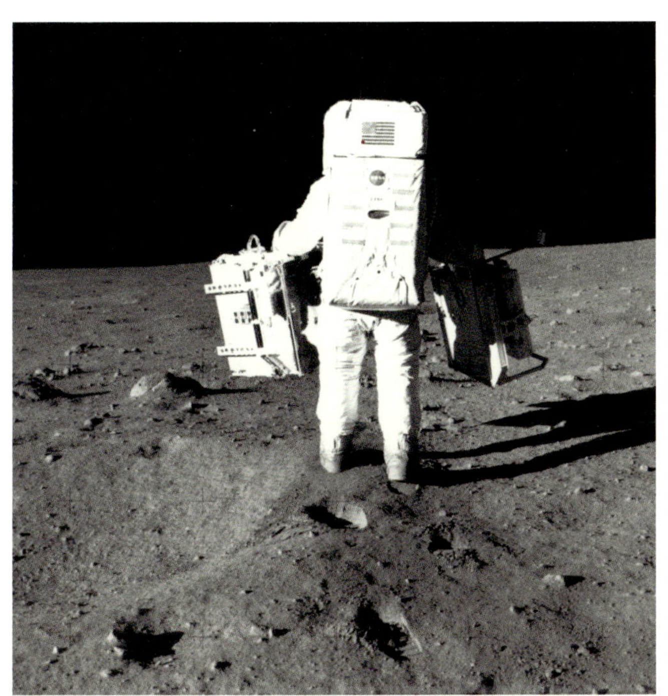

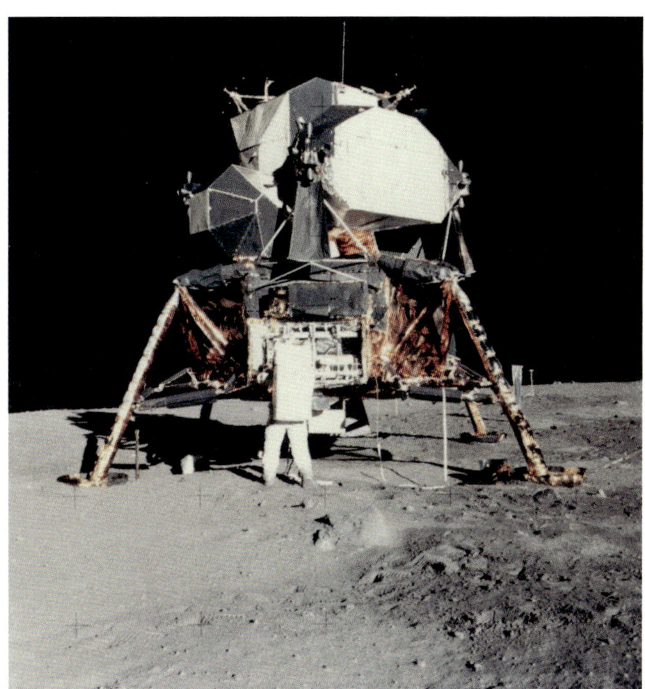

83, 84 | NEIL ARMSTRONG (NASA APOLLO 11) |

Buzz Aldrin on the Moon with Components of the Early Apollo Scientific Experiments Package

and *Buzz Aldrin with Apollo 11 Lunar Module on the Moon*, 1969

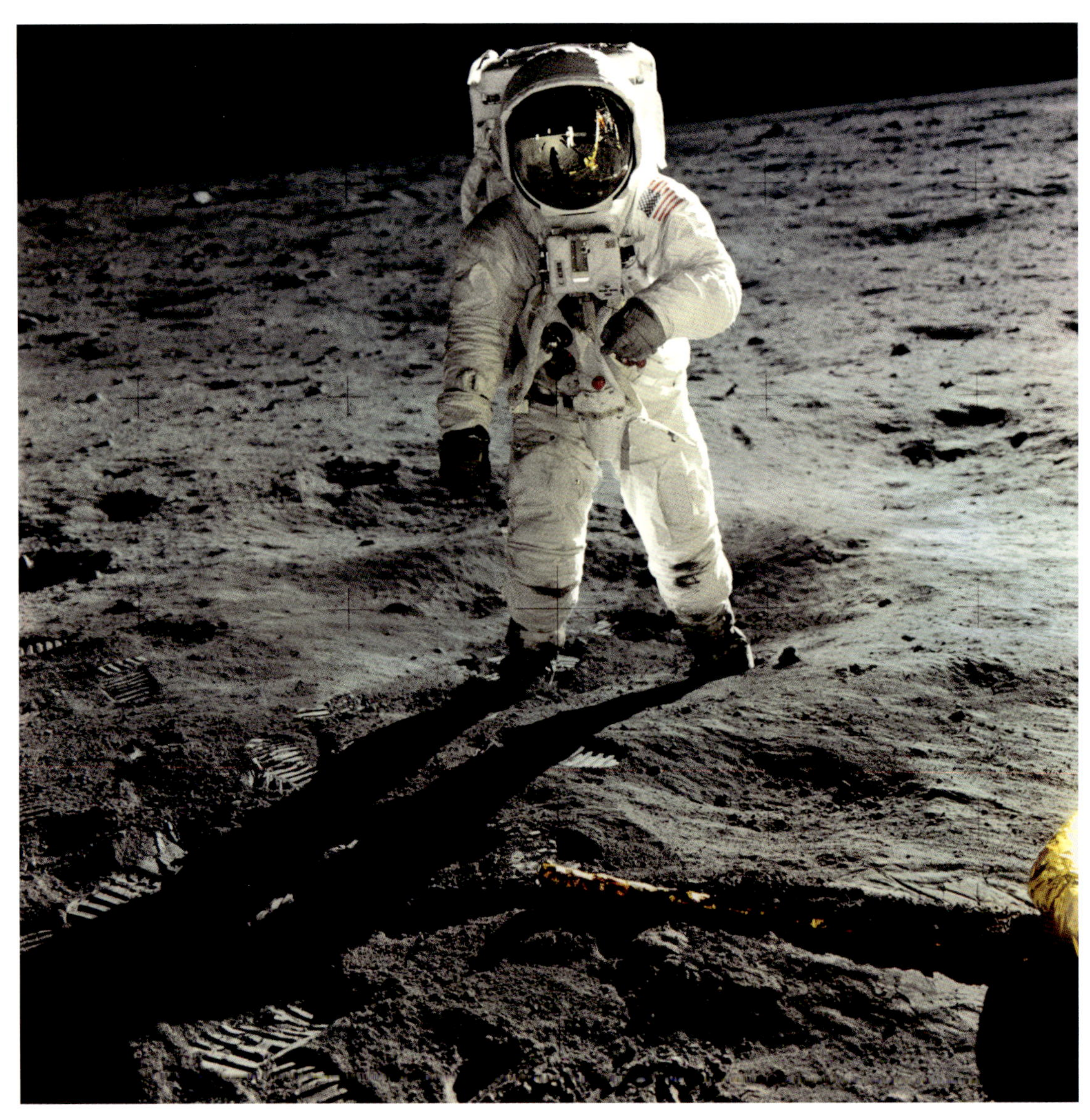

85 | NEIL ARMSTRONG (NASA APOLLO 11) |
Buzz Aldrin Walking on the Surface of the Moon near a Leg of the Lunar Module, 1969

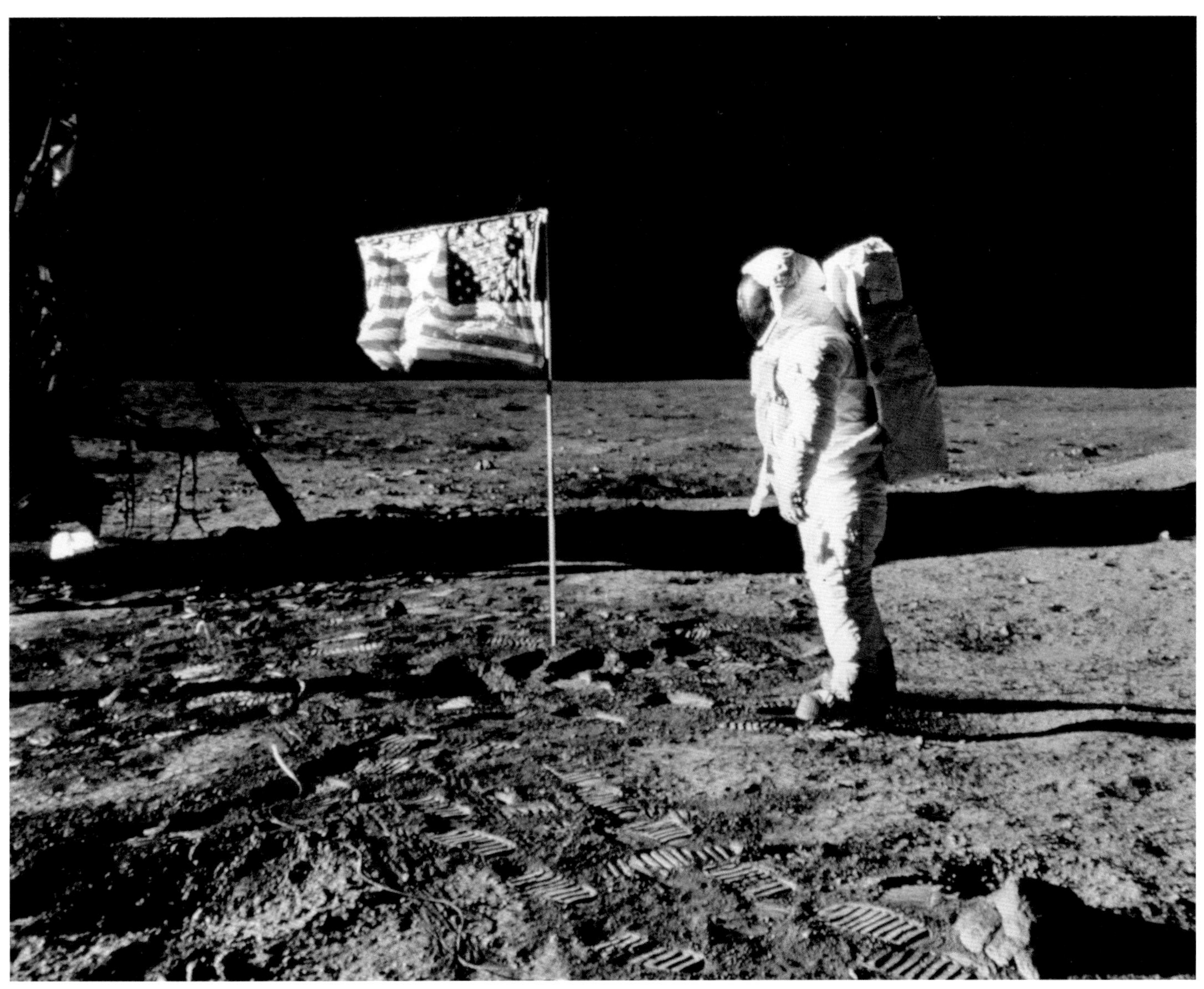

86 | NEIL ARMSTRONG (NASA APOLLO 11) | *Buzz Aldrin on the Moon with the American Flag*, 1969

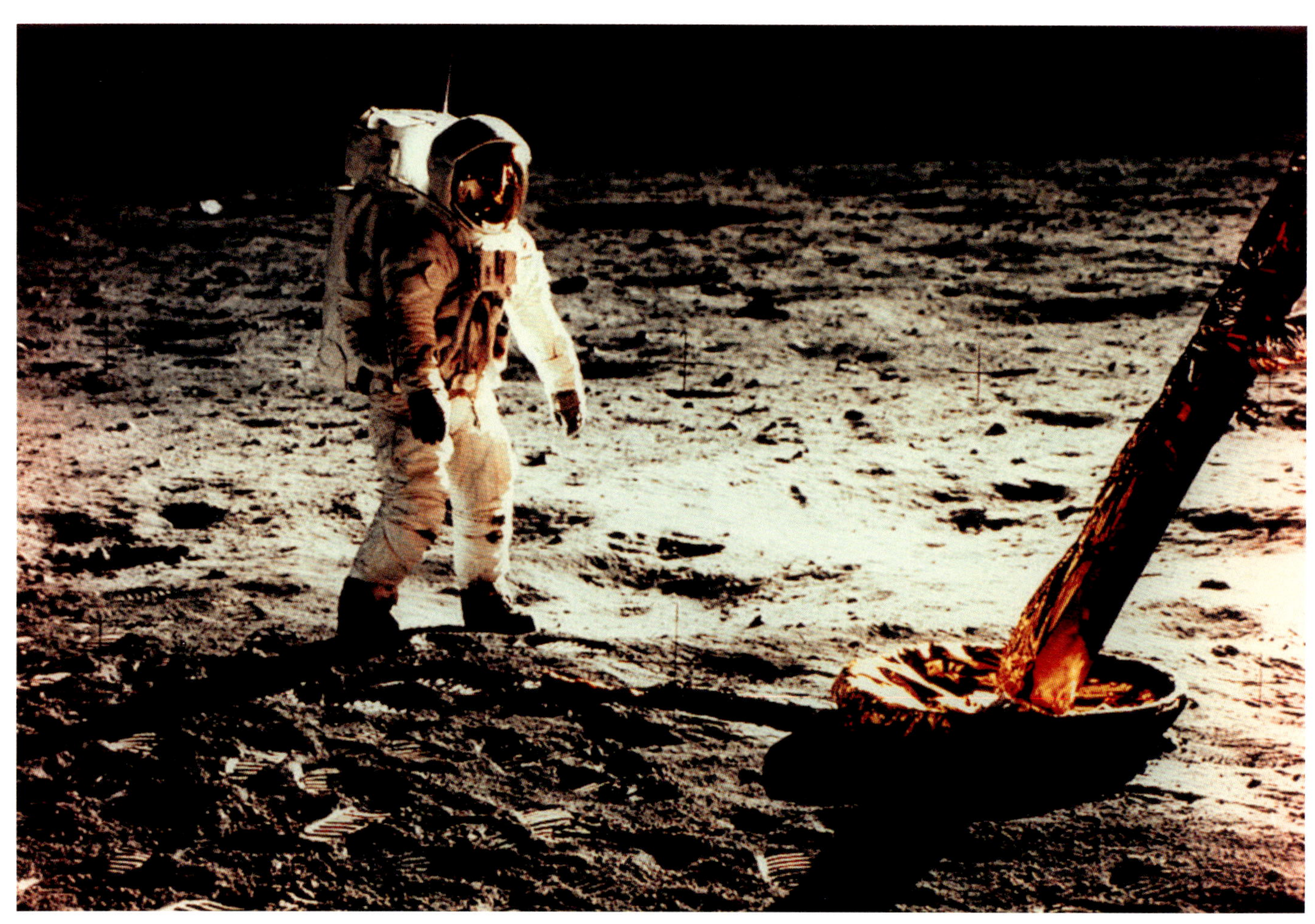

87 | NEIL ARMSTRONG (NASA APOLLO 11) |

Buzz Aldrin Walking on the Surface of the Moon near a Leg of the Lunar Module, 1969

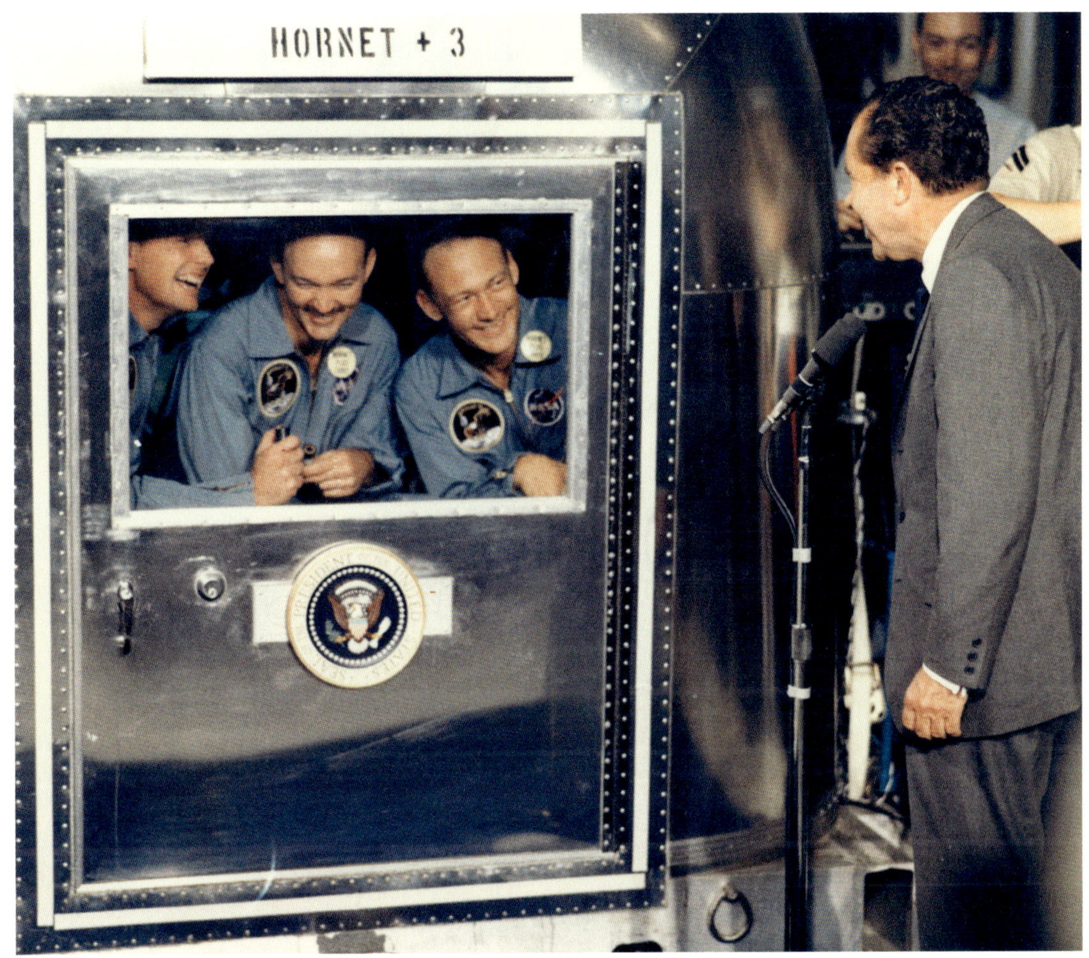

President Richard M. Nixon Welcomes the Apollo 11 Astronauts aboard Recovery Ship USS Hornet, 1969

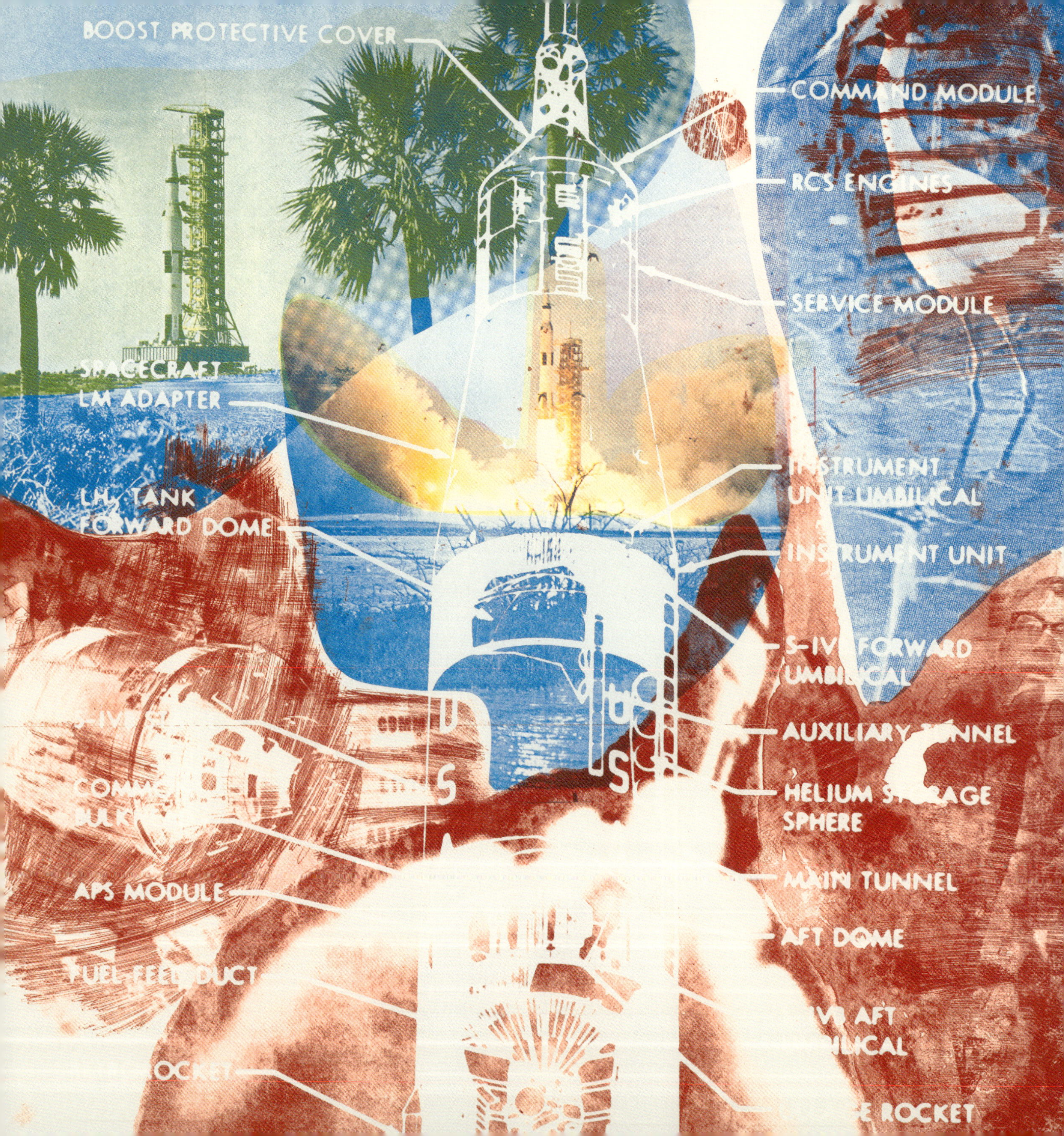
BOOST PROTECTIVE COVER
COMMAND MODULE
RCS ENGINES
SERVICE MODULE
SPACECRAFT
LM ADAPTER
FORWARD DOME
INSTRUMENT UNIT
INSTRUMENT UNIT UMBILICAL
AUXILIARY TUNNEL
HELIUM STORAGE SPHERE
MAIN TUNNEL
AFT DOME
APS MODULE
FUEL FEED DUCT
U
S
U
S

ART AFTER APOLLO

BETH SAUNDERS

On July 16, 1969, hundreds of thousands of people gathered in the muggy wetlands of Florida, having made pilgrimages from across the United States and beyond to watch the launch of Apollo 11. Garry Winogrand's photograph *Apollo 11 Moon Shot, Cape Kennedy, Florida* (1969, pl. 89) wryly captures the spectacle surrounding the occasion: a crowd wielding binoculars and cameras turns its gaze skyward, while a woman in the foreground aims her lens directly at Winogrand. As documentation of the historic milestone, the photograph lacks the decisive moment of takeoff; a mere plume of smoke lingers in the sky to suggest the rocket launch. Instead, the mirrorlike exchange at the heart of the picture points to the way photography mediates significant historical events and generates collective experiences. It also suggests the massive publicity machine that NASA mobilized to gain support for its enormously expensive endeavors. The moon landing, and especially its presence in popular culture, inspired many artists concerned with the impact of technology on art and the role of mediation in interpreting reality; in this context, photography played a particularly significant role. In the wake of Apollo 11, the subject of the moon in art took on new meaning and symbolism as artists responded to the cultural and metaphysical transformation of humanity's relationship to the cosmos.

Among the crowds gathered to witness Apollo 11's launch was Robert Rauschenberg, who had recently taken up residence in Florida on Captiva Island. In his recollections of the day, the artist mused on the strange harmony of nature and technology at the site, noting that as the rocket took off "all you could hear were frogs and alligators."[1] Rauschenberg had been invited as part of the NASA Artists' Cooperation Program. Established in 1962 under the supervision of James Dean, director of NASA's Educational Media Division, and Hereward Lester Cooke, curator of painting at the National Gallery of Art, the program commissioned artists to commemorate in sketches and painting the achievements of the space program.[2] They were meant to provide a subjective record of the agency's activities, creating realistic yet inspirational renderings of astronauts on space walks, gleaming machinery, and the majesty of the cosmos that complemented and humanized the cold optical gaze supplied by NASA's robust photography program. As Cooke described it:

> *When a major launch takes place at Cape Canaveral more than two hundred cameras record every split second of the blast off. The artist can add very little to this in the way of factual record. But, as [Honoré] Daumier pointed out a century ago, the camera sees everything and understands nothing. It is the emotional impact, interpretation, and hidden significance of these events which lie within the scope of the artist's vision.*[3]

Rauschenberg might seem an anomalous choice among the program's other artists, such as Paul Calle, Lamar Dodd, Peter Hurd, John McCoy, and John Meigs, who worked in expressionistic and realist modes (fig. 17). By 1969, however, Rauschenberg had been anointed with a grand prize at the Venice Biennale, signaling his institutional approval on a global stage, and he had cofounded Experiments in Art and Technology (E.A.T.), a collaborative program that paired artists and engineers on projects to "benefit society as a whole."[4] Furthermore, his collage style, which integrated realism through photographic images and abstraction by means of cutting, fragmentation, and layered graphic marks, offered a means by which NASA could court the avant-garde while remaining moored to recognizable imagery.

NASA gave Rauschenberg full access to the Kennedy Space Center and supplied him with photographs, maps, and press images. Using this source material, the artist produced a series of nineteen drawings and collages and thirty-four lithographs. The prints, made at the Gemini G. E. L. printmaking studio in Los Angeles, unfold as a series of impressions in black-and-white, monochrome, and color; engineering drawings and official images of astronauts commingle with Florida oranges and long-necked water birds. *Sky Garden (Stoned Moon)* (1969, pl. 90), the largest lithograph in the series, was a technical feat of printmaking to rival that of the moon landing. At a height of seven and a half feet, it required three lithographic stones laminated together and a specially built press that needed five people to pull the print.[5] The resulting composition layers multiple renderings of the Saturn V rocket used to launch Apollo 11—a schematic drawing, a photographic transfer of the device on the launchpad surrounded by palm trees, and another of the explosive launch—next to an image of a great white heron to create a visual pun that likens rocket to bird.[6] Rather than placing nature in opposition to technology, *Sky Garden* and the other prints in the Stoned Moon series propose a symbiotic and hopeful union.

Rauschenberg's optimism surrounding the moon landing is also expressed in an untitled drawing created during this period (1969, pl. 91). Photographic transfers culled from the popular press place the recognizable faces of John F. Kennedy and other politicians alongside a key to the Norman Rockwell painting *We Are All Astronauts* (1969) as reproduced in *Look* magazine, which identified and valorized the unsung heroes who worked alongside the famous astronauts to realize the dream of landing on the moon.[7] In Rauschenberg's version, those figures appear as mere outlines dwarfed by a large contour tracing of the artist's foot. The latter evokes Neil Armstrong's famous "one small step" statement and calls to mind Buzz Aldrin's iconic photograph of a boot print in moon dust (see fig. 16b). By inserting a trace of himself, Rauschenberg enacts Rockwell's declaration that "we are all astronauts."

FIG. 17. John Meigs (American, 1916–2003), *To the Moon*, 1969. Watercolor on paper. 10½ x 13 13/16 in. (26.7 x 35.1 cm). National Air and Space Museum, Washington, D.C., Transferred from the National Aeronautics and Space Administration

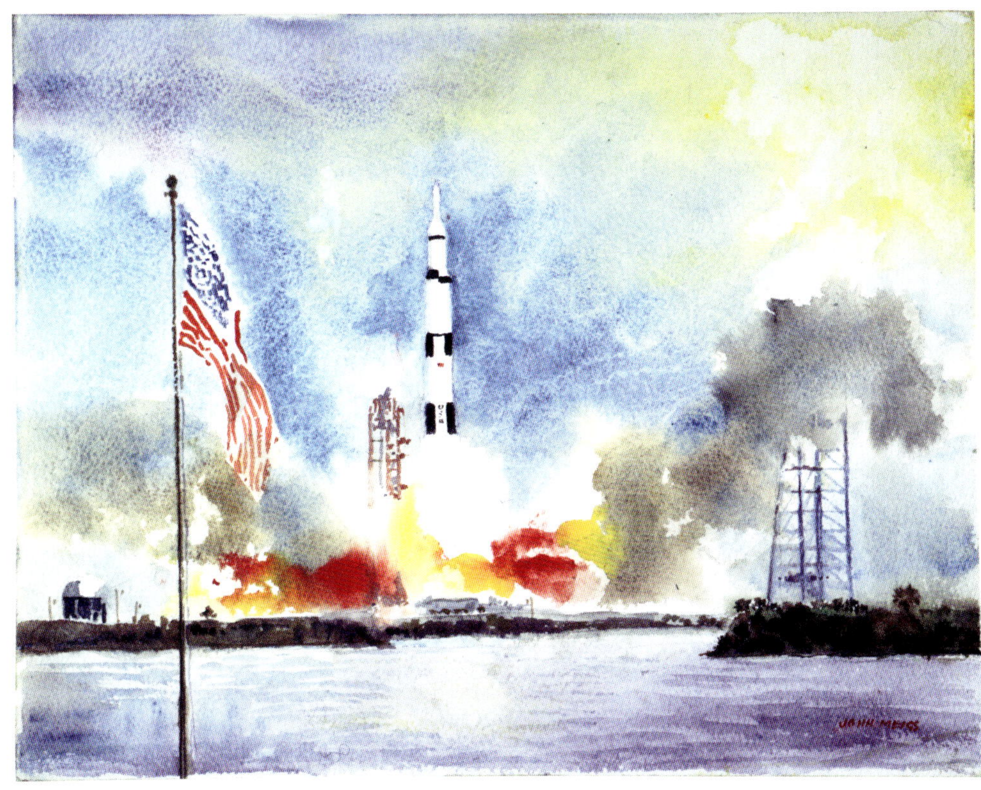

While Rauschenberg imagined standing on the moon, astronauts were actually leaving their mark on the satellite. Aldrin and Armstrong famously planted the American flag in an act of national pride (and global dominance) and left a small silicon disk etched with goodwill messages to the great beyond from four U.S. presidents and seventy-three worldwide heads of state. In addition to these NASA-sanctioned offerings to future space travelers and extraterrestrials, astronauts also deposited personal mementos and partnered with artists to install the first works of art exhibited intergalactically. On the Apollo 16 mission, one crew member left behind a family portrait bearing the inscription, "This is the family of astronaut Charlie Duke from planet Earth who landed on the moon on April 20, 1972." Duke then photographed the portrait nestled in moon dust to prove to his loved ones that it was there (pl. 92), producing perhaps the ultimate example of what the French semiotician Roland Barthes identified as a characteristic unique to photography, its insistent evidence of a subject's "having been there."[8]

The November 1969 Apollo 12 mission turned the moon into a museum when astronauts placed on the satellite a tiny ceramic wafer printed with drawings by John Chamberlain, Forrest Myers, David Novros, Claes Oldenburg, Rauschenberg, and Andy Warhol (pl. 93). The brainchild of Myers, the microchip was the product of a collaboration with engineers from Bell Laboratories arranged through E.A.T.[9] One can imagine an alien visitor regarding the elemental line drawings—which include abstract geometric forms, a crude phallic rocket ship, and a primitive Mickey

Mouse — much as we view prehistoric cave paintings, as echoes from an obscure civilization, familiar yet distinctly remote. In 1971 the moon's permanent collection grew to include a small aluminum sculpture of an astronaut designed by the Belgian artist Paul van Hoeydonck (pl. 94). Placed on the lunar surface by Apollo 15 crew member David Scott, it became a memorial to men who died in the line of duty attempting to reach the moon, and it was accompanied by a small plaque bearing their names. Scott snapped a photograph of the so-called Fallen Astronaut to commemorate the gesture (pl. 95). After the crew returned, at a time when interest in the Apollo program was waning, the sculpture began to draw negative press as an example of a frivolous use of government resources.

By the late 1960s, many on the political left viewed the moonshot as a distraction from the war in Vietnam and the struggles of the civil rights and feminist movements (fig. 18). They argued that economic expenditures on space travel were a waste of resources that could be better used to fight poverty back at home.[10] This critique was made with biting humor by Gil Scott-Heron in his 1970 spoken-word poem "Whitey on the Moon," which asks, "Was all that money I made last year for Whitey on the moon? How come there ain't no money here? Hm! Whitey's on the moon."[11] Despite the incredible excitement surrounding Apollo 11, once the goal of reaching the moon had been achieved, NASA found itself in the predicament of needing to validate its continued existence.

Publicity was integral to NASA's success from the start, and the organization deployed photography and video to capture not only scientific data but also heroic breakthroughs. NASA imagery permeated 1960s popular culture; this phenomenon is typified by Harry Gordon's *"Rocket" Dress* (1968, pl. 96), featuring a screenprinted view of Mercury-Atlas 8 blasting off into space. The mod garment, made of paper rather than fabric, was one of a series of disposable "poster dresses" featuring bold imagery that could be cut at the seams and hung on the wall. The vast array of photographic images disseminated by NASA in the

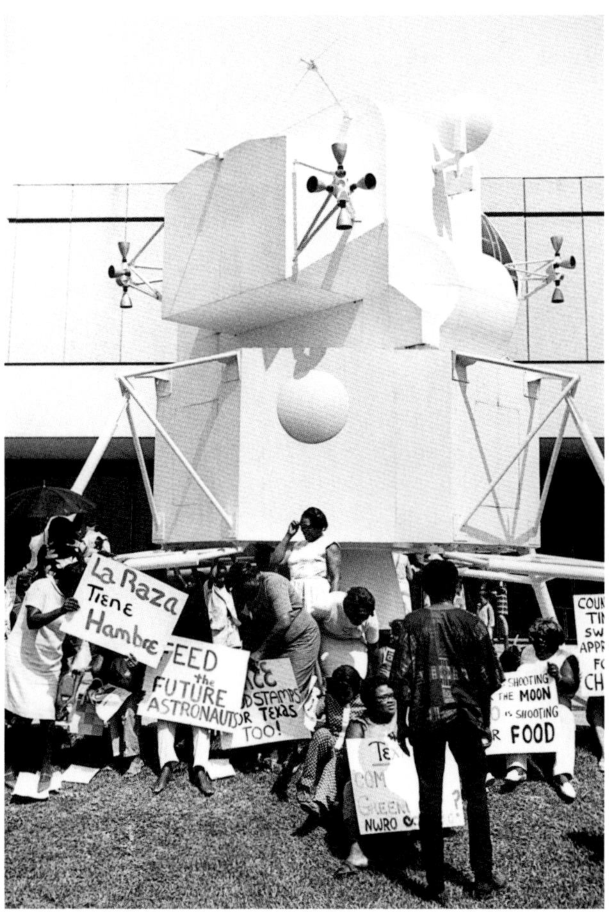

FIG. 18. Mario De Biasi, National Welfare Rights Organization protesters gathered in front of a mock-up of the Apollo lunar module at the Manned Spacecraft Center in Houston, 1969. Archivio Mario De Biasi

mass media also became source material for artists who incorporated strategies of appropriation into conceptual explorations of mediation itself.

It was through the relatively young medium of television that millions of people most famously experienced the moon landing, in the form of the video footage recorded by Neil Armstrong and broadcast around the world.[12] The link between lunar imagery and televisual transmission is especially potent in works by pioneering video artist Nam June Paik. His 1965 installation *Moon Is the Oldest TV* features a circle of television monitors that broadcast in static black-and-white the lunar cycle, orchestrated by the placement of magnets on the cathode-ray tubes. Paik subsequently collaborated with experimental film and video artist Jud Yalkut on works centered on lunar imagery. In *Electronic Moon No. 2* (1966–72, pl. 97), moonlight reflecting off rippling water fills the screen against the soundtrack of Claude Debussy's "Moonlight Sonata." An evanescent multicolored orb appears, flickering and soon pulsing, stretching the boundaries of its spherical shape before reconstituting itself. The final silhouettes of a lit match, a fork and knife, a man's face, and his hand grabbing a woman's breast evoke a new kind of techno-eroticism bathed in the light of a television screen, one inflected with the same masculine energy associated with the space program's self-presentation.

New technologies and space travel also offered opportunities for reimagining modern life, as in the work of the radical architecture firm Superstudio, founded in Florence in 1966. The collective's project Interplanetary Architecture (1970–71, pls. 98, 99) developed as a response to a magazine competition calling for space architecture in the wake of Apollo 11.[13] Superstudio member Alessandro Poli made collages and drawings as preparatory material and visuals for a film inspired by the moon landing, which envisions life on the satellite as an expanded and improved version of that on our home planet. The imagined settlement includes a soccer field, space-age balloon structures, and an earth-moon superhighway conceived as a malleable Cartesian plane on which astronauts careen. Poli's utopian vision—clipped from popular magazines like Italy's illustrated weekly *Epoca* that documented the American space program—was matched by a real political critique. Among his research materials for the project was an issue of *Quindici*, the journal of the avant-garde Gruppo 63, containing a full-page response to the moon landing by a communist workers' group from outside Venice that stated, "The conquest of the moon has demonstrated the immense potential of man, but it has also shown how the master uses science and technology to increase his power and our exploitation."[14] Poli offers an alternative future in which science and technology are harnessed toward a free society.

In addition to appropriating imagery found in mass-media publications, many artists in the 1970s turned their attention to the format of the space program's scientific data. Long interested in natural history and the visualization of science, cross-disciplinary artist Nancy Graves pored over the photographs, charts, maps, and publications issued by NASA, recasting them as complex systems of her own graphic markings. In 1972 she created a series of ten lithographs (pls. 100, 101) inspired by topographical maps of the moon issued by the U.S. Geological Survey, which display photographic data gathered during the Lunar Orbiter and Apollo missions overlaid with bright colors designating the various geological features of the moon (fig. 19). In Graves's interpretation of Fra Mauro, site of the Apollo 14 landing, bright pinks, purples, yellows, and blues echo the colors of the official document, but the lunar topography dissolves in a thrumming rhythm of pointillist dots. The resulting composition is faithful to its source material and yet arbitrary in its formal vocabulary—as are, Graves suggests, all mapping projects. Christina Hunter notes that for Graves, lunar charts were especially compelling because their accuracy could not be verified with the naked eye: "Graves utilizes the *factual* physical remoteness of these locales to indicate the *conceptual* distance between a map and any place, and by extension between an artistic representation and physical reality."[15]

Other artists responding to Apollo 11 were particularly interested in making visible this traversal of physical and conceptual distance. California-based conceptualist Michelle Stuart's earliest pieces inspired by the moon landing are painstaking photorealist drawings of the lunar surface based on NASA photographs, similar to contemporaneous works made by Vija Celmins (fig. 20). Whereas the technical virtuosity of Celmins's photorealistic drawings

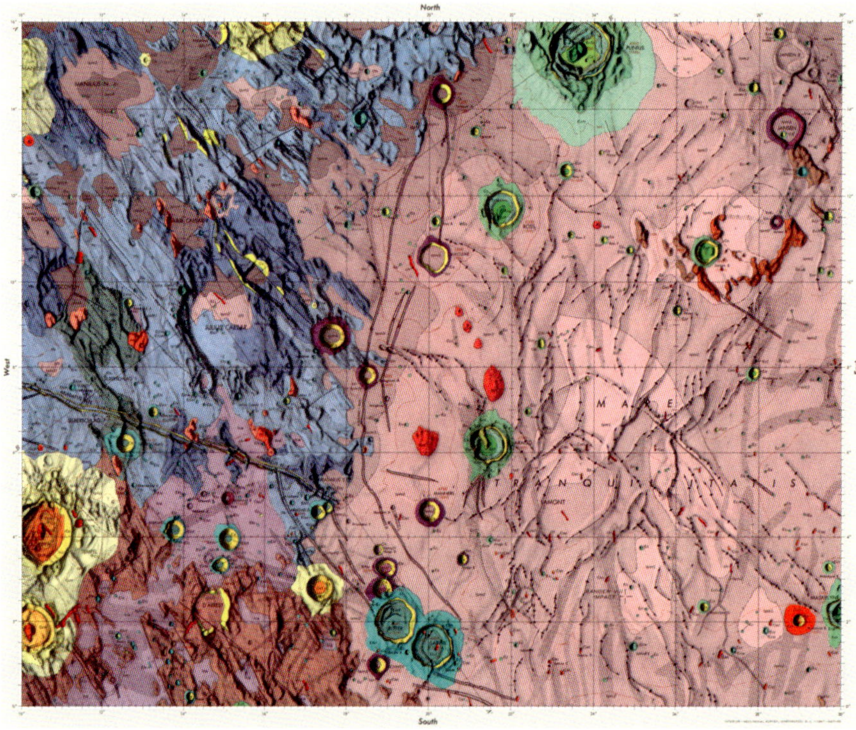

FIG. 19. United States Geological Survey, Geologic Atlas of the Moon, Julius Caesar Quadrangle, I-510 (LAC-60), 1967

challenged the limits of human perception in a machine age,[16] Stuart's concerns were more firmly rooted in the natural world. Stuart's *#7 Moon Tide* (1969, pl. 102) represents an investigation of our planet's relationship to its moon, charted across a sequence of squares. Within those defined boundaries, Stuart included cut pieces of photographs ordered from NASA, her own drawings based on those images, and graphite rubbings of the earth's surface. Tiny pinpricked holes and pencil lines create subtle waves representing the magnetic forces that bind the two celestial bodies and the tidal rhythm that results. The performative act of rubbing the earth links *#7 Moon Tide* to Stuart's later works of Land Art, a movement heavily influenced by the widening of eco-consciousness in the wake of human space travel.

The midcentury rhetoric of space exploration frequently linked the moon's desolate landscape with that of the American West, a comparison drawn in California-based photographer Judy Dater's *Self-Portrait at Craters of the Moon* (1981, pl. 103). As "the final frontier," outer space became the last uncharted territory staked out for American progress, and missions to the moon focused on potential geological resources that could benefit science and industry. The government's vast financial support for space travel and the photographic recording of lunar exploration call to mind the great survey expeditions to the American West documented by the likes of Timothy O'Sullivan and Eadweard Muybridge in the latter half of the nineteenth century. Dater's photograph, part of a series made during ten trips to national parks in the western United States between 1980 and 1983, is a highly personal feminist exploration of the same territory. The small, stark-white figure of the artist appears otherworldly in the barren volcanic fields of Idaho known as Craters of the Moon. The shadow of a

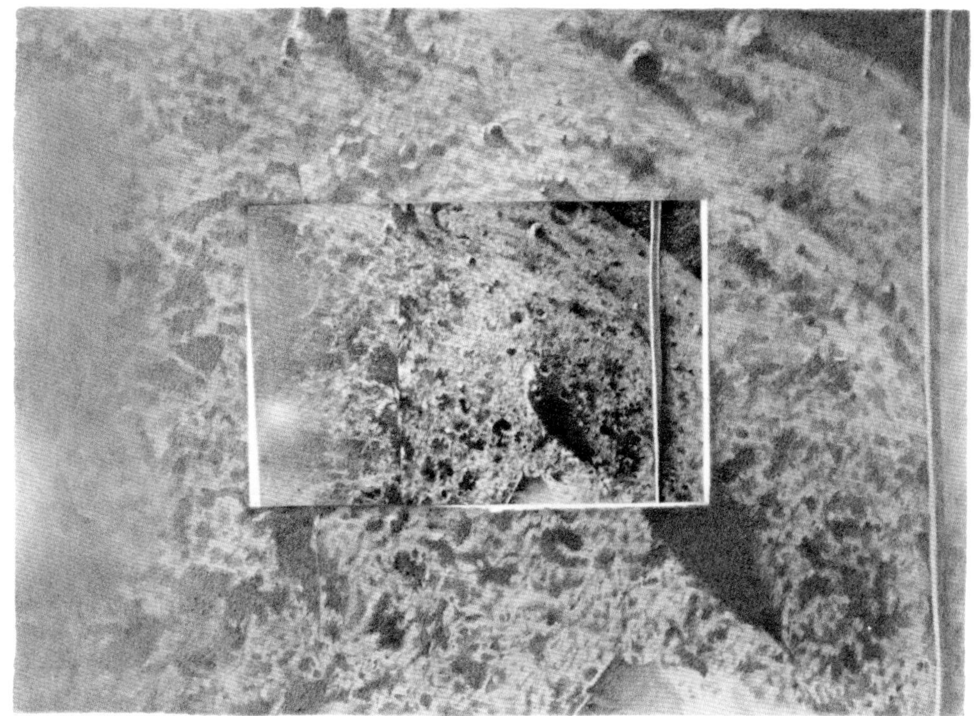

FIG. 20. Vija Celmins (American, b. Riga, Latvia, 1938), *Moon Surface (Luna 9) #1*, 1969. Graphite on acrylic ground on paper, 13¾ x 18½ in. (35 x 47.2 cm). Museum of Modern Art, New York, Mrs. Florene M. Schoenborn Fund (584.1970)

camera tripod in the foreground heightens the impression of an alien landscape and evokes the iconic photographs made during the Apollo 11 moon landing.[17] Dater's performative intervention into the landscape is a potent reminder that no woman has set foot on the moon.

Addressing this historical exclusion, on August 28, 1999, the Poland-born American artist Aleksandra Mir became *First Woman on the Moon* (pl. 104), staging the triumphant moment on a beach in the Netherlands. Over the course of a day, workers transformed the shoreline in the small town of Wijk aan Zee into a hilly and crater-filled lunar landscape, as sunbathers watched in amusement. At sunset Mir climbed a hill and planted an American flag, accompanied by the celebratory drumming of a band. She then invited the exuberant crowds to join her on the "moon" and claim their own slice of history, shouting, "I am the first gay man, first black man, first German, first pregnant woman, on the moon!"[18] before bulldozers returned the beach to its natural state. Mir grew up in Gothenburg, Sweden, home to the Hasselblad firm that had supplied photographic equipment to NASA. The company donated cameras to document her performance, which was also broadcast live on Dutch television in imitation of the media spectacle surrounding Apollo 11. Mir's feminist and transnational recasting of the moon landing joyously reasserts a sentiment conveyed by photographer Stephen Shames, who, while documenting the activities of the Black Panther Party three decades earlier, had highlighted the graffiti message "The moon belongs to the people!!!" (1971, pl. 105).

A different sort of restaging, in this case an imagined reenactment of Zambia's short-lived space program, is the subject of The Afronauts, a series by Spanish photographer Cristina De Middel. In *Bambuit* (2012, pl. 106), a figure wearing a space suit of boldly printed African fabrics and a globular glass helmet fashioned from a streetlight cover wanders incongruously through the tall grasses of a terrestrial environment. When Zambia became an independent nation in 1964, Edward Mukaka Nkoloso — a political revolutionary, science teacher, and self-appointed director of the country's National Academy of Science, Space Research, and Philosophy — recruited teenagers and prepared them to endure the hardships of space travel, with the goal of beating the Americans and Soviets by landing Africans on the moon and Mars. The sincerity of Nkoloso's project is unclear; some have interpreted it as a satirization of imperialism, a cover for his activities as a freedom fighter, or a fanciful yet hopeful dream for his country's future.[19] De Middel's photographs use photography's status as evidentiary proof to envision Nkoloso's fantasies as a reality.

Mir's and De Middel's projects nod to the conspiracy theory that the Apollo 11 landing was faked in a Hollywood studio (possibly under the direction of Stanley Kubrick) so that the United States could claim to have won the space race. The Swiss artists Jojakim Cortis and Adrian Sonderegger also play on this idea in their tongue-in-cheek examination of how iconic images are constructed. Following in the spiritual footsteps of another mischievous Swiss duo, Peter Fischli and David Weiss, the artists build painstakingly accurate models of famous photographs and then rephotograph them with the camera pulled back to reveal the tools of their artifice. In *Making of AS11-40-5878 (by Edwin Aldrin, 1969)* (2014, pl. 108), a square pile of gray dust bears the recognizable mark of a boot print left on the lunar surface. Strewn around that familiar composition, however, are paintbrushes, a saw, bags of cement, and a print of Aldrin's original photograph. Cortis and Sonderegger's work calls attention to the way certain images form a collective cultural experience, a principle that also figures in Brazilian sculptor and photographer Vik Muniz's *Memory Rendering of the Man on the Moon* (1990, pl. 107). To produce it, Muniz first drew Armstrong's iconic photograph of Aldrin entirely from memory, then photographed his drawing and printed it through a halftone screen, thus mimicking the format of the mass-media publications that disseminated the original image. Compared side by side with Armstrong's photograph, Muniz's creation has the same basic contours, although the details are lacking. The resulting image appears a bit hazy, as though materialized from a cloudy recollection, yet it is nonetheless instantly identifiable.

Similarly, the British conceptual artist Lenka Clayton's *Moon 25/12/2015 (& 1977)* (2015, pl. 109), a reinterpretation of an image produced by the Lunar Reconnaissance Orbiter, uses a distinct format to call attention to the transmission of photographs across media and memory. While composing her work over the course of several months on an aptly named Smith Corona Skyriter typewriter, Clayton repeatedly folded and reinserted the paper to create a variety of graphic markings that form subtle shadows across the surface of a full moon. Despite its cumbersome means of creation, the resulting image is faithful to the photographic source material. As in earlier works by Celmins, Graves, and Stuart, the artist's laborious process unites technology and handicraft. For Clayton, the method renders a machine-made image personal; it represents the first full moon to occur on Christmas since 1977, the year of her birth. Canadian photographer and conceptualist Penelope Umbrico's *Everyone's Moon 2015-11-04 14:22:59* (2015, pl. 110) likewise examines the relationship of photography and technology to individual experience. In the video, 1.1 million images of the full moon made by both professional and amateur photographers and captured by the artist as screenshots from the photo-sharing site Flickr scroll by at lightning speed. Emphasizing the accumulation of images on the internet, the work also expresses how looking at (and photographing) the moon is both a universal impulse and subjective experience.

The intertwined relationship of looking and longing is suggested in *Moon Watch* (2012, pl. 111) by Sarah Charlesworth, an American artist known for appropriating photographic images as a means of investigating how the medium functions as a visual language. Her sparse diptych explores the metaphysical and even magical association of photography with the cosmos. On the left, a small image of the moon hovers at the top of a vertical field of pure black, while on the right, a telescope on a tripod tilts longingly toward it against a solid white background. The paired images of moon and telescope suggest an experience suspended between the intimate and the infinite, conjuring visions of a lone astronomer in an observatory or a solitary viewer inside a camera obscura. While Charlesworth's diptych evokes a sense of wonder that harks back to the dawn of astronomical photography and the Romantic period, Kiki Smith's *Tidal* (1998, pl. 113) was directly inspired by nineteenth-century lunar photographs produced at the Harvard College Observatory. An accordion-folded artist's book comprising thirteen photogravures of the full moon — from photographs made using the telescope at Columbia University's observatory — *Tidal* is as sumptuous as a medieval book of hours. The lunar orbs float on the page above a delicate skirt of Japanese paper printed with a continuous expanse of ocean. Their repetition, echoed by the folded-paper format, evokes cyclical time (there are thirteen full moons in a year) and the female body's monthly cycle, and suggests a devotional practice.

Photography's relationship to temporality and to nature are central themes in Darren Almond's *Fifteen Minute Moon* (2000, pl. 112). Although the photograph recalls the picturesque landscapes of nineteenth-century British artists in its seeming depiction of a misty morning view, it is in fact the product of a fifteen-minute exposure made by the light of a full moon, and what appears to be a gleaming stream is actually car headlights on a nearby road. The resulting image challenges the eye's (and by extension, the camera's) capacity to interpret reality accurately. In Sharon Harper's similarly enchanting *Moon Studies and Star Scratches, No. 6* (2004, pl. 114), multiple exposures of the night sky made at different locations seem to bend time and space into an ethereal abstraction. Stars appear as streaking lines of light and the moon registers as multiple shining crescents and dots puncturing an expanse of night sky composed of glowing orange and inky purple and black tones. Her photographs make visible a sense of time's passing, suggesting both singular moments captured by a camera and the unfathomable time of light-years.

Duration is also key to the process behind Kikuji Kawada's *Artificial Moon Trail, Tokyo* (1969, pl. 115), made in the immediate wake of the Apollo 11 landing. The Japanese artist pointed his camera at the full moon, rotating it during a long exposure to create an abstract geometric pattern that traces an impossible trajectory through the night sky. Kawada's image is a literal enacting of the meaning of "photography"— writing with light — and, like Harper's, it distills the medium to its essential elements to sublime effect. Works such as these that refer to the earliest mechanics of photography demonstrate that the moon still holds a primordial power to astound. They suggest that although from its inception photography has been most closely associated with the sun, it is the moon that has remained its steadfast muse.

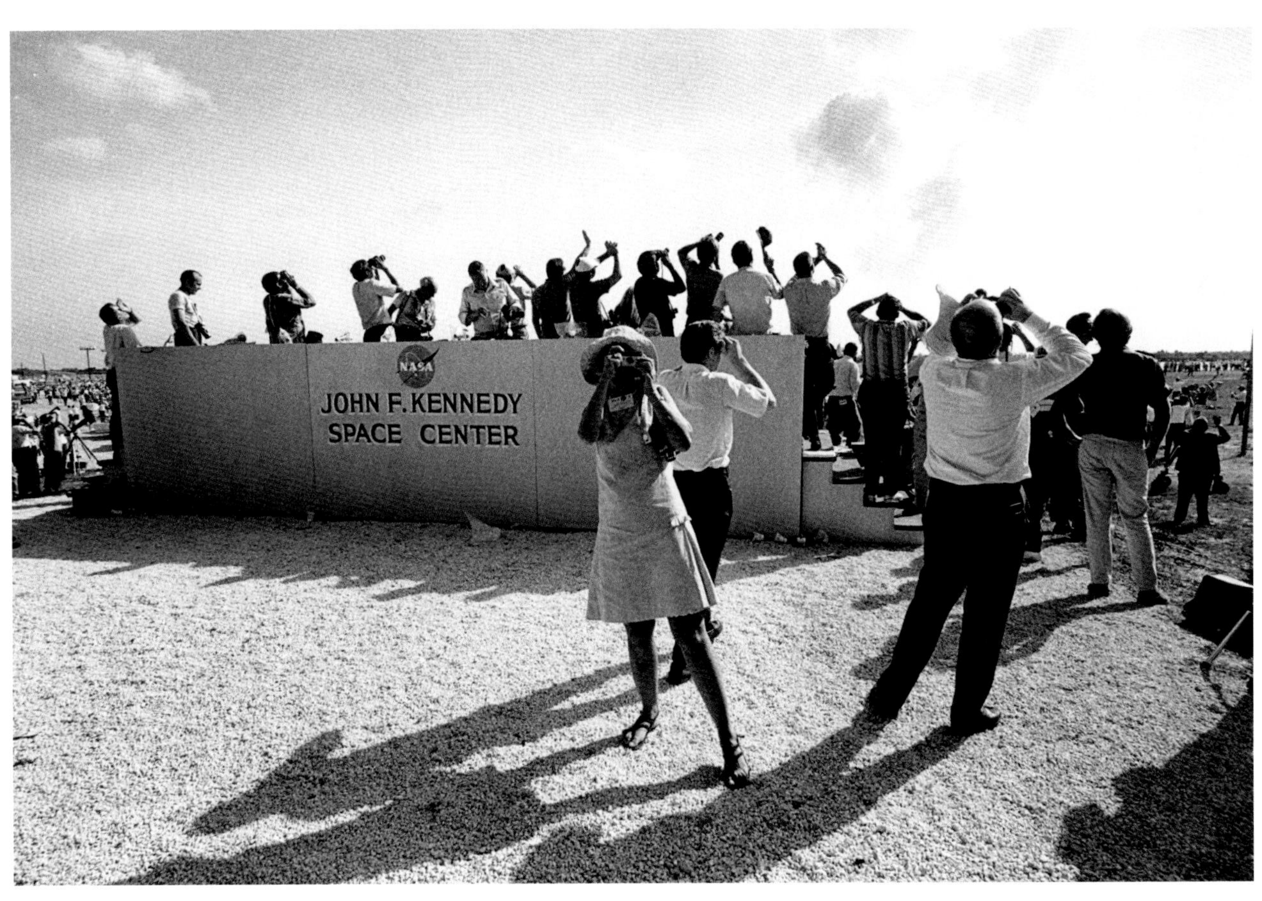

89 | GARRY WINOGRAND | *Apollo 11 Moon Shot, Cape Kennedy, Florida*, 1969

90 | ROBERT RAUSCHENBERG | *Sky Garden (Stoned Moon)*, 1969

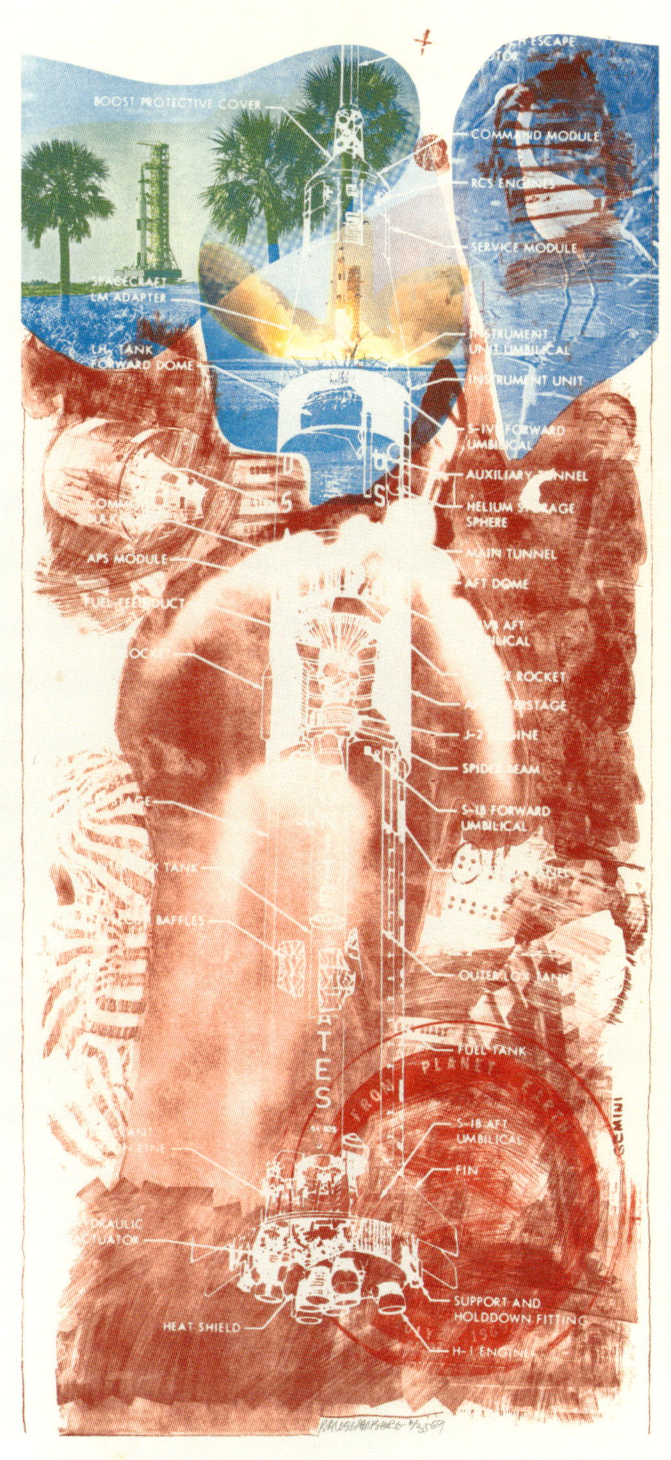
BOOST PROTECTIVE COVER
COMMAND MODULE
SERVICE MODULE
SPACECRAFT LM ADAPTER
INSTRUMENT UNIT UMBILICAL
INSTRUMENT UNIT
APS MODULE
MAIN TUNNEL
AFT DOME
SPIDER BEAM
S-IB FORWARD UMBILICAL
FUEL TANK
S-IB AFT UMBILICAL
FIN
HEAT SHIELD
SUPPORT AND HOLDDOWN FITTING
H-1 ENGINE
FROM PLANET EARTH
GEMINI

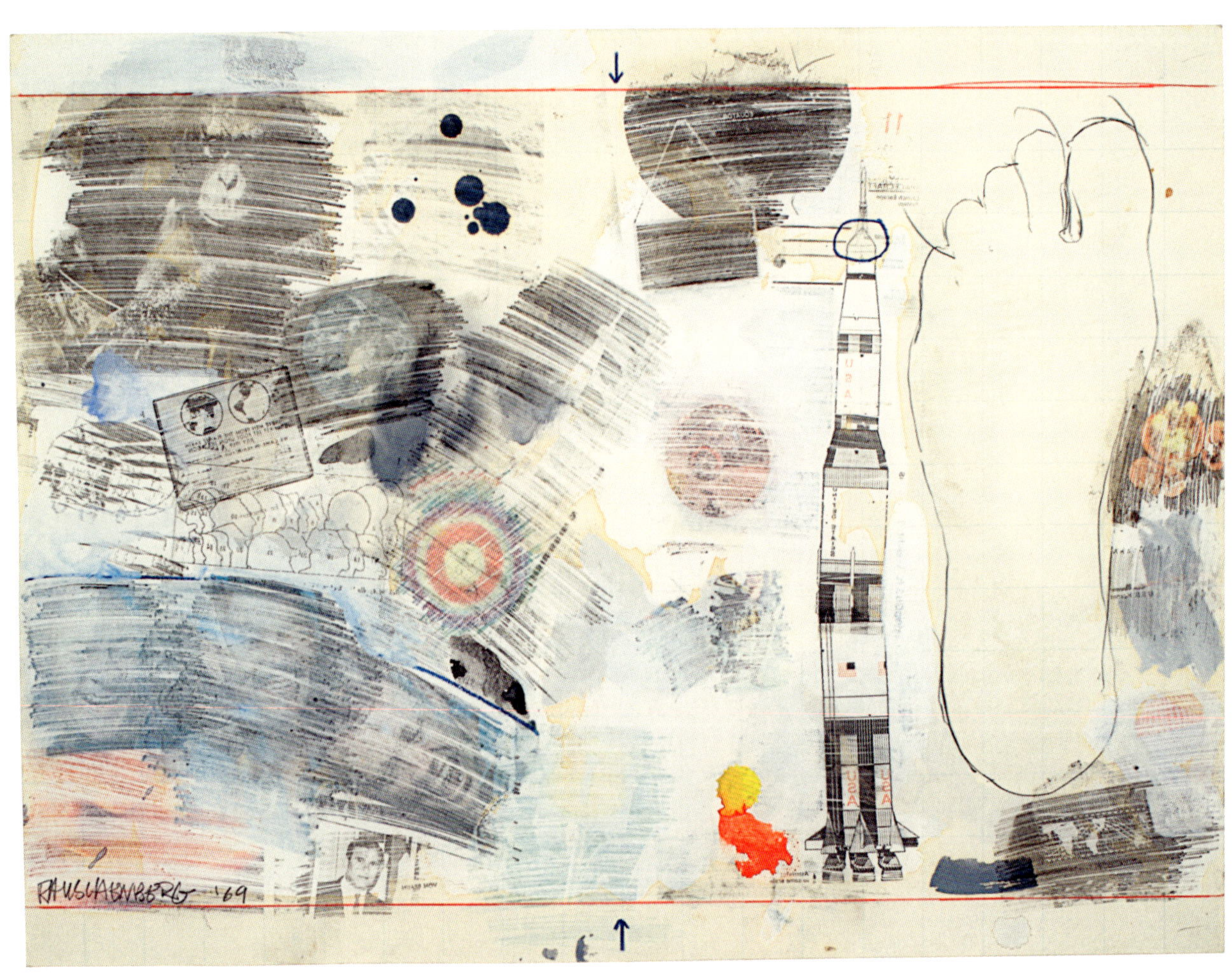

91 | ROBERT RAUSCHENBERG | *Untitled*, 1969

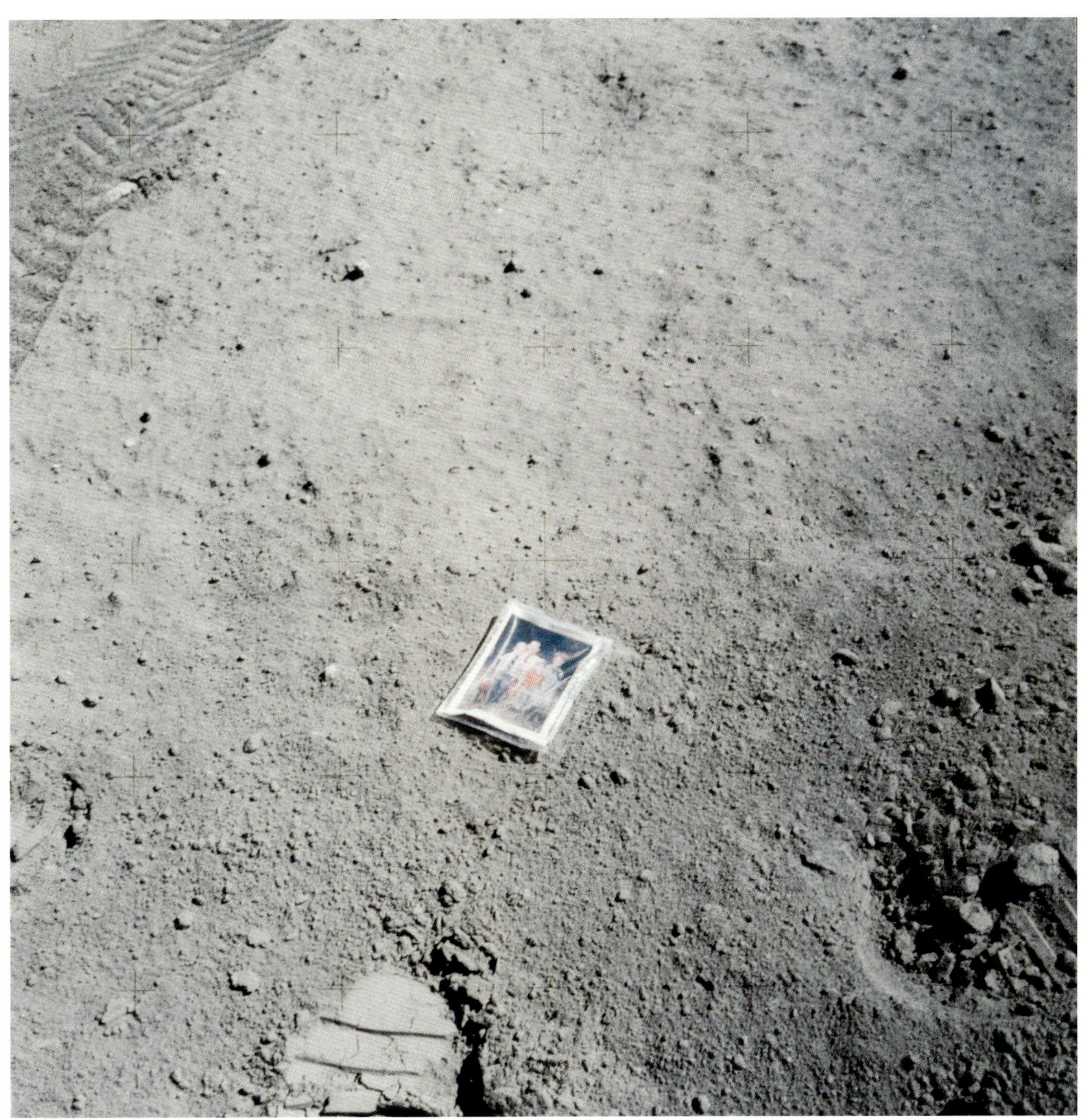

92 | CHARLES DUKE (NASA APOLLO 16) | *Duke Family Photograph on the Lunar Surface*, 1972

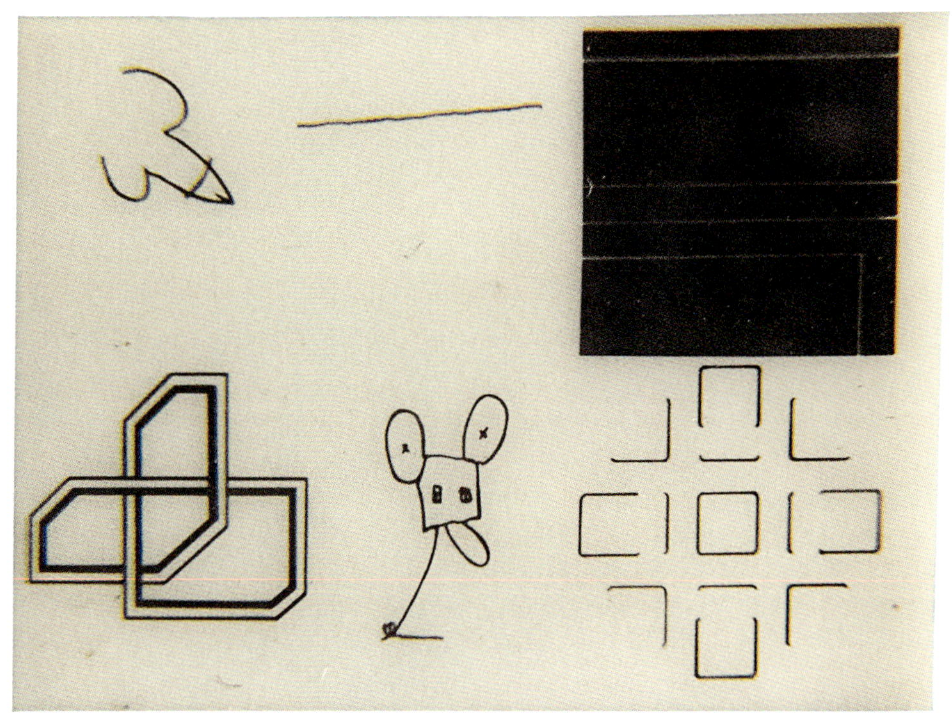

93 | JOHN CHAMBERLAIN, FORREST MYERS, DAVID NOVROS, CLAES OLDENBURG, ROBERT RAUSCHENBERG, AND ANDY WARHOL | *The Moon Museum*, 1969

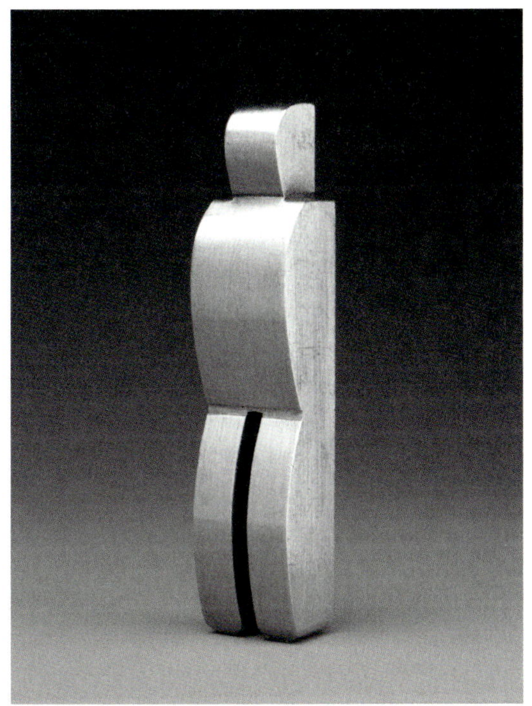

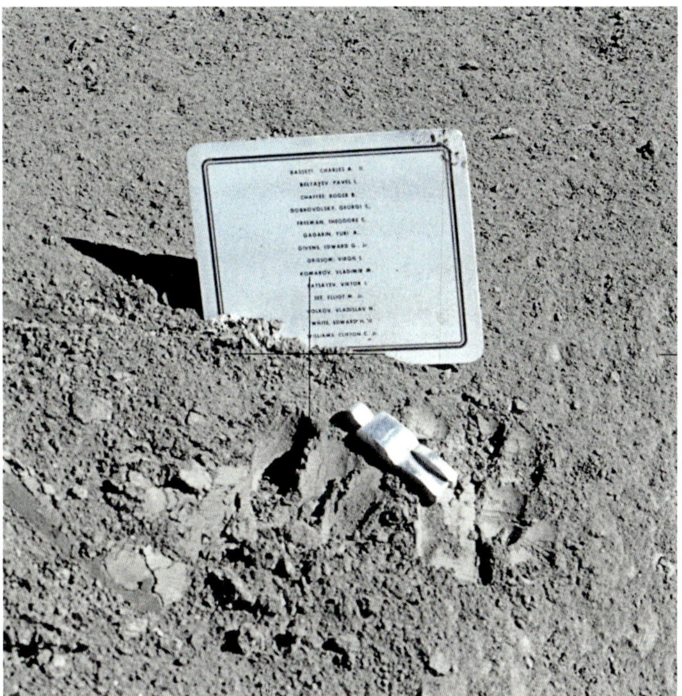

94 | PAUL VAN HOEYDONCK | *Fallen Astronaut* replica, 1971

95 | DAVID SCOTT (NASA APOLLO 15) | *Paul van Hoeydonck's "Fallen Astronaut" Sculpture on the Lunar Surface*, 1971

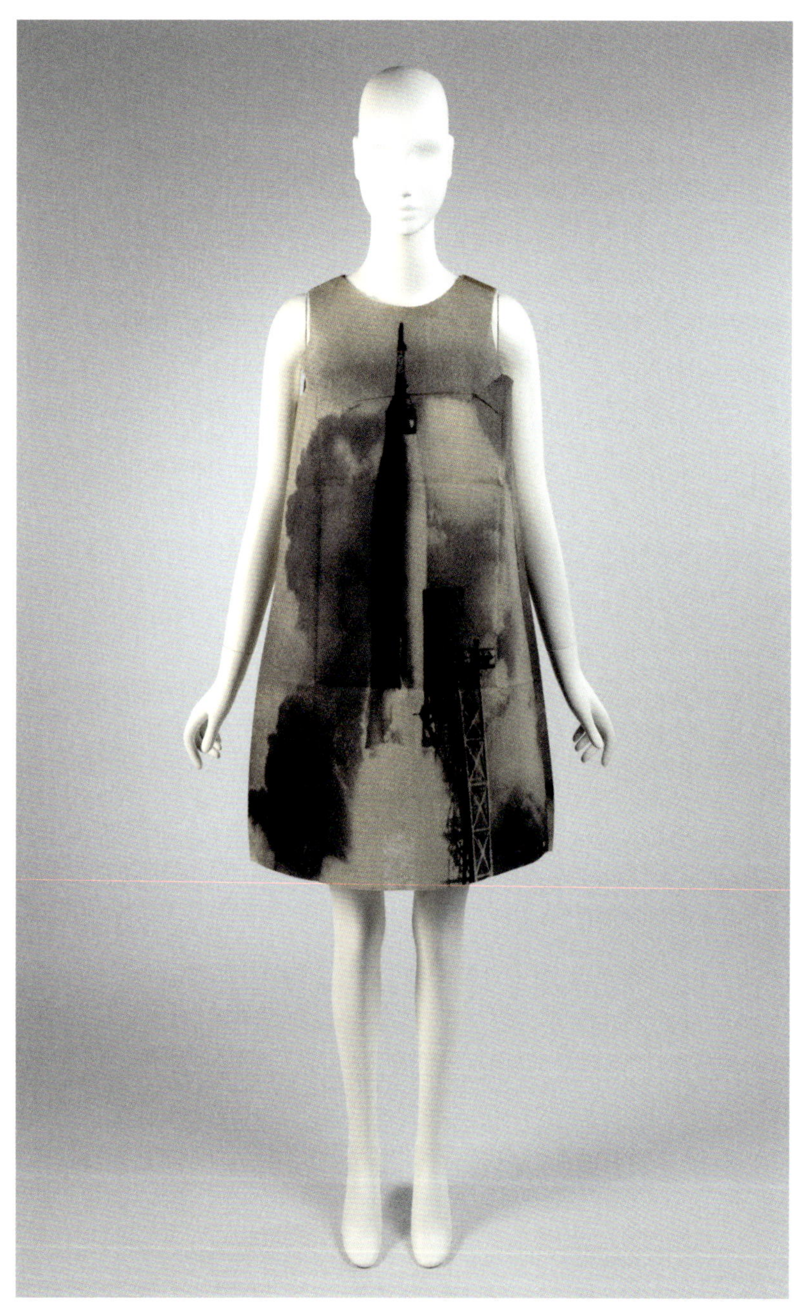

96 | HARRY GORDON | *"Rocket" Dress*, 1968

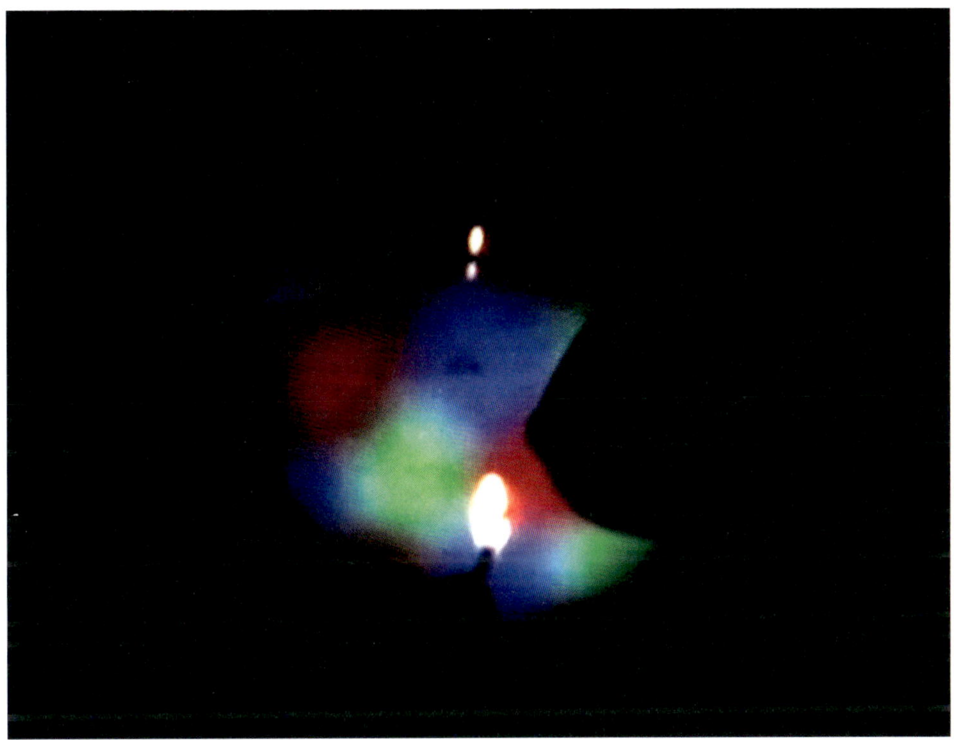

97 | NAM JUNE PAIK AND JUD YALKUT | Still from *Electronic Moon No. 2*, 1966–72

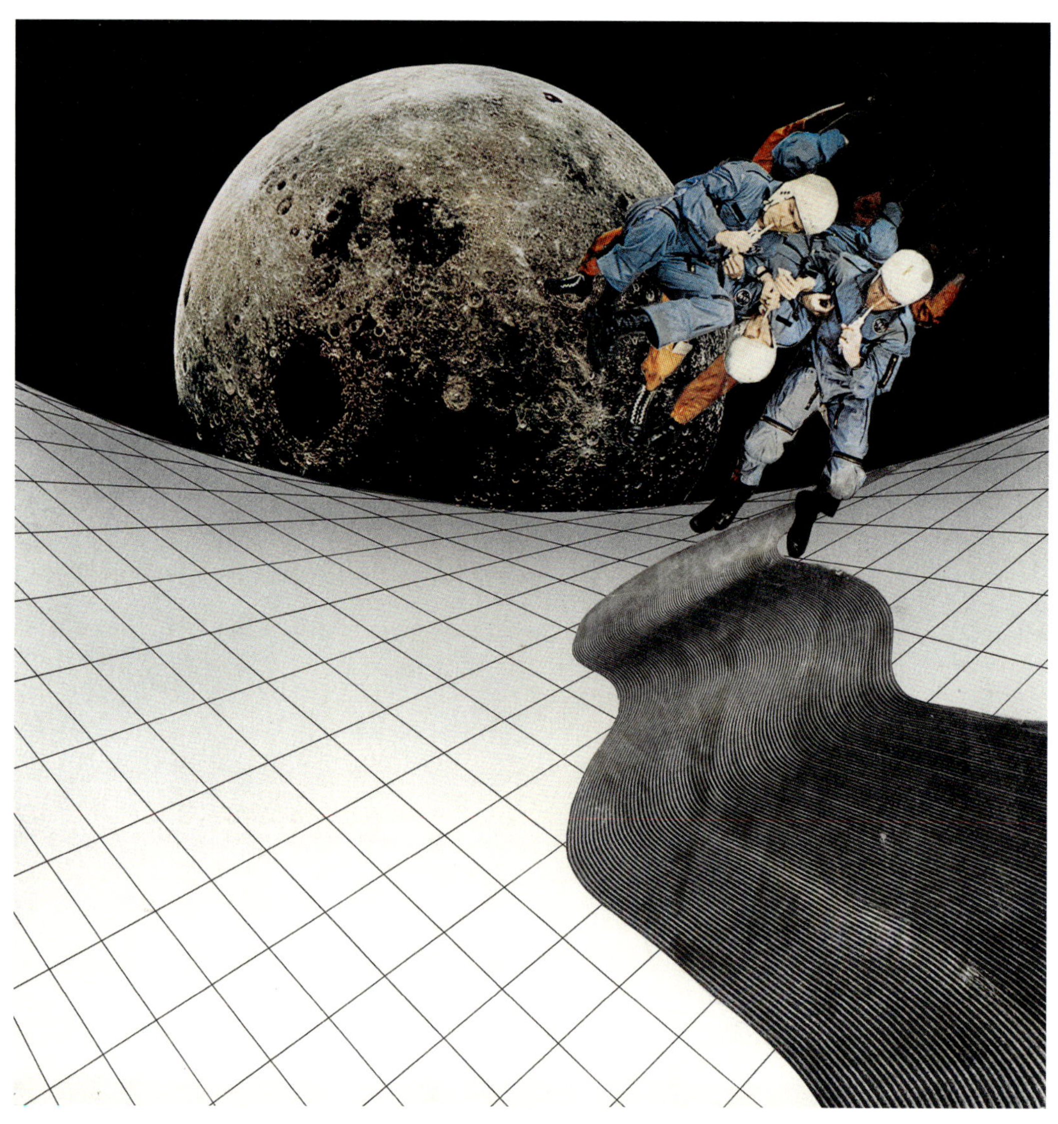

98 | ALESSANDRO POLI AND SUPERSTUDIO | *Highway with Sine Wave of Energy,* from the series Interplanetary Architecture, 1970–71

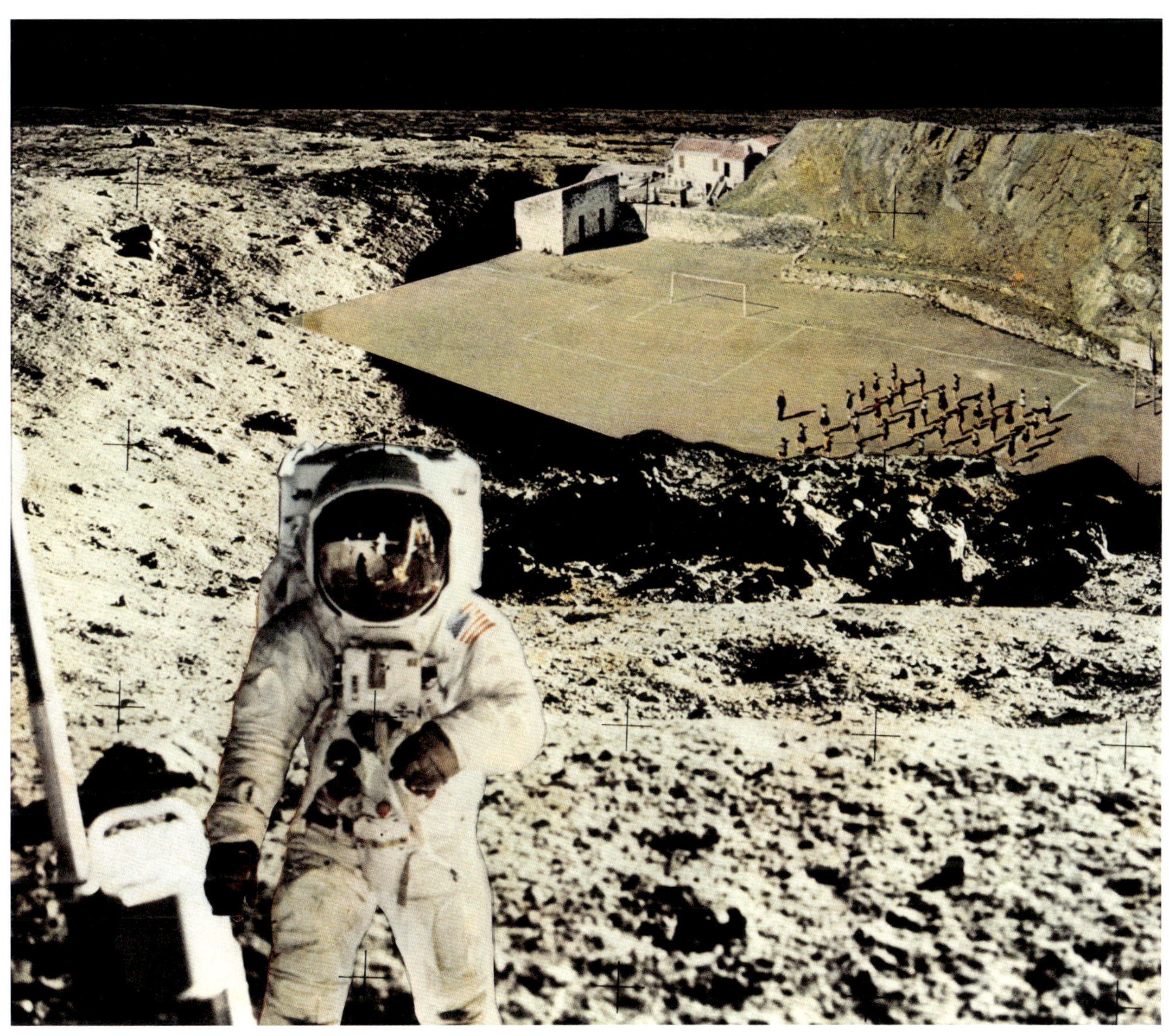

99 | ALESSANDRO POLI AND SUPERSTUDIO | *New Rural Landscapes,* from the series Interplanetary Architecture, 1970–71

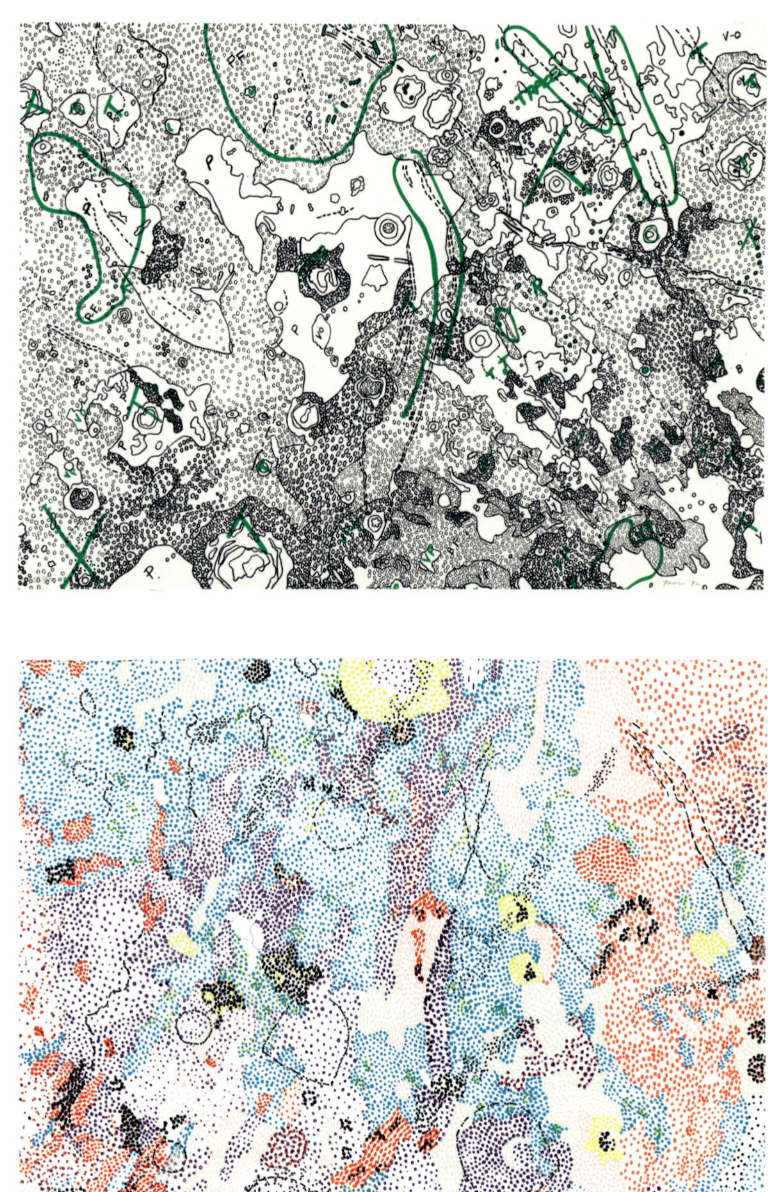

100, 101 | NANCY GRAVES | *IV Julius Caesar Quadrangle of the Moon* and *II Fra Mauro Region of the Moon*, from the series Lithographs Based on Geologic Maps of Lunar Orbiter and Apollo Landing Sites, 1972

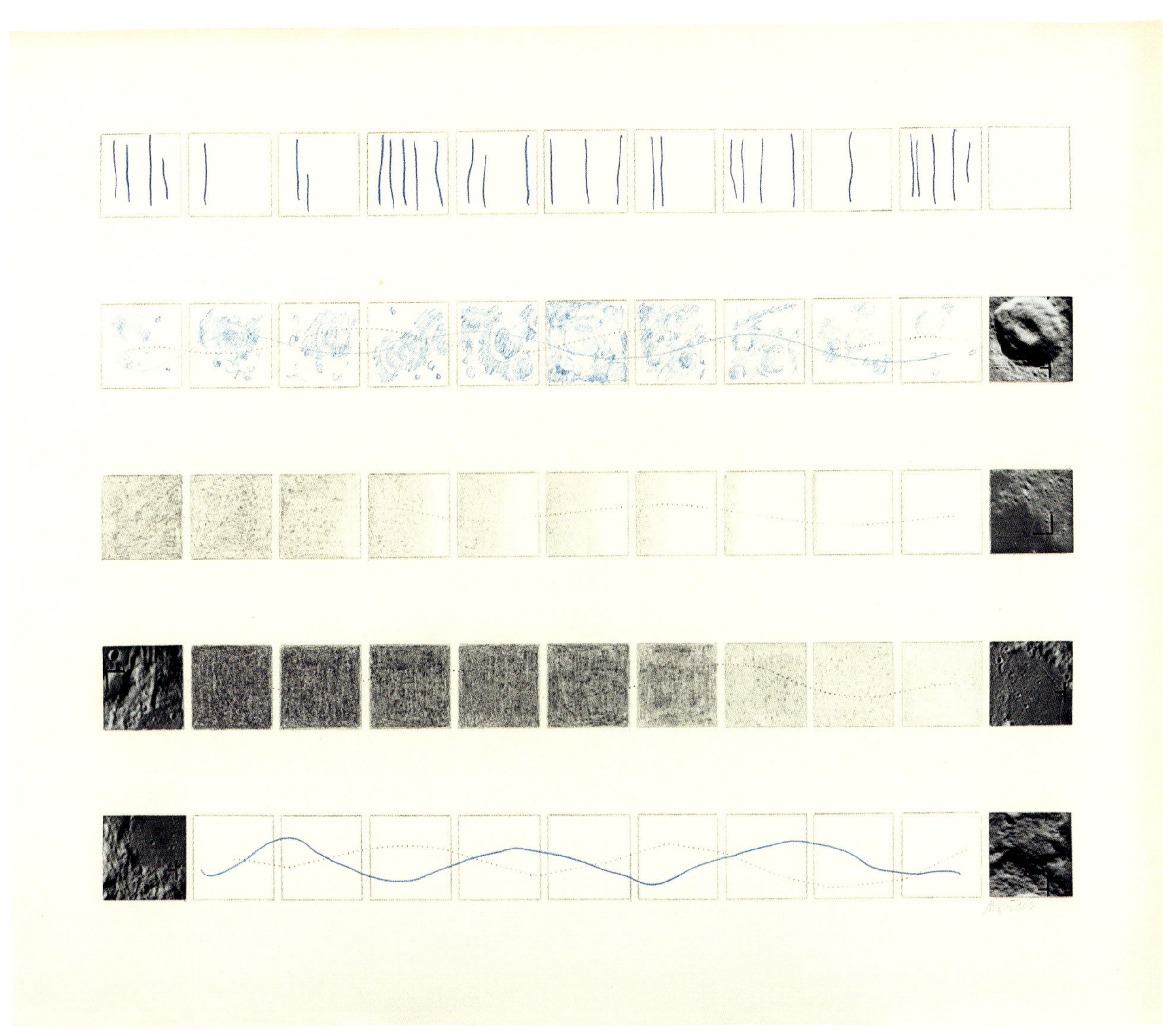

 | MICHELLE STUART | #7 *Moon Tide*, 1969

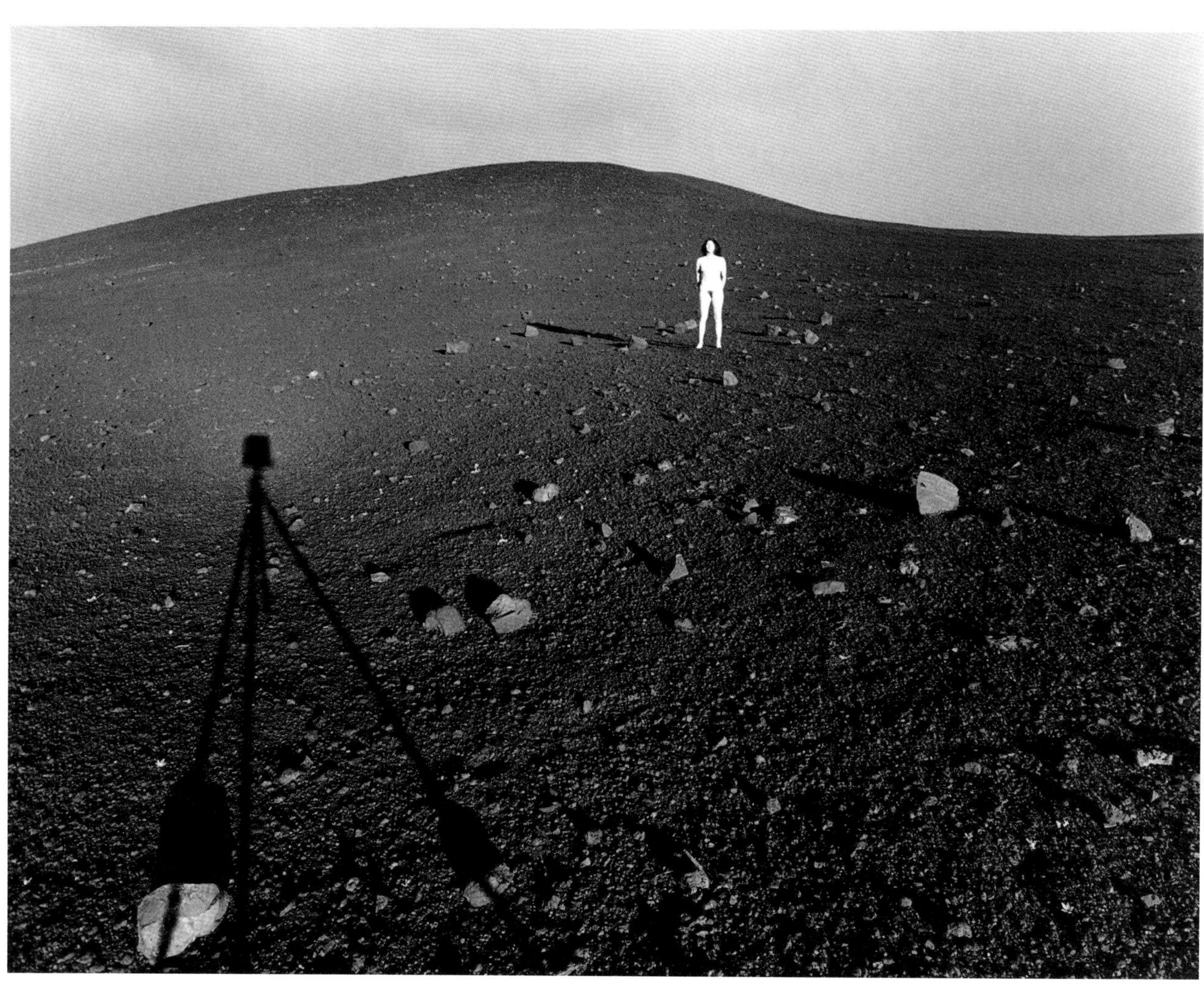

103 | JUDY DATER | *Self-Portrait at Craters of the Moon*, 1981

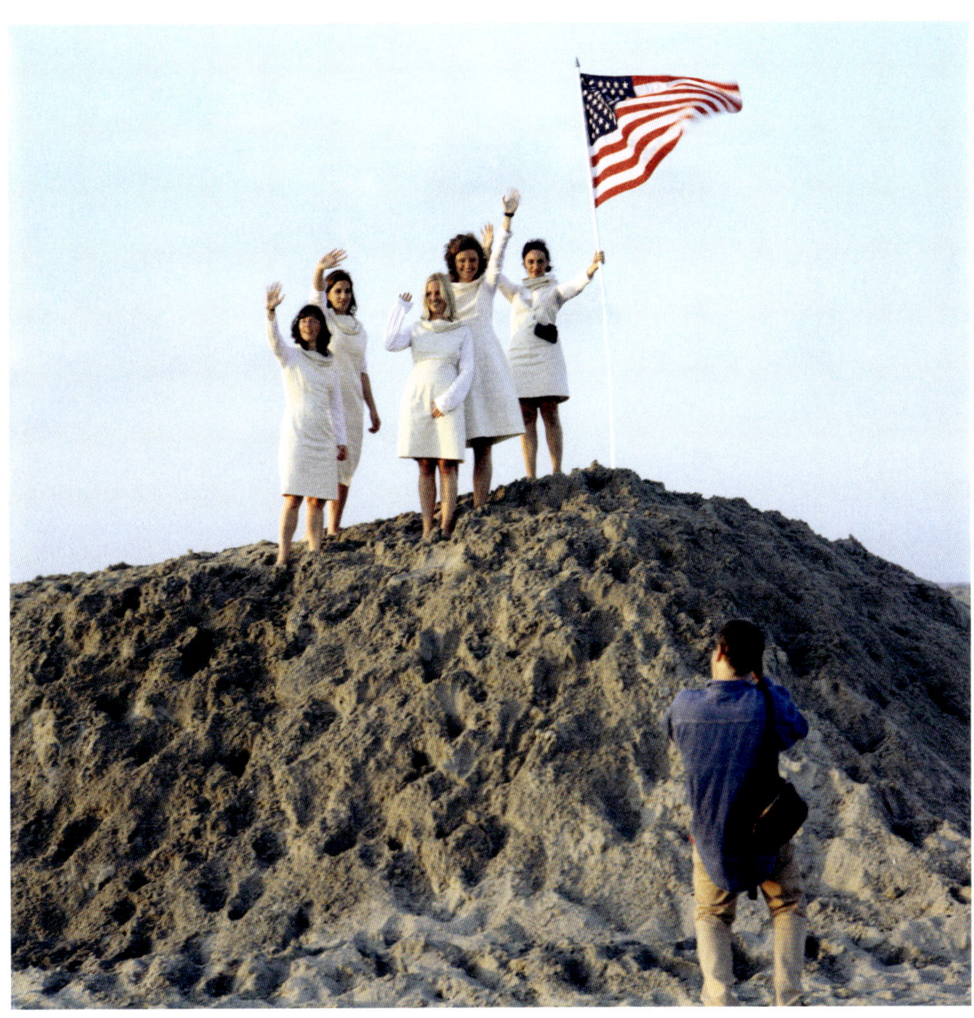

 | ALEKSANDRA MIR | *First Woman on the Moon*, 1999

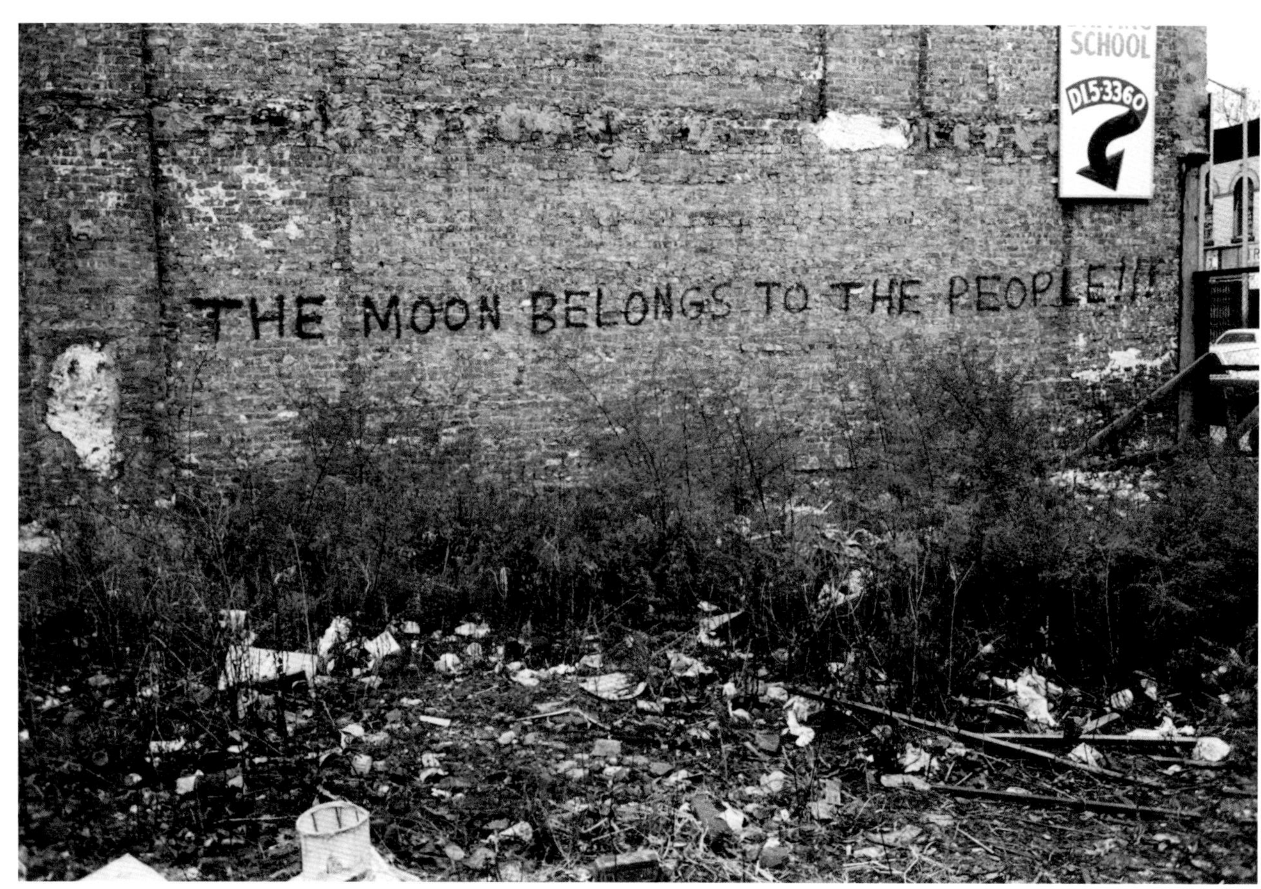

105 | STEPHEN SHAMES | *The Moon Belongs to the People*, 1971

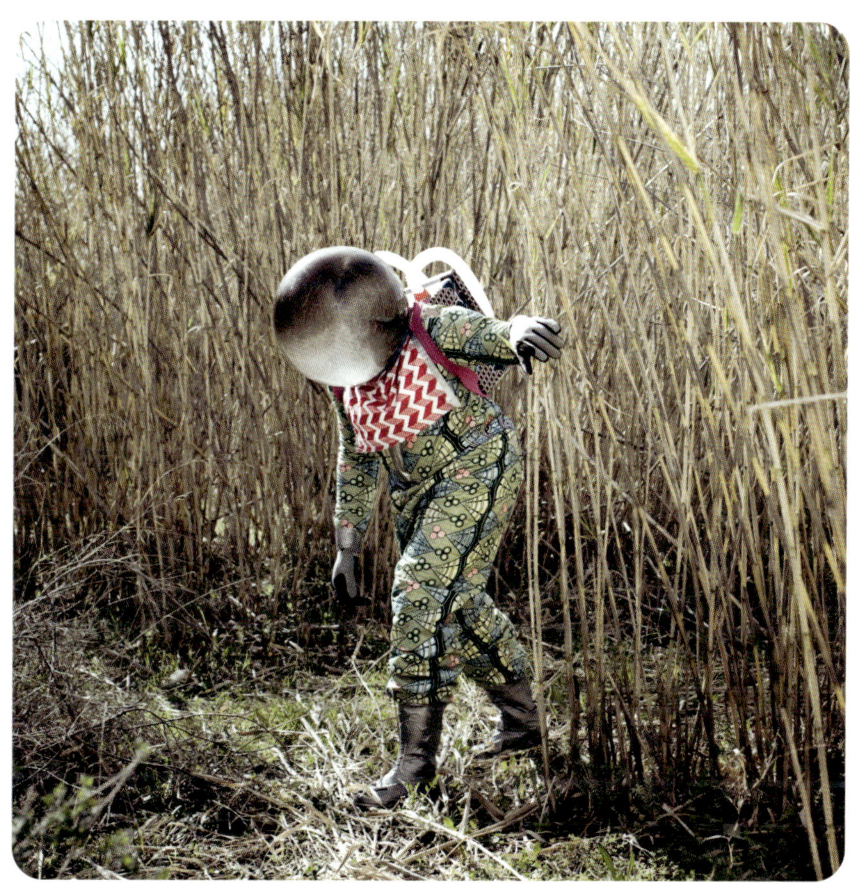

 | CRISTINA DE MIDDEL | *Bambuit*, from the series The Afronauts, 2012

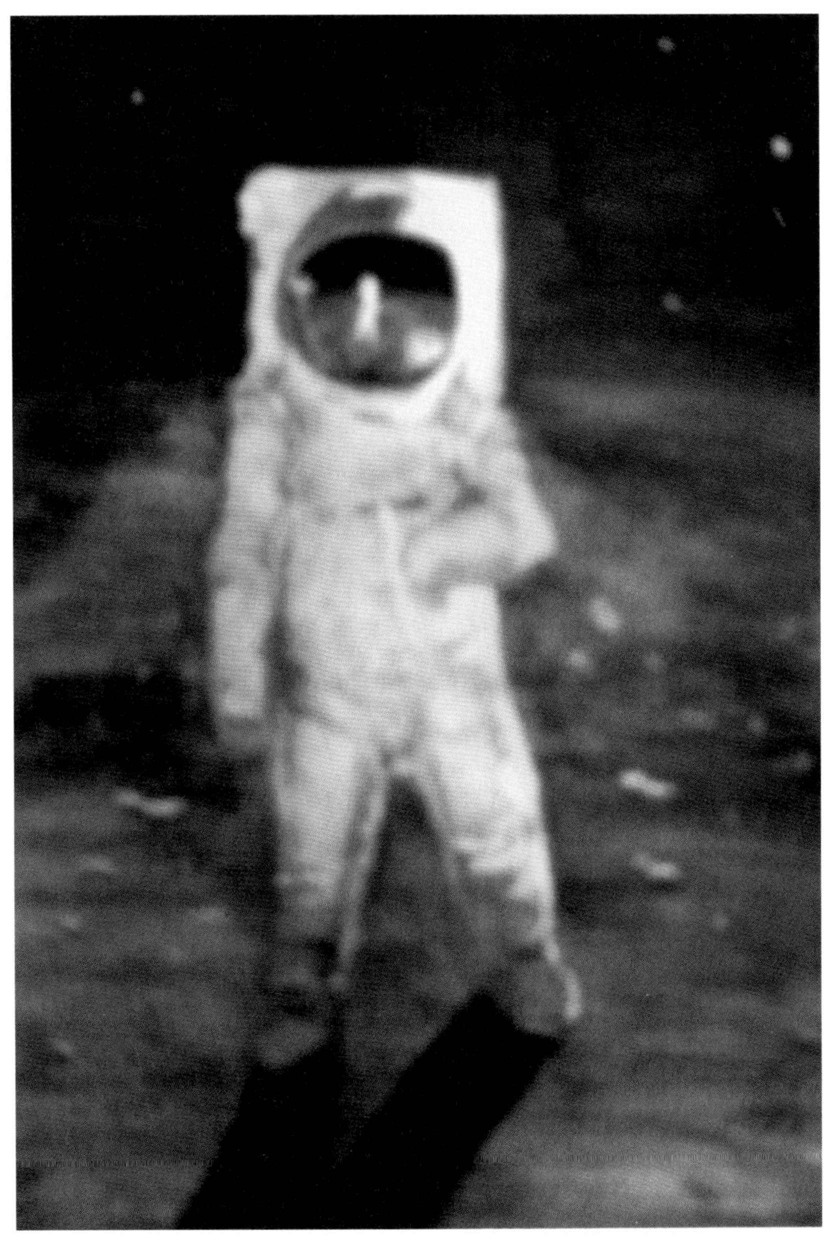

107 | VIK MUNIZ | *Memory Rendering of the Man on the Moon*, 1990

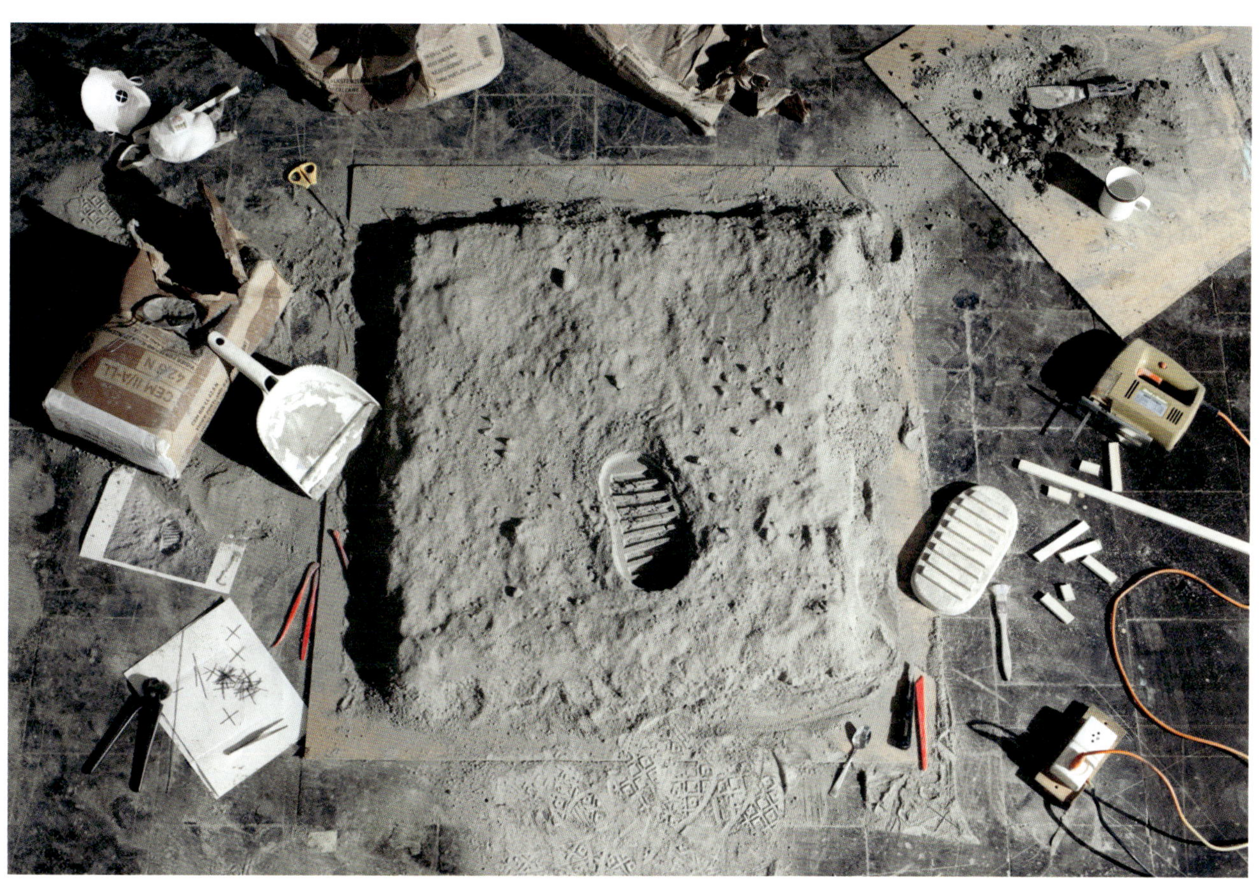

 | JOJAKIM CORTIS AND ADRIAN SONDEREGGER | *Making of AS11-40-5878 (by Edwin Aldrin, 1969)*, 2014

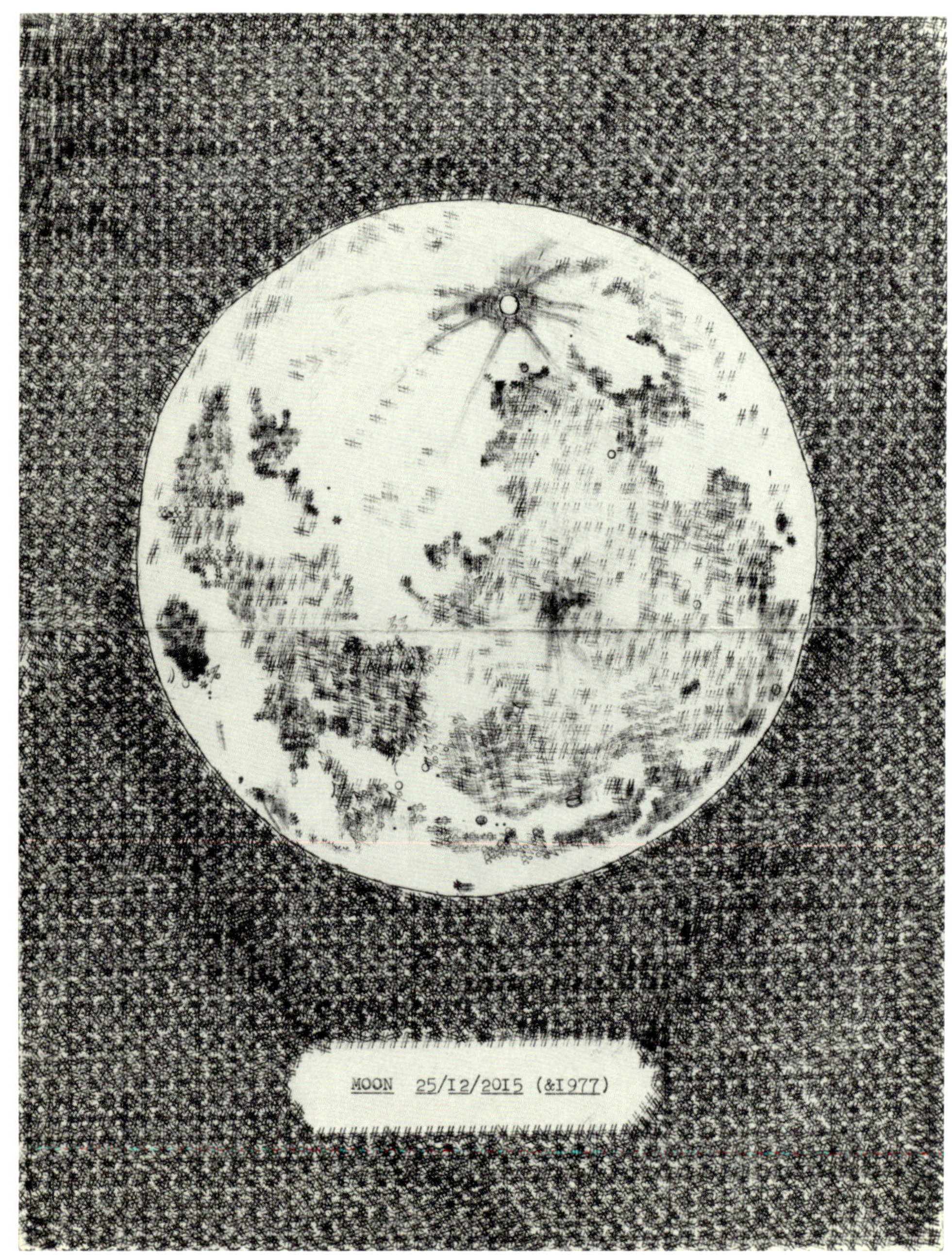

109 | LENKA CLAYTON | *Moon 25/12/2015 (& 1977)*, 2015

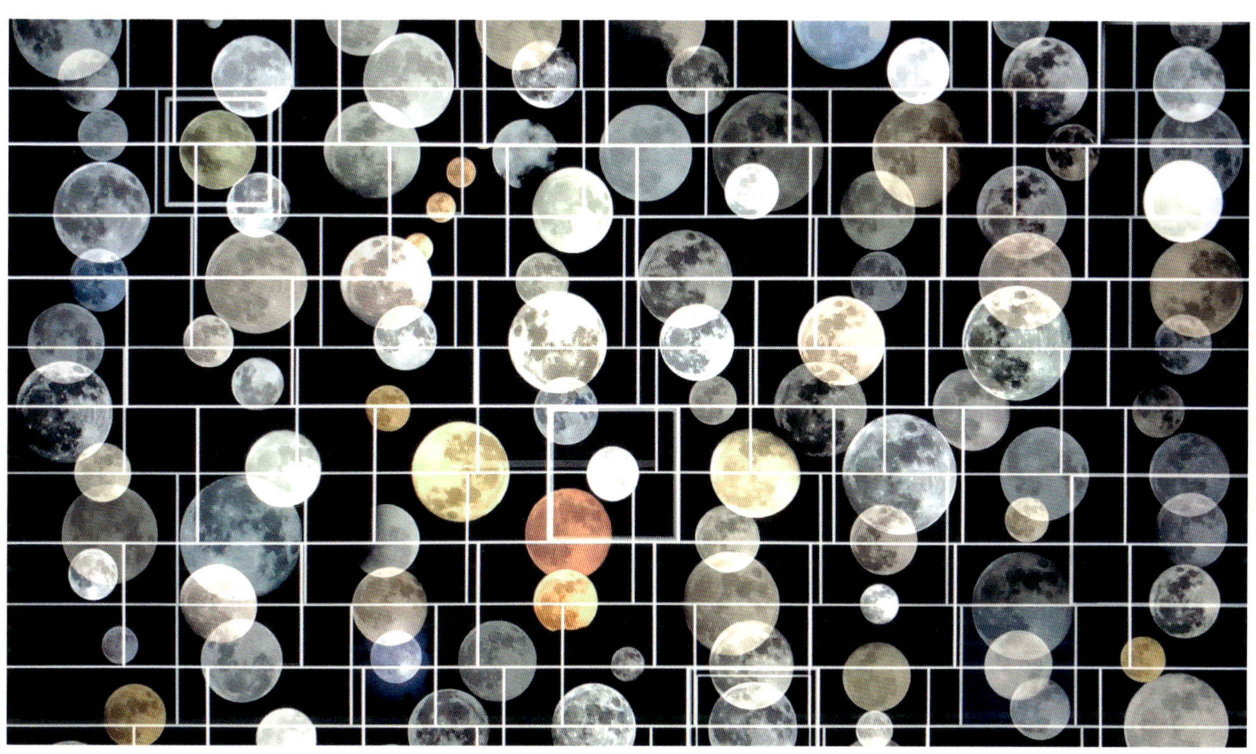

 Still from *Everyone's Moon 2015-11-04 14:22:59*, 2015

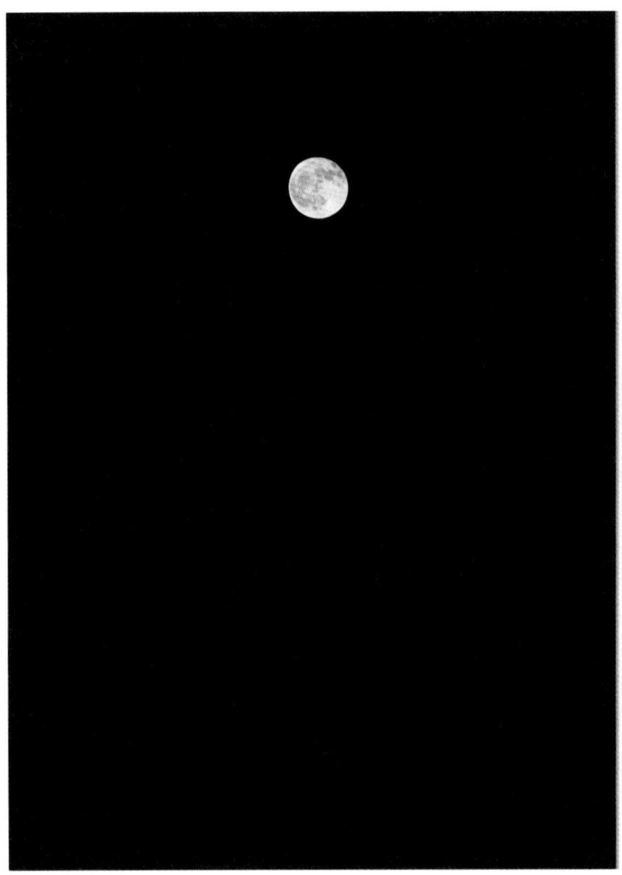

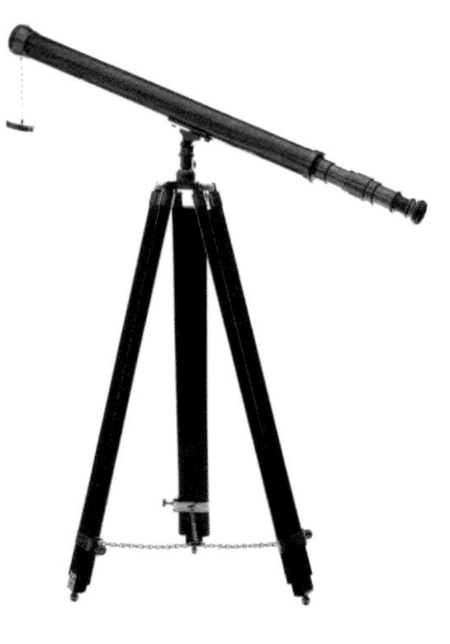

 | SARAH CHARLESWORTH | *Moon Watch*, 2012

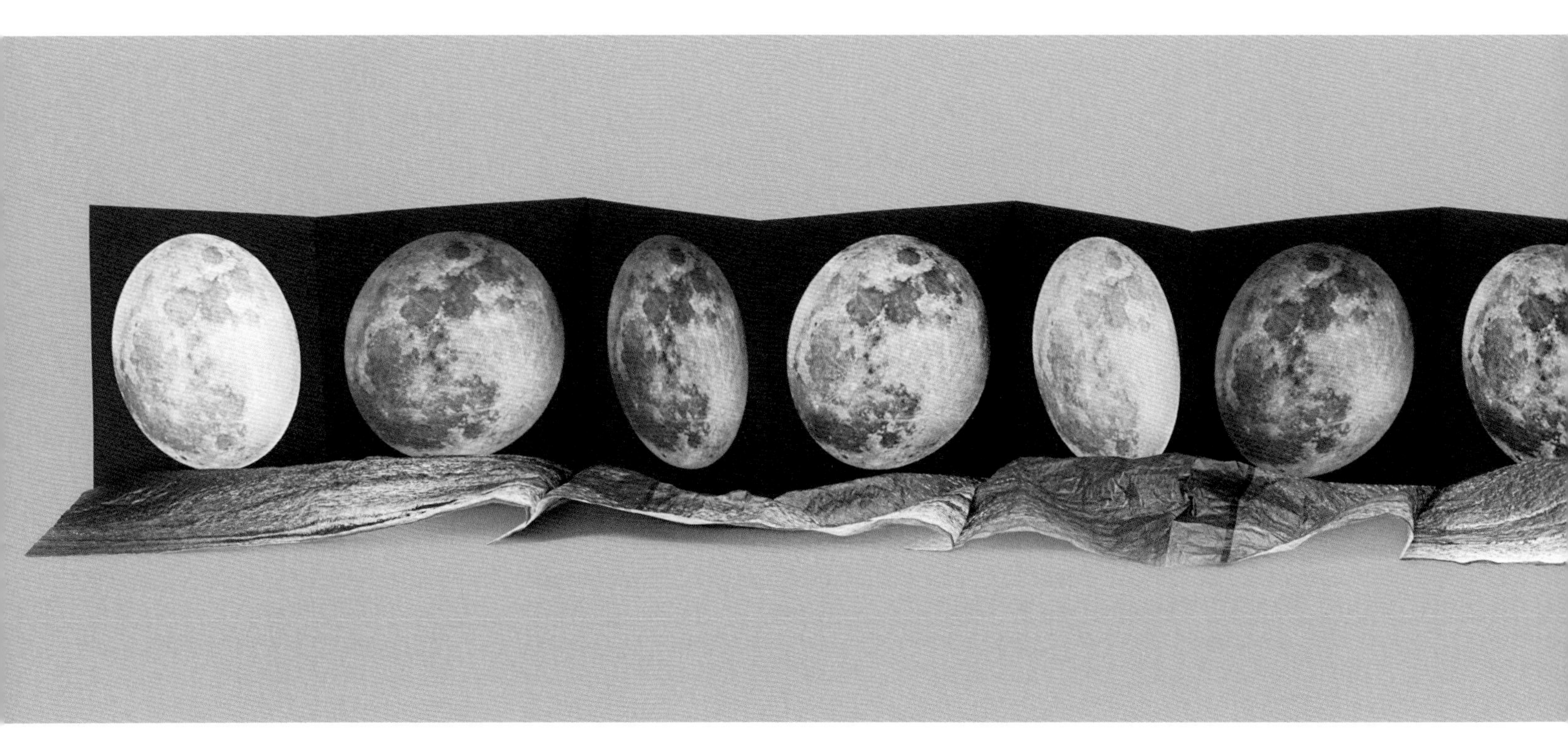

113 | KIKI SMITH | *Tidal*, 1998

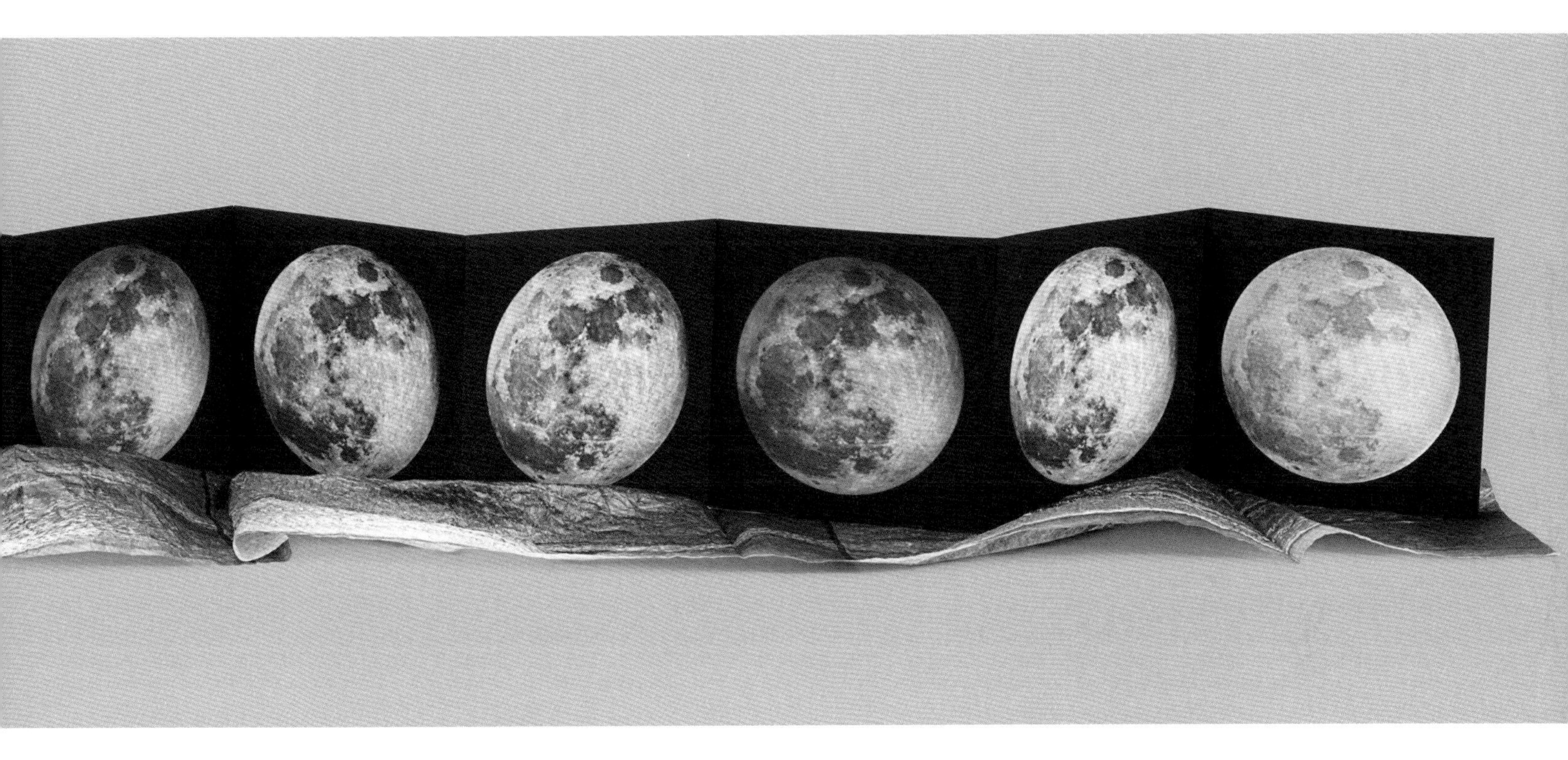

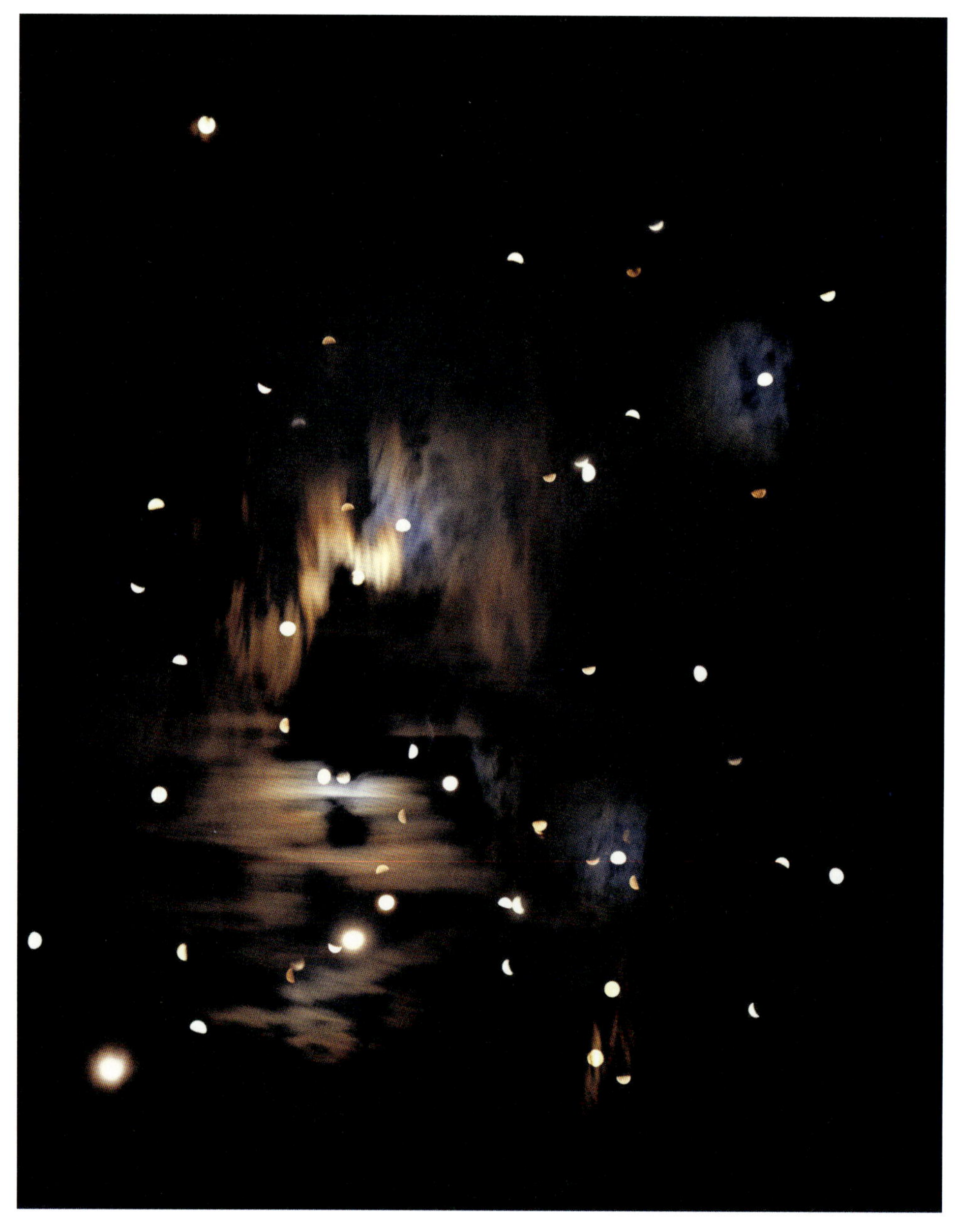

 Moon Studies and Star Scratches, No. 6. June–September 2004. Saratoga Springs, New York; Middlesex, Vermont; Johnson, Vermont; Eden Mills, Vermont; Greensboro, North Carolina, 2004

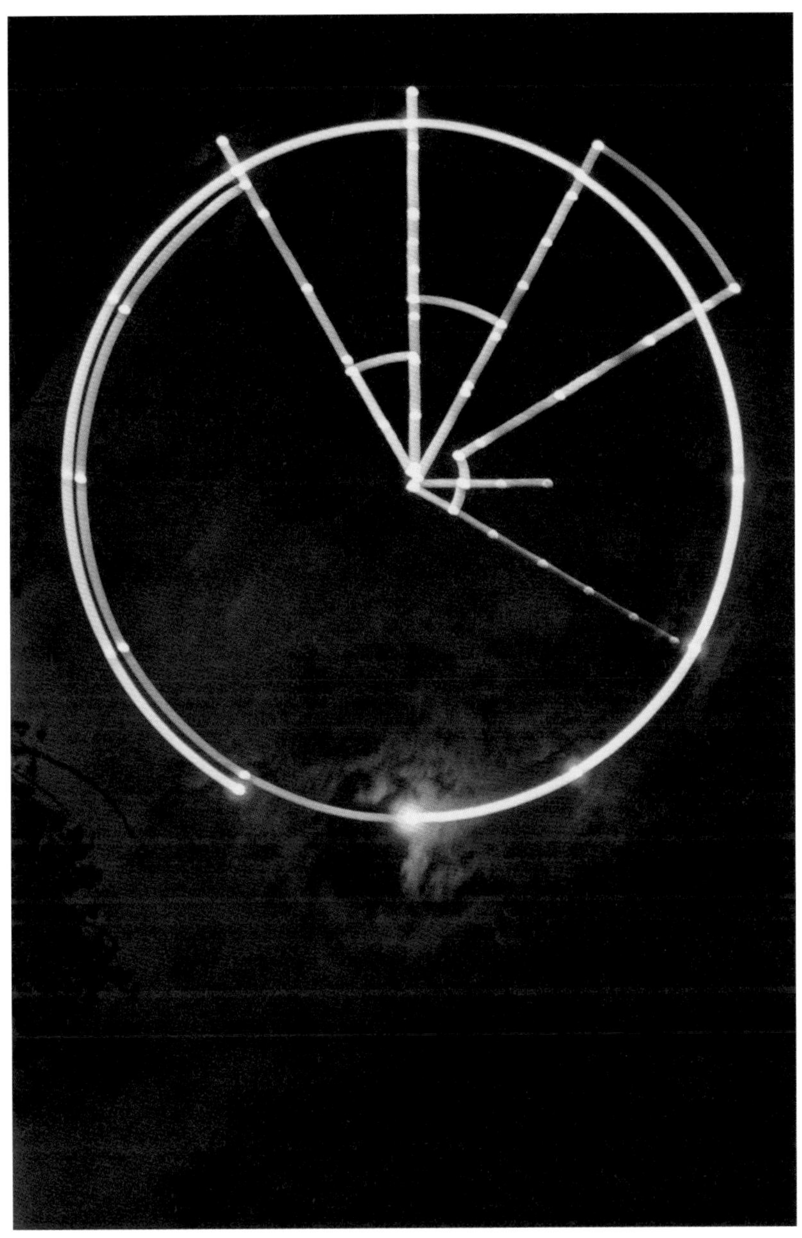

115 | KIKUJI KAWADA | *Artificial Moon Trail, Tokyo*, 1969

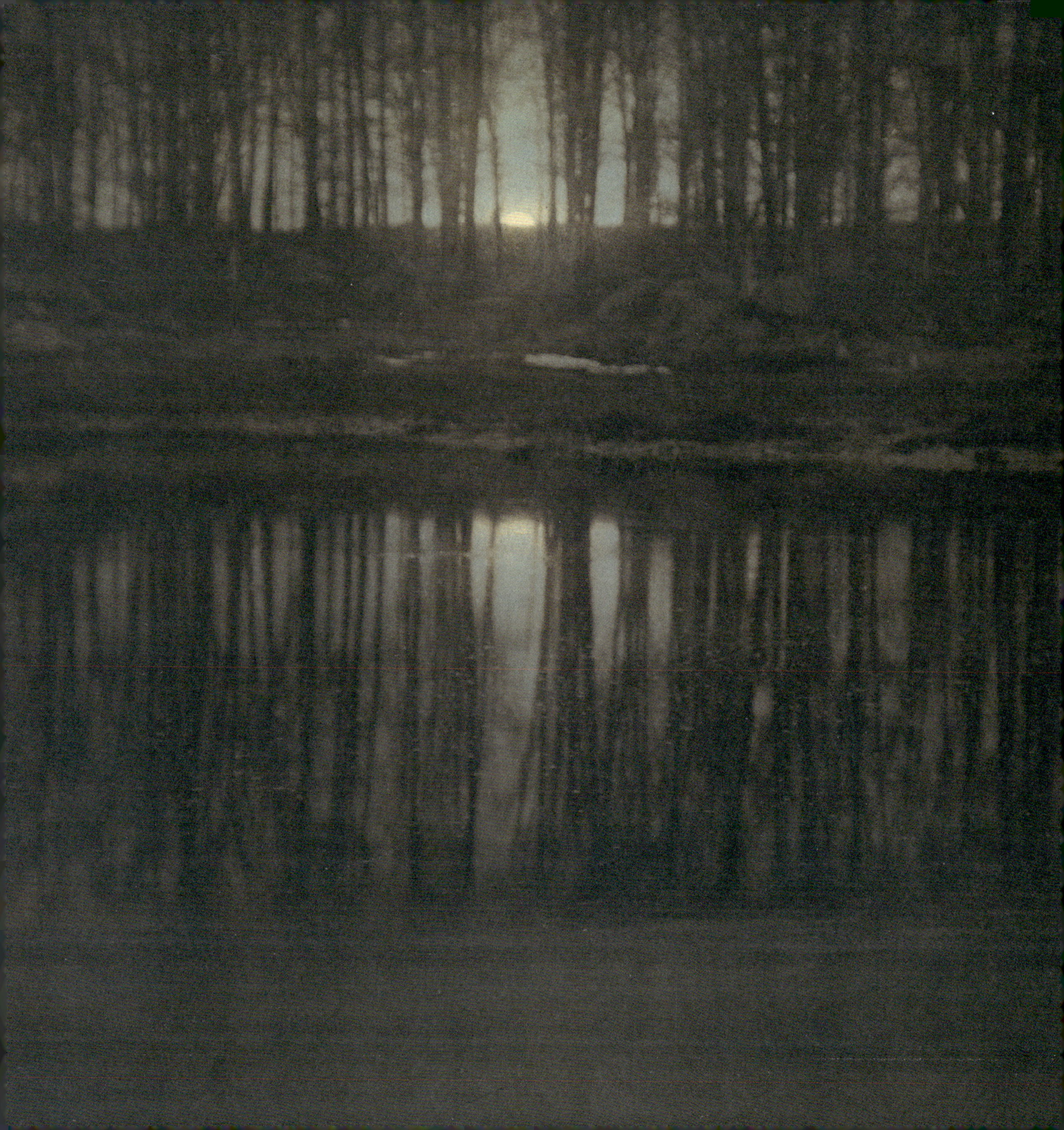

WORKS IN THE EXHIBITION

This list is organized by plate number; additional works included in the exhibition but not illustrated or illustrated as figures appear unnumbered at the end of each section, in chronological order. Unless otherwise indicated, dimensions are for the image and are given with height preceding width (preceding depth, where applicable). A sheet bound in a book is indicated as "in" the volume, whereas an unbound sheet is "from" the volume.

MAPPING THE MOON

1 | Galileo Galilei
Italian, 1564–1642
Two Drawings of Waxing Moon, in *Starry Messenger* (*Sidereus Nuncius*), 1610
Published by Thomas Baglioni, Venice
Engraving
Smithsonian Institution Libraries, Washington, D.C. (QB41 .G15 1610b)

2 | Francesco Fontana
Italian, 1580–1656
Lunar Map, in *New Observations of Heavenly and Earthly Objects* (*Novae coelestium, terrestriumque rerum observationes*), 1646
Published by Gaffarum, Naples
Engraving
The Metropolitan Museum of Art, The Elisha Whittelsey Collection, The Elisha Whittelsey Fund, 1948 (48.144)

3 | Claude Mellan
French, 1598–1688
Full Moon, 1635
Engraving; first state of two
Sheet: 9¾ x 8⁹⁄₁₆ in. (24.8 x 21.7 cm)
The Metropolitan Museum of Art, The Elisha Whittelsey Collection, The Elisha Whittelsey Fund, 1960 (60.634.36)

4 | Claude Mellan
French, 1598–1688
The Moon in Its First Quarter, 1635
Engraving; first state of two
Sheet: 9³⁄₁₆ x 6⅞ in. (23.3 x 17.5 cm)
The Metropolitan Museum of Art, The Elisha Whittelsey Collection, The Elisha Whittelsey Fund, 1960 (60.634.38)

5 | Claude Mellan
French, 1598–1688
The Moon in Its Final Quarter, 1635
Engraving; first state of two
Sheet: 9⅛ x 5⅜ in. (23.1 x 13.7 cm)
The Metropolitan Museum of Art, The Elisha Whittelsey Collection, The Elisha Whittelsey Fund, 1960 (60.634.37)

6 | Johannes Hevelius
Polish, 1611–1687
Lunar Map, in *Selenography, or the Description of the Moon* (*Selenographia, sive lunae descriptio*), 1647
Published by Andreas Hënefeld, Danzig
Engraving
Collection of Bethany and Robert B. Millard

7 | John Russell
British, 1745–1806
A Drawing of a Part for the Map of the Moon, 1794
Graphite
Sheet: 8 in. x 6⅛ in. (20.3 x 15.6 cm)
Yale Center for British Art, New Haven, Conn., Paul Mellon Collection (B1975.4.920)

8 | John Russell
British, 1745–1806
Lunar Planisphere, Flat Light, 1805
Engraving
41⅝ x 41 in. (105.7 x 104.1 cm)
Yale Center for British Art, New Haven, Conn., Paul Mellon Fund (B2016.39.1)

9 | Charles F. Blunt
British, active early 19th century
The Moon's Phases, in *Lecture on Astronomy: Beauty of the Heavens; a Pictorial Display of the Astronomical Phenomena of the Universe*, 1842
Published by Tilt and Bogue, London
Lithograph
Smithsonian Institution Libraries, Washington, D.C. (QB68 .B65 1842)

10 | John William Draper
American, b. England, 1811–1882
Moon, 1840s
Daguerreotype
3¾ x 2¾ in. (9.5 x 7 cm)
New York University Archives (MC298, object 192)

11 | Gustavus W. Pach
American, b. Germany, 1845–1904
The Great Refractor, ca. 1880
Albumen silver print
9 x 6⅞ in. (22.8 x 17.5 cm)
Harvard University Archives, Cambridge, Mass. (UAV630.271 [222])

12 | Samuel Dwight Humphrey
American, 1823–1883
Multiple Exposures of the Moon: Nine Exposures Ranging from Two Minutes to Half a Second, 1849
Daguerreotype
1⁹⁄₁₆ x 1³⁄₁₆ in. (4 x 3 cm)
John G. Wolbach Library, Harvard College Observatory, Cambridge, Mass. (OB-1)

13 | Antoine-François-Jean Claudet
French, active Great Britain,

1797–1867
Multiple Exposures of the Moon, 1846–52
Daguerreotype
2½ x 2 in. (6.4 x 5.1 cm)
The Metropolitan Museum of Art, The Horace W. Goldsmith Foundation Fund, through Joyce and Robert Menschel, 2019 (2019.47)

14 | John Adams Whipple
American, 1822–1891
View of the Moon, 1852
Daguerreotype
3⅞ x 3 in. (9.8 x 7.6 cm)
John G. Wolbach Library, Harvard College Observatory, Cambridge, Mass. (OB-8)

15 | John Adams Whipple
American, 1822–1891
James Wallace Black
American, 1825–1896
The Moon, 1857–60
Salted paper print from glass negative
8 5/16 x 6¼ in. (21.1 x 15.8 cm)
The Metropolitan Museum of Art, Robert O. Dougan Collection, Gift of Warner Communications Inc., 1978 (1978.649.7)

16 | Henry Draper
American, 1837–1882
Lunar Transparency, 1863
Albumen silver print in original wood and glass mount
Overall: 33¼ x 22½ x 8 in. (84.5 x 57.2 x 20.3 cm)
Collection of James and Abigail Draper

17 | Warren De La Rue
British, 1815–1889
The Moon, ca. 1856
Albumen silver print
6⅝ x 6 7/16 in. (16.8 x 16.4 cm)
Randy and Yulia G. Liebermann Lunar and Planetary Exploration Collection

18 | Austin Augustus Turner
American, ca. 1813–1866
After Warren De La Rue
British, 1815–1889
Twelve Photographs of the Moon, 1863
Published by D. Appleton and Company, New York
Albumen silver prints
Each 2½ x 2 1/16 in. (6.3 x 5.2 cm)
The Metropolitan Museum of Art, Purchase, Susan and Thomas Dunn Gift and funds from various donors, 2014 (2014.753.1–.14)

19 | Warren De La Rue
British, 1815–1889
Robert Howlett
British, 1831–1858
The Moon, ca. 1858
Published by Smith, Beck and Beck, London
Albumen silver transparencies on glass; stereograph
3¼ x 6 13/16 in. (8.3 x 17.3 cm)
Hans P. Kraus Jr. Inc., New York

20 | William Henry Fox Talbot
British, 1800–1877
Warren De La Rue
British, 1815–1889
The Half Moon, ca. 1864
Photoglythic engraving
5 x 4 in. (12.7 x 10.1 cm)
Hans P. Kraus Jr. Inc., New York

21 | Lewis Morris Rutherfurd, American, 1816–1892
The Moon, New York, 1865
Albumen silver print
27 15/16 x 17 11/16 in. (71 x 45 cm)
Collection of Alexander W. Rutherfurd

22 | Camille Flammarion
French, 1842–1925
Adolphe-Alexandre Martin
French, 1824–1896
After Warren De La Rue
British, 1815–1889
The Moon (*La Lune*), in *Astronomical Gallery, Photographs of Science and Art* (*Galerie astronomique, photographie des sciences et des art*), 1867
Albumen silver print
Photography Collection, Miriam and Ira D. Wallach Division of Art, Prints and Photographs, The New York Public Library, Astor, Lenox, and Tilden Foundations (115PH016)

23 | James Nasmyth
British, 1808–1890
An Ideal Sketch of "Pico," in *The Moon: Considered as a Planet, a World, and a Satellite*, by Nasmyth and James Carpenter, 3rd edition, 1885
Published by Scribner and Welford, New York
Woodburytype
The Metropolitan Museum of Art, Joyce F. Menschel Photography Library (QB581 .N210 1885 Rare Books)

24 | James Nasmyth
British, 1808–1890
Normal Lunar Crater, in *The Moon: Considered as a Planet, a World, and a Satellite*, by Nasmyth and James Carpenter, 1st edition, 1874
Published by John Murray, London
Heliotype
Private collection

25 | John Brett
British, 1831–1902
Gassendi's Crater on the Moon, 1884
Black chalk heightened with gouache
16⅞ x 14⅞ in. (42.8 x 37.8 cm)
Collection of Roberta J. M. Olson and Alexander B. V. Johnson

26 | Etienne Léopold Trouvelot
French, 1827–1895
Interior of Crater, in *Astronomical Notebook*, 1872–73
Ink, gouache, and graphite
4¾ x 8¼ in. (12 x 21 cm)
Houghton Library, Harvard University, Cambridge, Mass., transferred from the Harvard College Observatory to Widener Library, Harvard (MS Am 2081)

27 | Etienne Léopold Trouvelot
French, 1827–1895
Mare Humorum, from *The Trouvelot Astronomical Drawings Manual*, 1882
Published by Charles Scribner's Sons, New York
Chromolithograph
Sheet: 16 x 20 in. (40.6 x 50.8 cm)
Rare Book Division, The New York Public Library, Astor, Lenox and Tilden Foundations

28 | Etienne Léopold Trouvelot
French, 1827–1895
Total Eclipse of the Sun, Observed July 29, 1878, at Creston, Wyoming Territory, from *The Trouvelot Astronomical Drawings Manual*, 1882
Published by Charles Scribner's Sons, New York
Chromolithograph
Sheet: 16 x 20 in. (40.6 x 50.8 cm)
Rare Book Division, The New York Public Library, Astor, Lenox and Tilden Foundations

29 | John Adams Whipple
American, 1822–1891
Partial Eclipse of the Sun, 1851
Daguerreotype
4¼ x 3½ in. (10.2 x 8.9 cm)
John G. Wolbach Library, Harvard College Observatory, Cambridge, Mass. (OB-2)

30 | William Langenheim
American, b. Germany, 1807–1874
Frederick Langenheim
American, b. Germany, 1809–1879
Eclipse of the Sun, 1854
7 daguerreotypes
From 1¼ x 1 in. (3.2 x 2.5 cm) to

2 13/16 x 2 5/16 in. (7.2 x 5.9 cm)
The Metropolitan Museum of Art, Gilman Collection, Gift of The Howard Gilman Foundation, 2005 (2005.100.614a–g)

31 | H. A. Lawrence
British, active 1880s
Charles Ray Woods
British, active 1880s
Solar Eclipse from Caroline Island, 1883
Gelatin silver print
5 1/4 x 5 3/16 in. (13.3 x 13.1 cm)
The Metropolitan Museum of Art, Gift of Donald Lokuta and Melissa Tomich, 2018 (2018.382)

32 | Paul Henry
French, 1848–1905
Prosper Henry
French, 1849–1903
Lunar Photograph, South Pole, 1890
Albumen silver print from glass negative
9 x 6 7/16 in. (22.9 x 16.4 cm)
The Metropolitan Museum of Art, Gift of Rolf Mayer, 1995 (1995.125)

33 | George W. Ritchey
American, 1864–1945
Mirror for Mount Wilson Observatory Telescope, early 20th century
Gelatin silver print
8 x 10 in. (20.3 x 25.4 cm)
Randy and Yulia G. Liebermann Lunar and Planetary Exploration Collection

34 | George W. Ritchey
American, 1864–1945
The Moon (10 Days Old), ca. 1901
Gelatin silver print
10 x 8 in. (25.4 x 20.3 cm)
Randy and Yulia G. Liebermann Lunar and Planetary Exploration Collection

35 | William H. Pickering
American, 1858–1938
Lunar Fancies, in *The Moon: A Summary of the Existing Knowledge of Our Satellite, with a Complete Photographic Atlas*, 1903
Published by Doubleday, Page, and Company, New York
Halftone
The Metropolitan Museum of Art, Joyce F. Menschel Photography Library (QB581 .P6 1903 Rare Books)

36 | *Transparency of the Moon from Negatives Made at the Lick Observatory, Mount Hamilton, California*, ca. 1896
Gelatin silver transparency on glass
Overall: 18 11/16 x 15 3/4 in. (47.5 x 40 cm)
Miriam and Ira D. Wallach Division of Art, Prints and Photographs, The New York Public Library, Astor, Lenox and Tilden Foundations (92PH041.001)

37, 38 | Maurice Loewy
French, b. Czech Republic, 1833–1907
Pierre Puiseux
French, 1855–1928
Photographic Atlas of the Moon, Published by the Paris Observatory (Atlas photographique de la lune, publié par l'Observatoire de Paris), 1896–1910
Published by Imprimerie Nationale, Paris
83 photogravures (71 plates with tissue overlays and 12 frontispiece images)
Each sheet: 28 x 24 in. (71.1 x 61 cm)
Private collection

39 | Charles Le Morvan
French, 1865–1933
Systematic Photographic Map of the Moon, Increasing and Decreasing Phases, 1899–1909
Published by the French Academy of Sciences, Paris, 1914
48 photogravures
Each sheet: 19 3/4 x 15 1/4 in. (50.2 x 38.7 cm)
The Metropolitan Museum of Art, Funds from various donors, 2016 (2016.321.1.1–.50a, b, 2016.321.2.1–.26a, b)

Four-Inch Refracting Telescope, ca. 1800
Manufactured by Watkins of Charing Cross, London
Brass and glass
28 x 16 x 38 in. (71.1 x 40.6 x 96.5 cm)
Collection of Bethany and Robert B. Millard

John Adams Whipple
American, 1822–1891
James Wallace Black
American, 1825–1896
The Moon, 1857–60
Salted paper print from glass negative
8 7/16 x 6 5/16 in. (21.4 x 16 cm)
The Metropolitan Museum of Art, Robert O. Dougan Collection, Gift of Warner Communications Inc., 1981 (1981.1229.1)

John Adams Whipple
American, 1822–1891
James Wallace Black
American, 1825–1896
The Moon, 1857–60
Salted paper print from glass negative
8 1/4 x 6 3/16 in. (21 x 15.7 cm)
The Metropolitan Museum of Art, Robert O. Dougan Collection, Gift of Warner Communications Inc., 1981 (1981.1229.19)

John Adams Whipple
American, 1822–1891
James Wallace Black
American, 1825–1896
The Moon, 1857–60
Salted paper print from glass negative
8 1/4 x 6 3/8 in. (21 x 16.2 cm)
The Metropolitan Museum of Art, Robert O. Dougan Collection, Gift of Warner Communications Inc., 1981 (1981.1229.55)

Lewis Morris Rutherfurd
American, 1816–1892
Full Moon: The Left Hand Moon Was Photographed June 2d, 1871. The Right Hand Moon Was Photographed Aug. 29, 1871, 1870s
Published by E. and H. T. Anthony, New York
Albumen silver prints; stereograph
3 3/8 x 6 7/8 in. (8.6 x 17.5 cm)
The Metropolitan Museum of Art, Gift of Weston J. Naef, in memory of Kathleen W. Naef and Weston J. Naef Sr., 1982 (1982.1182.291)

The Wonderful Universe Explorer, The Great 36-Inch Equatorial Telescope, Lick Observatory, Mount Hamilton, California, 1902
Published by Underwood and Underwood, New York
Albumen silver prints; stereograph
3 7/16 x 7 in. (8.8 x 17.8 cm)
Gift of Weston J. Naef, in memory of Kathleen W. Naef and Weston J. Naef Sr., 1982 (1982.1182.266)

DAYDREAMS BY MOONLIGHT

40 | Filippo Morghen
Italian, 1730–after 1807
Pumpkins Used as Dwellings to Secure against Wild Beasts, from *The Collection of the Most Notable Things Seen by John Wilkins, Erudite English Bishop, on His Famous Trip from the Earth to the Moon*, 1766–67
Etching and aquatint
10 13⁄16 x 15 3⁄16 in. (27.4 x 38.6 cm)
The Metropolitan Museum of Art, Harris Brisbane Dick Fund, 1932 (32.74.8)

41 | Leopoldo Galluzzo
Italian, active early 19th century
Plate from *Other Discoveries Made on the Moon by Sigr. Herschell* [*sic*], ca. 1836
Published by Litr. Fergolic, Naples
Lithograph with applied color
Sheet: 18½ x 13 5⁄16 in. (47 x 33.8 cm)
Smithsonian Institution Libraries, Washington, D.C. (NX650.M6S42 folio)

42 | Gustave Doré
French, 1832–1883
It Looked Round and Shining Like a Glittering Island, in *Aventures du Baron de Munchhausen*, by Rudolf Erich Raspe, 1862
Published by Furne, Jouvet et Cie, Paris
Engraving
The Metropolitan Museum of Art, Gift of Lincoln Kirstein, 1970 (1970.565.275)

43 | Man Ray
American, 1890–1976
The World (*Le Monde*), in *Electricité*, 1931
Commissioned by La Compagnie Parisienne de Distribution d'Electricité, Paris
Photogravure
10¼ x 8 1⁄16 in. (26 x 20.5 cm)
The Metropolitan Museum of Art, Gift of Joyce F. Menschel, 2013 (2013.1098.13.9)

44 | Caspar David Friedrich
German, 1774–1840
Two Men Contemplating the Moon, ca. 1825–30
Oil on canvas
13¾ x 17¼ in. (34.9 x 43.8 cm)
The Metropolitan Museum of Art, Wrightsman Fund, 2000 (2000.51)

45 | Thomas Rowlandson
British, 1757–1827
The Assignation, 1799
Aquatint with applied color
Sheet: 20¼ x 14⅜ in. (51.4 x 36.5 cm)
Yale Center for British Art, New Haven, Conn., Paul Mellon Collection (Folio A 2018 39)

46 | Carlo Naya
Italian, 1816–1882
Night View of the Grand Canal, Venice, ca. 1875
Albumen silver print from glass negative
16⅜ x 21⅛ in. (41.6 x 53.7 cm)
The Metropolitan Museum of Art, Gift of Estate of Emily Delafield Floyd, 1961 (61.568.9)

47 | Edward J. Steichen
American, b. Luxembourg, 1879–1973
The Pond — Moonrise, 1904
Platinum print with applied color
15⅝ x 19 in. (39.7 x 48.2 cm)
The Metropolitan Museum of Art, Alfred Stieglitz Collection, 1933 (33.43.40)

48 | Georges Méliès
French, 1861–1938
The Giant Cannon, preparatory drawing for the film *A Trip to the Moon* (*Le Voyage dans la lune*, 1902), re-created 1930
Ink
9 11⁄16 x 12½ in. (24.6 x 31.8 cm)
Cinémathèque Française, Paris (D015/73)

49 | Georges Méliès
French, 1861–1938
Square in the Eye, preparatory drawing for the film *A Trip to the Moon* (*Le Voyage dans la lune*, 1902), re-created 1930
Ink
9½ x 12½ in. (24.1 x 31.7 cm)
Cinémathèque Française, Paris (D082/20)

50 | Georges Méliès
French, 1861–1938
Earthlight, preparatory drawing for the film *A Trip to the Moon* (*Le Voyage dans la lune*, 1902), re-created 1930
Ink
9½ x 12 7⁄16 in. (24.2 x 31.6 cm)
Cinémathèque Française, Paris (D095/46)

51 | Georges Méliès
French, 1861–1938
The Grotto of Giant Mushrooms, preparatory drawing for the film *A Trip to the Moon* (*Le Voyage dans la lune*, 1902), re-created 1930
Ink and gouache
9½ x 12 7⁄16 in. (24.2 x 31.6 cm)
Cinémathèque Française, Paris (D015/69)

52 | Bliss Brothers Studio
American, active 1880–1910
Gorge, A Trip to the Moon, Pan-American Exposition, Buffalo, New York, 1901
Gelatin silver print
9 7⁄16 x 7½ in. (24 x 19 cm)
Miriam and Ira D. Wallach Division of Art, Prints and Photographs, The New York Public Library, Astor, Lenox and Tilden Foundations (106PH226.004)

53 | Georges Méliès
French, 1861–1938
A Trip to the Moon (*Le Voyage dans la lune*), 1902
Digital video transferred from restored 35 mm film, hand-colored, silent (with new score by Jeff Mills), 14 min.
Courtesy MK2 Films

54 | *"Man in the Moon" Postcards*, 1900s–1940s
68 gelatin silver prints and chromolithographs
Approximately 5½ x 3½ in. (14 x 8.9 cm)
The Metropolitan Museum of Art, Gift of Peter J. Cohen, 2018 (2018.369.1–.347)

55 | *Drinking with the Moon*, 1910s
Gelatin silver print with applied color
5 3⁄16 x 3⅜ in. (13.2 x 8.6 cm)
The Metropolitan Museum of Art, Twentieth-Century Photography Fund, 2010 (2010.194)

56 | Emile-Antoine Bayard
French, 1837–1891
Alphonse-Marie-Adolphe de Neuville
French, 1835–1885
Illustrations in *From the Earth to the Moon followed by Around the Moon* (*De la terre à la lune suivi de Autour de la lune*), by Jules Verne, 1872
Published by J. Hetzel, Paris
Rotogravures
Houghton Library, Harvard University, Cambridge, Mass. Gift of the Institute of Aeronautical Sciences New York, 1955 (*FC8 V5946 865dm)

57 | Nadar
French, 1820–1910
Nadar with His Wife, Ernestine, in a Balloon, ca. 1865, printed 1890s
Gelatin silver print from glass negative
3 9/16 x 3 1/16 in. (9 x 7.8 cm)
The Metropolitan Museum of Art, Gilman Collection, Museum Purchase, 2005 (2005.100.313)

58 | Otto Hunte
German, 1881–1960
Production sketch for the film *Woman in the Moon* (*Frau im Mond*), 1929
Ink and gouache
18 7/8 x 25 1/4 in. (48 x 64.2 cm)
Cinémathèque Française, Paris (D021/12)

59 | *Woman in the Moon* (*Frau im Mond*), 1929
Directed by Fritz Lang
American, b. Austria, 1890–1976
Digital video transferred from 35 mm film, black-and-white, silent (with new score by Javier Peréz Azpeitia), 162 min.
Courtesy Kino Lorber Inc.

60 | *Constructing the Model for Germany's "Moon Rocket,"* 1929
Gelatin silver print
5 5/8 x 7 9/16 in. (14.3 x 19.2 cm)
The Metropolitan Museum of Art, Gift of Mary and Dan Solomon, 2016 (2016.796.29)

61 | *Destination Moon*, 1950
Directed by Irving Pichel
American, 1891–1954
Digital video transferred from Technicolor film, sound, 91 min.
Wade Williams Collection, licensed through Corinth Films Inc.

62 | Chesley Bonestell
American, 1888–1986
Study for *A Lunar Landscape*, 1957
Acrylic over photomontage
7 7/8 x 30 11/16 in. (20 x 78 cm)
Randy and Yulia G. Liebermann Lunar and Planetary Exploration Collection

Bliss Brothers Studio
American, active 1880–1910
A Street in Moon City, A Trip to the Moon, Pan-American Exposition, Buffalo, New York, 1901
Gelatin silver print
9 7/16 x 7 1/2 in. (24 x 19 cm)
Miriam and Ira D. Wallach Division of Art, Prints and Photographs, The New York Public Library, Astor, Lenox and Tilden Foundations (106PH226.005)

MOONSHOT

63–66 | Vasily M. Baturin
Russian, active 1960s
Brothers of the Cosmos, 1969
Album of 117 gelatin silver prints by various artists, including Baturin, and notes in ink
Each sheet: 9 7/16 x 12 13/16 in. (24 x 32.5 cm)
Stephen White Collection II

67 | TASS (Telegraph Agency of the Soviet Union)
USSR Luna 3
The Far Side of the Moon, 1959
Distributed by United Press International
Gelatin silver print
6 13/16 x 6 5/8 in. (17.3 x 16.9 cm)
The Metropolitan Museum of Art, Gift of Mary and Dan Solomon, 2016 (2016.796.1)

68 | Aerojet-General Corporation
Mock-Up of Moon Suit in Action, 1962
Gelatin silver print
9 7/16 x 7 9/16 in. (24 x 19.2 cm)
The Metropolitan Museum of Art, Purchase, Vital Projects Fund Inc. Gift, through Joyce and Robert Menschel, 2018 (2018.268)

69 | Aerojet-General Corporation
Moon Mobile Carries Two Explorers, Life-Support Systems, 1962
Gelatin silver print
7 5/8 x 9 9/16 in. (19.3 x 24.3 cm)
The Metropolitan Museum of Art, Purchase, Vital Projects Fund Inc. Gift, through Joyce and Robert Menschel, 2018 (2018.270)

70 | Ralph Turner
American, b. 1935
High Relief of Alphonsus Peak, Made at the Lunar and Planetary Laboratory, Tucson, 1966
Epoxy
50 3/8 x 48 1/16 in. (128 x 122 cm)
Randy and Yulia G. Liebermann Lunar and Planetary Exploration Collection

71 | NASA Surveyor 6
Day 322, Survey U, Sectors 15 and 16, 1967
Instant black-and-white prints
29 5/16 x 11 7/8 in. (74.4 x 30.1 cm)
The Metropolitan Museum of Art, Purchase, Nancy and Edwin Marks Gift, 1992 (1992.5153)

72 | NASA Lunar Orbiter 2
Close-Up of Crater Copernicus, 1966
Gelatin silver print
14 x 13 7/8 in. (35.5 x 35.2 cm)
The Metropolitan Museum of Art, The Horace W. Goldsmith Foundation Fund, through Joyce and Robert Menschel, 2016 (2016.513)

73 | NASA Lunar Orbiter 3
Far Side of the Moon at Apolune, 1967
Gelatin silver print
16 3/16 x 13 3/4 in. (41.1 x 34.9 cm)
The Metropolitan Museum of Art, The Horace W. Goldsmith Foundation Fund, through Joyce and Robert Menschel, 2016 (2016.516)

74 | NASA Lunar Orbiter 3
Crater Kepler and Vicinity, 1967
Gelatin silver print
11 7/8 x 13 13/16 in. (30.2 x 35.1 cm)
The Metropolitan Museum of Art, The Horace W. Goldsmith Foundation Fund, through Joyce and Robert Menschel, 2016 (2016.515)

75 | NASA Lunar Orbiter 4
Crater Aristarchus, Schroter's Valley, and Vicinity, 1967
Gelatin silver print
13 3/4 x 12 1/4 in. (34.9 x 31.1 cm)
The Metropolitan Museum of Art, The Horace W. Goldsmith Foundation Fund, through Joyce and Robert Menschel, 2016 (2016.514)

76 | NASA Lunar Orbiter 5
Crater Aristarchus, 1967
Gelatin silver print
51 x 57 in. (129.5 x 144.8 cm)
Randy and Yulia G. Liebermann Lunar and Planetary Exploration Collection

77 | NASA Lunar Orbiter 5
Lunar Panorama #158, 1967
Gelatin silver print
18 7/8 x 64 3/16 in. (48 x 163 cm)
Randy and Yulia G. Liebermann Lunar and Planetary Exploration Collection

78 | William Anders
American, b. 1933
NASA Apollo 8
Earthrise, 1968

Color laser print
20½ x 20¼ in. (52 × 51.5 cm)
Gift of Jules Bergman, 1984, Miriam and Ira D. Wallach Division of Art, Prints and Photographs, The New York Public Library, Astor, Lenox and Tilden Foundations (84PH002.060)

79 | Harrison Schmitt
American, b. 1935
NASA Apollo 17
Blue Marble, 1972
Color laser print
11 x 10 1/16 in. (28 x 25.5 cm)
Gift of Jules Bergman, 1984, Miriam and Ira D. Wallach Division of Art, Prints and Photographs, The New York Public Library, Astor, Lenox and Tilden Foundations (84PH002.001A)

80 | James McDivitt
American, b. 1929
NASA Gemini 4
Ed White Extravehicular Activity (EVA), 1965
Color laser print
19⅞ x 20 11/16 in. (50.5 x 52.5 cm)
Gift of Jules Bergman, 1984, Miriam and Ira D. Wallach Division of Art, Prints and Photographs, The New York Public Library, Astor, Lenox and Tilden Foundations (84PH002.13)

81 | NASA Apollo 11
Apollo 11 Command and Service Modules Photographed from the Lunar Module in Orbit, 1969
Chromogenic print
9 1/16 x 7⅝ in. (23 x 19.3 cm)
The Metropolitan Museum of Art, Gift of Mary and Dan Solomon, 2016 (2016.796.26)

82 | *Apollo 11 Moon Landing on Television Screens*, 1969
3 gelatin silver prints
Each 2⅝ x 4 3/16 in. (6.6 x 10.7 cm)
The Metropolitan Museum of Art, Gift of Jeffrey Fraenkel, 2014 (2014.494.1–.3)

83 | Neil Armstrong
American, 1930–2012
NASA Apollo 11
Buzz Aldrin on the Moon with Components of the Early Apollo Scientific Experiments Package, 1969
Chromogenic print
7¼ x 7⅛ in. (18.4 x 18.1 cm)
The Metropolitan Museum of Art, Gift of Mary and Dan Solomon, 2016 (2016.796.21)

84 | Neil Armstrong
American, 1930–2012
NASA Apollo 11
Buzz Aldrin with Apollo 11 Lunar Module on the Moon, 1969
Chromogenic print
7¼ x 7 in. (18.2 x 17.8 cm)
The Metropolitan Museum of Art, Gift of Mary and Dan Solomon, 2016 (2016.796.27)

85 | Neil Armstrong
American, 1930–2012
NASA Apollo 11
Buzz Aldrin Walking on the Surface of the Moon near a Leg of the Lunar Module, 1969, printed later
Dye transfer print
16⅛ x 16⅜ in. (41 x 41.6 cm)
The Metropolitan Museum of Art, Purchase, Alfred Stieglitz Society Gifts, 2017 (2017.421)

86 | Neil Armstrong
American, 1930–2012
NASA Apollo 11
Buzz Aldrin on the Moon with the American Flag, 1969
Gelatin silver print
7½ x 9½ in. (19 x 24.1 cm)
The Metropolitan Museum of Art, Gift of Mary and Dan Solomon, 2016 (2016.796.20)

87 | Neil Armstrong
American, 1930–2012
NASA Apollo 11
Buzz Aldrin Walking on the Surface of the Moon near a Leg of the Lunar Module, 1969
Chromogenic print
6¾ x 10 1/16 in. (17.1 x 25.6 cm)
The Metropolitan Museum of Art, Gift of Mary and Dan Solomon, 2016 (2016.796.24)

88 | NASA Apollo 11
President Richard M. Nixon Welcomes the Apollo 11 Astronauts aboard Recovery Ship USS Hornet, 1969
Chromogenic print
7⅝ x 9 1/16 in. (19.4 x 23 cm)
The Metropolitan Museum of Art, Gift of Mary and Dan Solomon, 2016 (2016.796.22)

Dmitri Baltermants
Russian, 1912–1990
Soviet Space Dogs Mishka and Tsigan, 1951
Gelatin silver print
13⅜ x 15⅜ in. (34 x 39.1 cm)
The Metropolitan Museum of Art, Twentieth-Century Photography Fund, 2018 (2018.511)

Aerojet-General Corporation
12 Man Lunar Expedition, 1962
Gelatin silver print
7 3/16 x 9⅜ in. (18.3 x 23.8 cm)
The Metropolitan Museum of Art, Purchase, Vital Projects Fund Inc. Gift, through Joyce and Robert Menschel, 2018 (2018.269)

Lunar Globe, ca. 1963
Produced by Paul Räth Verlag for Pergamon Press, London
Offset lithography on board with plastic stand
18 x 13 x 13 in. (45.7 x 33 x 33 cm)
Randy and Yulia G. Liebermann Lunar and Planetary Exploration Collection

NASA Ranger 9
Alphonsus Peak, 1965
Gelatin silver print
16 11/16 x 15 9/16 in. (42.4 x 39.6 cm)
Randy and Yulia G. Liebermann Lunar and Planetary Exploration Collection

Edwin "Buzz" Aldrin
American, b. 1930
NASA Apollo 11
Buzz Aldrin's Footprint on the Surface of the Moon, 1969
Chromogenic print
7⅜ x 7 9/16 in. (18.7 x 19.2 cm)
The Metropolitan Museum of Art, Funds from various donors, 2006 (2006.47)
Fig. 16b, p. 112

CBS News Television Coverage of Apollo 11 Moon Landing, 1969
Video, color, sound, 1 min., 45 sec.
Courtesy CBS News / Wazee Digital

Hasselblad 70 mm camera used for astronaut training, 1969
Metal and glass
5⅞ x 4 5/16 x 8 7/16 in. (15 x 11 x 21.5 cm)
Smithsonian National Air and Space Museum, Washington, D.C., transferred from NASA (A19790875000)
Fig. 15a, p. 111

Hasselblad 70 mm film magazine used on Apollo 11, 1969
Steel, velcro, and duct tape
3½ x 3½ x 3½ in. (8.9 x 8.9 x 8.9 cm)
Smithsonian National Air and Space Museum, Washington, D.C., transferred from NASA (A19791524000)
Fig. 15b, p. 111

Lunar Globe, 1969
Published by Rand McNally, Skokie, Ill.
Offset lithography on board with metal half-meridian ring and wood stand
14½ x 12 x 12 in.
(36.8 x 30.5 x 30.5 cm)
Private collection

NASA Apollo 11
Apollo 11 Blast Off, Kennedy Space Center, Florida, 1969
Distributed by United Press International
Gelatin silver print
7 x 9⅛ in. (17.8 x 23.2 cm)
The Metropolitan Museum of Art, Gift of Mary and Dan Solomon, 2016 (2016.796.14)

NASA Apollo 11
Apollo 11 Broadcast at Mission Control at the Manned Spacecraft Center, Houston, 1969
Gelatin silver print
6⅞ x 8⅜ in. (17.5 x 21.2 cm)
The Metropolitan Museum of Art, Gift of Mary and Dan Solomon, 2016 (2016.796.13)

NASA Apollo 11
Astronauts in Lifeboat after Apollo 11 Splashdown, 1969
Gelatin silver print
7 5⁄16 x 9 9⁄16 in. (18.5 x 24.3 cm)
The Metropolitan Museum of Art, Gift of Mary and Dan Solomon, 2016 (2016.796.12)

NASA Apollo 11
Television Transmission of Neil Armstrong's First Steps on the Moon, 1969
Video, black-and-white, sound, 1 min.
Courtesy NASA TV

ART AFTER APOLLO

89 | Garry Winogrand
American, 1928–1984
Apollo 11 Moon Shot, Cape Kennedy, Florida, 1969
Gelatin silver print
10¾ x 15⅞ in. (27.3 x 40.4 cm)
The Metropolitan Museum of Art, Purchase, Vital Projects Fund Inc. Gift, through Joyce and Robert Menschel, 2014 (2014.570)

90 | Robert Rauschenberg
American, 1925–2008
Sky Garden (Stoned Moon), 1969
Lithograph and screenprint
89¼ x 41 in. (226.7 x 104.1 cm)
Collection of Marc and Laura Andreessen

91 | Robert Rauschenberg
American, 1925–2008
Untitled, 1969
Solvent transfer with watercolor, gouache, pencil, and colored pencil on paperboard
15 x 20 in. (38.1 x 50.8 cm)
Robert Rauschenberg Foundation, New York (69.D014)

92 | Charles Duke
American, b. 1935
NASA Apollo 16
Duke Family Photograph on the Lunar Surface, 1972
Color laser print
16⅛ x 15 15⁄16 in. (41 x 40.5 cm)
Gift of Jules Bergman, 1984, Miriam and Ira D. Wallach Division of Art, Prints and Photographs, The New York Public Library, Astor, Lenox and Tilden Foundations (84PH002.076)

93 | John Chamberlain
American, 1927–2011
Forrest Myers
American, b. 1941
David Novros
American, b. 1941
Claes Oldenburg
American, b. 1929
Robert Rauschenberg
American, 1925–2008
Andy Warhol
American, 1928–1987
The Moon Museum, 1969
Lithograph of tantalum nitride on ceramic wafer
½ × ¾ in. (1.3 x 1.9 cm)
Collection of Forrest W. Myers

94 | Paul van Hoeydonck
Belgian, b. 1925
Fallen Astronaut replica, 1971
Aluminum
3½ × ¾ × 11⁄16 in.
(8.9 x 1.9 x 1.7 cm)
Collection of Roberta W. Waddell

95 | David Scott
American, b. 1932
NASA Apollo 15
Paul van Hoeydonck's "Fallen Astronaut" Sculpture on the Lunar Surface, 1971, printed 2019
Inkjet exhibition print from digital file

96 | Harry Gordon
American, 1930–2007
"Rocket" Dress, 1968
Screenprinted tissue, wood pulp, and rayon mesh
34 x 24 in. (86.4 x 61 cm)
The Metropolitan Museum of Art, Purchase, Gould Family Foundation Gift, in memory of Jo Copeland, 2009 (2009.73)

97 | Nam June Paik
American, b. Korea, 1932–2006
Jud Yalkut
American, 1938–2013
Electronic Moon No. 2, 1966–72
Digital video, transferred from 16 mm film, color, sound, 4 min., 52 sec.
Electronic Arts Intermix, New York

98 | Alessandro Poli
Italian, b. 1941
Superstudio
Italian, active 1966–86
Highway with Sine Wave of Energy, from the series Interplanetary Architecture, 1970–71
Photomontage
19 5⁄16 x 19 5⁄16 in. (49.1 x 49.1 cm)
Alessandro Poli fonds, Canadian Centre for Architecture, Montreal; Don de Alessandro Poli / Gift of Alessandro Poli (ARCH400644)

99 | Alessandro Poli
Italian, b. 1941
Superstudio
Italian, active 1966–86
New Rural Landscapes, from the series Interplanetary Architecture, 1970–71
Photomontage
10 15⁄16 x 12 15⁄16 in. (27.8 x 32.9 cm)
Alessandro Poli fonds, Canadian Centre for Architecture, Montreal; Don de Alessandro Poli / Gift of Alessandro Poli (ARCH280640)

100 | Nancy Graves
American, 1940–1995
IV Julius Caesar Quadrangle of the Moon, from the series Lithographs Based on Geologic Maps of Lunar Orbiter and Apollo Landing Sites, 1972
Lithograph
Sheet: 22⅜ in. x 30 in.
(56.8 x 76.2 cm)
Nancy Graves Foundation, New York

101 | Nancy Graves
American, 1940–1995
II Fra Mauro Region of the Moon, from the series Lithographs Based on Geologic Maps of Lunar Orbiter and Apollo Landing Sites, 1972
Lithograph
Sheet: 22⅜ in. x 30 in.

(56.8 x 76.2 cm)
Nancy Graves Foundation,
New York

102 | Michelle Stuart
American, b. 1933
#7 Moon Tide, 1969
Graphite, colored pencil, and gelatin silver prints
23¼ x 27¹³⁄₁₆ in. (59.1 x 70.6 cm)
The Metropolitan Museum of Art, Purchase, Vital Projects Fund Inc. Gift, through Joyce and Robert Menschel, 2018 (2018.267)

103 | Judy Dater
American, b. 1941
Self-Portrait at Craters of the Moon, 1981
Gelatin silver print
14½ × 18½ in. (36.8 x 47 cm)
The Metropolitan Museum of Art, Twentieth-Century Photography Fund, 2018 (2018.473)

104 | Aleksandra Mir
Swedish-American, b. Poland, 1967
First Woman on the Moon, 1999
Digital video of live event on August 28, 1999, produced by Casco Projects, Utrecht, on location in Wijk aan Zee, Netherlands, color, sound, 14 min.
Courtesy of the artist

105 | Stephen Shames
American, b. 1947
The Moon Belongs to the People, 1971, printed later
Gelatin silver print
12¹³⁄₁₆ x 18¹¹⁄₁₆ in. (32.5 x 47.5 cm)
The Metropolitan Museum of Art, Purchase, Vital Projects Fund Inc. Gift, through Joyce and Robert Menschel, 2018 (2018.271)

106 | Cristina De Middel
Spanish, b. 1975
Bambuit, from the series The Afronauts, 2012
Inkjet print
39⅜ x 39⅜ in. (100 x 100 cm)
Collection of Valerie Dillon

107 | Vik Muniz
Brazilian, b. 1961
Memory Rendering of the Man on the Moon, 1990
Gelatin silver print
15⁷⁄₁₆ x 10¹¹⁄₁₆ in. (39.2 x 27.2 cm)
The Metropolitan Museum of Art, Purchase, Anonymous Gift, 1995 (1995.323.1)

108 | Jojakim Cortis
Swiss, b. 1978
Adrian Sonderegger
Swiss, b. 1980
Making of AS11-40-5878 (by Edwin Aldrin, 1969), 2014
Chromogenic print
47¼ x 70⅞ in. (120 x 180 cm)
The Metropolitan Museum of Art, Purchase, Vital Projects Fund Inc. Gift, through Joyce and Robert Menschel, 2019 (2019.54)

109 | Lenka Clayton
British, b. 1977
Moon 25/12/2015 (& 1977), 2015
Ink
Sheet: 11 x 8½ in. (27.9 x 21.6 cm)
The Metropolitan Museum of Art, Funds from various donors, 2018 (2018.199)

110 | Penelope Umbrico
Canadian, b. America, 1957
Everyone's Moon 2015-11-04 14:22:59, 2015
Single-channel video, color, sound, 16 min.
The Metropolitan Museum of Art, Purchase, Henry Nias Foundation Inc. Gift, 2016 (2016.365)

111 | Sarah Charlesworth
American, 1947–2013
Moon Watch, 2012
2 chromogenic prints with lacquered wood frames
Overall: 41 x 63 in. (104.1 x 160 cm)
Paula Cooper Gallery, New York (SCHA-17-PH)

112 | Darren Almond
British, b. 1971
Fifteen Minute Moon, 2000
Chromogenic print
49⅝ x 49⅜ in. (126 x 125.4 cm)
The Metropolitan Museum of Art, Purchase, The Horace W. Goldsmith Foundation Gift, through Joyce and Robert Menschel, 2001 (2001.212)

113 | Kiki Smith
American, b. 1954
Tidal, 1998
Published by the LeRoy Neiman Center for Print Studies, Columbia University, New York
Accordion-fold book with photogravure, offset lithography, and screenprint
19¼ in. x 10 ft. 6¼ in. (48.9 x 320.7 cm)
The Metropolitan Museum of Art, John B. Turner Fund, 2018 (2018.942a, b)

114 | Sharon Harper
American, b. 1966
Moon Studies and Star Scratches, No. 6. June–September 2004. Saratoga Springs, New York; Middlesex, Vermont; Johnson, Vermont; Eden Mills, Vermont; Greensboro, North Carolina, 2004
Chromogenic print
50 x 40 in. (127 x 101.6 cm)
The Metropolitan Museum of Art, Purchase, Vital Projects Fund Inc. Gift, through Joyce and Robert Menschel, 2019 (2019.53)

115 | Kikuji Kawada
Japanese, b. 1933
Artificial Moon Trail, Tokyo, 1969, printed 1989
Gelatin silver print
12¾ x 8⁷⁄₁₆ in. (32.4 x 21.4 cm)
The Metropolitan Museum of Art, Twentieth-Century Photography Fund, 2015 (2015.288)

Alessandro Poli
Italian, b. 1941
Superstudio
Italian, active 1966–86
Constructing an Inflatable Building, from the series Interplanetary Architecture, 1970–71
Photomontage
13¼ in. x 11 in. (33.7 x 28 cm)
Alessandro Poli fonds, Canadian Centre for Architecture, Montreal; Don de Alessandro Poli/Gift of Alessandro Poli (ARCH280639)

Alessandro Poli
Italian, b. 1941
Superstudio
Italian, active 1966–86
Planetary Transportation for a Larger Earth, from the series Interplanetary Architecture, 1970–71
Photomontage
27¼ x 19½ in. (69.2 x 49.5 cm)
Alessandro Poli fonds, Canadian Centre for Architecture, Montreal; Don de Alessandro Poli/Gift of Alessandro Poli (ARCH280638)

NOTES

MAPPING THE MOON

Epigraph: John Adams Whipple, letter to the editor, *Photographic Art-Journal* 6, no. 1 (July 1853), p. 66.

1. The British astronomer Thomas Hariot, for one, recorded gazing at the moon through a telescope six months before Galileo; Samuel Y. Edgerton, *The Mirror, the Window, and the Telescope: How Renaissance Linear Perspective Changed Our Vision of the Universe* (Ithaca, N.Y.: Cornell University Press, 2009), p. 154.
2. On photography, objectivity, and science in the nineteenth century, see Corey Keller, "Sight Unseen: Picturing the Invisible," in *Brought to Light: Photography and the Invisible, 1840–1900*, ed. Corey Keller, exh. cat. (San Francisco: San Francisco Museum of Modern Art, in association with Yale University Press, 2008), pp. 21–23.
3. Scott L. Montgomery, *The Moon and the Western Imagination* (Tucson: University of Arizona Press, 1999), pp. 153–55.
4. Ibid., pp. 172–84.
5. Frances Terpak, "Imaging the Moon," in Barbara Maria Stafford and Frances Terpak, *Devices of Wonder: From the World in a Box to Images on a Screen* (Los Angeles: Getty Research Institute, 2001), pp. 197–200.
6. "'[T]ruth to nature' had little to do with objectivity," as Peter Galison put it in "Judgment against Objectivity," in *Picturing Science, Producing Art*, ed. Caroline Jones and Peter Galison (New York: Routledge, 1998), p. 328.
7. These ideas are explored with great nuance in Keller, "Sight Unseen," p. 22.
8. Richard Adams Locke, "Great Astronomical Discoveries Lately Made by Sir John Herschel . . . at the Cape of Good Hope," *New York Sun*, August 25–31, 1835; reprinted in *Moon Walk 1835: Was Neil Armstrong Really the First Man on the Moon?*, ed. C. W. Tazewell, 5th ed. (Virginia Beach, Va.: W. S. Dawson Co., 1990); Christopher Phillips, "'Magnificent Desolation': The Moon Photographed," in *Cosmos: From Romanticism to the Avant-Garde*, ed. Jean Clair, exh. cat. (Montreal: Montreal Museum of Fine Arts; Munich: Prestel, 1999), p. 144.
9. Larry J. Schaaf, *Out of the Shadows: Herschel, Talbot, and the Invention of Photography* (New Haven: Yale University Press, 1992), pp. 48–49.
10. Dominique François Arago, "Report [of the Commission of the Chamber of Deputies . . . July 3, 1839]," in *Classic Essays on Photography*, ed. Alan Trachtenberg (New Haven: Leete's Island Books, 1989), pp. 20–21.
11. Wolfgang Baier, *Quellendarstellungen zur Geschichte der Fotografie* (Leipzig: VEB Fotokinoverlag, 1966), p. 396; quoted in Ann Thomas, "Capturing Light: Photographing the Universe," in *Beauty of Another Order: Photography in Science*, ed. Ann Thomas (New Haven: Yale University Press, in association with National Gallery of Canada, Ottawa, 1997), p. 196.
12. On Draper and photography, see Deborah Jean Warner, "The Draper Family Material: National Museum of American History," *History of Photography* 24, no. 1 (Spring 2000), pp. 16–23; Sarah Kate Gillespie, *The Early American Daguerreotype: Cross-Currents in Art and Technology* (Cambridge, Mass.: MIT Press, 2016), pp. 114–17.
13. A heliostat is a device that tracks the sun from a fixed point. John William Draper, "Chemistry and Physics Experiments: Journal 1836–1842," Smithsonian Libraries, Special Collection (Dibner), MS 001672 B quarto, entry for March 16, 1840.
14. Draper also published a short report on his experiments with daguerreotypes: John W. Draper, "Remarks on the Daguerreotype," *American Repertory of Arts, Sciences, and Manufactures* 1, no. 6 (July 1840), pp. 401–4.
15. This ninth-plate daguerreotype, from the collection of early photography pioneer Alfred Swaine Taylor, is housed in a case bearing the 1846–52 London studio address of John Jabez Edwin Mayall, who worked as an assistant to Claudet in 1846. Claudet's well-documented scientific experiments with photography — he exhibited a lunar daguerreotype at the 1851 Great Exhibition — make him the more likely author of this image. "Photographic Exhibitions in Britain 1839–1865: Claudet, Antoine François Jean (1797–1867)," http://peib.dmu.ac.uk; Karen Hellman, "Antoine Claudet, a Figure of Photography, 1839–1867" (PhD diss., CUNY Graduate Center, 2010), pp. 109–13.

    On Humphrey's work, see Melissa Banta, *A Curious and Ingenious Art:*

Reflections on Daguerreotypes at Harvard (Iowa City: Published for Harvard University Library by University of Iowa Press, 2000), pp. 35–40.

16. Ibid.; M. Susan Barger, "The Moon, 6 August 1851," in Christian A. Johnson Memorial Gallery, *Annual Report for the Year 1989* (Middlebury, Vt.: Christian A. Johnson Memorial Gallery, 1990), pp. 7–16.
17. On activities in Liverpool, see Frances Robertson, "Science and Fiction: James Nasmyth's Photographic Images of the Moon," *Victorian Studies* 48, no. 4 (Summer 2006), pp. 605–7; Thomas, "Capturing Light," pp. 201–2.
18. "Meeting of the British Association of Science at Liverpool," *Illustrated London News*, September 30, 1854, pp. 300–301.
19. On popularization of the sciences in the nineteenth century, see Anne Secord, "Botany on a Plate: Pleasure and the Power of Pictures in Promoting Early Nineteenth-Century Scientific Knowledge," *Isis* 93, no. 1 (March 2002), pp. 28–57.
20. On De La Rue, see Dorrit Hoffleit, *Some Firsts in Astronomical Photography* (Cambridge, Mass: Harvard College Observatory, 1950), pp. 21–24; Thomas, "Capturing Light," pp. 201–6; Holly Rothermel, "Images of the Sun: Warren De la Rue, George Biddell Airy and Celestial Photography," *British Journal for the History of Science* 26, no. 2 (June 1993), pp. 137–69.
21. This association is attested to by labels on the stereoview housings. It was also discussed in a self-published pamphlet by Rose Teanby, "Robert Howlett, 1831–1858: Service of Rededication St Peter & St Paul, Wendling, October 14th 2017." I thank Hans Kraus for sharing information about this pamphlet.
22. De La Rue remarked to Talbot, "I have instructed Mr. Howlitt [*sic*] to take a magnified positive of one of my best lunar negatives and to send it to you for your acceptance—and if your leisure permits of your making an engraved plate as you propose to do I shall highly prize it"; Warren De La Rue to William Henry Fox Talbot, October 5, 1858, The Correspondence of William Henry Fox Talbot, doc. 7702, http://foxtalbot.dmu.ac.uk/letters/name.php?bcode=Dela-W&pageNo=0, no. 3.
23. De La Rue wrote, "If it would be possible to obtain plates that would print copies of the size of the original negative 1 in. diameter by your process I would like to use the prints in our diary which sells to the extent of about 25000 annually"; Warren De La Rue to William Henry Fox Talbot, May 19, 1859, ibid., doc. 7883, http://foxtalbot.dmu.ac.uk/letters/name.php?bcode=Dela-W&pageNo=0, no. 4. De La Rue also reported on his correspondence with Talbot at an annual meeting of the Royal Astronomical Society: "Mr. De La Rue . . . informs the Council that Mr. Fox Talbot has kindly proposed to apply his newly-invented art of heliographic engraving to the reproduction of one of the lunar photographs: should this be successful, it is hoped that the distribution of these beautiful objects will become more general"; "Report of the Council to the Thirty-Ninth Annual General Meeting of the Society," *Monthly Notices of the Royal Astronomical Society* 19, no. 4 (February 11, 1859), p. 138.
24. Larry J. Schaaf, *Sun Pictures: Talbot and Photogravure*, cat. 12 (New York: Hans P. Kraus Jr. Fine Photographs, 2003), no. 29, p. 43.
25. On Rutherfurd, see Samuel Devons, "Lewis Morris Rutherfurd 1816–1892," *Applied Optics* 15, no. 7 (July 1976), pp. 1731–40.
26. On Nasymth, see Robertson, "Science and Fiction: James Nasmyth's Photographic Images of the Moon," pp. 595–623.
27. For an analysis of Nasmyth's book, see Carol Armstrong, *Scenes in a Library: Reading the Photograph in the Book, 1843–1875* (Cambridge, Mass.: MIT Press, 1998), pp. 63–79.
28. J. Norman Lockyer, "The Moon," *Nature* 9, no. 226 (March 12, 1874), p. 358.
29. Camille Flammarion, *Mémoires biographiques et philosophiques d'un astronome* (Paris: Ernest Flammarion, Editeur, 1911), p. 405.
30. Camille Flammarion, *Les terres du ciel: Description astronomique, physique, climatologique, géographique des planètes qui gravitent avec la Terre autour du Soleil et de l'état probable de la vie à leur surface* (Paris: Didier, 1877).
31. On this subject, see Simon Schaffer, "On Astronomical Drawing," in *Picturing Science, Producing Art*, pp. 441–73.
32. Christiana Payne, *John Brett: Pre-Raphaelite Landscape Painter* (New Haven: Yale University Press, 2010), p. 111. In 1870 Brett traveled to Sicily on a Royal Astronomical Society mission to document the sun's corona during a total eclipse.
33. For a brief biography of Trouvelot, see "Minor Contributions and Notes: Etienne-Léopold Trouvelot," *Astrophysical Journal* 2, no. 2 (August 1895), pp. 166–67; DeWayne A. Backhus and Elizabeth K. Fitch, "Nineteenth Century E. L. Trouvelot Astronomical Prints at Emporia State University," *Transactions of the Kansas Academy of Science* 109, nos. 1, 2 (April 2006), pp. 11–12.
34. E. L. Trouvelot, *The Trouvelot Astronomical Drawings Manual* (New York: Charles Scribner's Sons, 1882), p. vi.
35. Jules Janssen, "Le progrès de l'astronomie physique" (1883), in Janssen, *Oeuvres scientifiques*, vol. 1 (Paris: Société d'Editions Géographiques, Maritimes et Coloniales, 1929), p. 483; quoted in Phillips, "Magnificent Desolation," p. 146.
36. Hoffleit, *Some Firsts in Astronomical Photography*, p. 17.
37. Thomas, "Capturing Light," p. 197, fig. 131; Rothermel, "Images of the Sun," pp. 148–51.
38. Thomas, "Capturing Light," p. 189, fig. 124; Jeff L. Rosenheim, cat. 118a–g, in Maria Morris Hambourg et al., *The Waking Dream: Photography's First Century. Selections from the Gilman Paper Company Collection*, exh. cat. (New York: The Metropolitan Museum of Art, 1993), p. 311.
39. On Rosse's telescope, see Schaffer, "On Astronomical Drawing," pp. 457–61; Rupert Cole, "The Remarkable Mary Rosse," *Science Museum Blog*, March 28,

2017, https://blog.sciencemuseum.org.uk/the-remarkable-mary-rosse/.

40. On Ritchey, see Deborah J. Mills, "George Willis Ritchey and the Development of Celestial Photography," *American Scientist* 54, no. 1 (March 1966), pp. 64–93.
41. Phillips, "Magnificent Desolation," p. 146.
42. S. A. Saunder, "A Photographic Atlas of the Moon," *Observatory* 27, no. 341 (February 1904), p. 95.
43. Alex Soojung-Kim Pang, "'Stars Should Henceforth Register Themselves': Astrophotography at the Early Lick Observatory," *British Journal for the History of Science* 30, no. 2 (June 1997), pp. 177–202.
44. Phillips, "Magnificent Desolation," p. 146; Martin Parr and Gerry Badger, *The Photobook: A History*, vol. 1 (London: Phaidon, 2004), pp. 54–55.
45. Phillips, "Magnificent Desolation," p. 146.

DAYDREAMS BY MOONLIGHT

1. Maurice Loewy and Pierre Puiseux, *Atlas photographique de la lune, publié par l'Observatoire de Paris*, 8 pts. (Paris: Imprimerie Nationale, 1896–1904), pt. 1, p. 2; quoted in Christopher Phillips, "Magnificent Desolation: The Moon Photographed," in *Cosmos: From Romanticism to Avant-Garde*, ed. Jean Clair, exh. cat. (Montreal: Montreal Museum of Fine Arts; Munich: Prestel, 1999), p. 146.
2. David Cressy, "Early Modern Space Travel and the English Man in the Moon," *American Historical Review* 111, no. 4 (October 2006), p. 981.
3. See Steven J. Dick, *Plurality of Worlds: The Origins of the Extraterrestrial Life Debate from Democritus to Kant* (Cambridge: Cambridge University Press, 1982).
4. Richard Adams Locke, *The Celebrated "Moon Story": Its Origins and Incidents, with a Memoir by the Author and an Appendix . . . by William N. Griggs* (New York: Bunnell and Price, 1852), pp. 78–95.
5. Ibid., p. 20.
6. Edgar Allan Poe, "The Literati of New York City — No. VI: Richard Adams Locke," *Godey's Lady's Book* 33 (October 1846), p. 161, https://www.eapoe.org/works/misc/litratb6.htm.
7. István Kornél Vida, "The 'Great Moon Hoax' of 1835," *Hungarian Journal of English and American Studies* 18, no. 1–2 (Spring–Fall 2012), p. 439.
8. For a nuanced analysis of Friedrich's moon paintings, see Sabine Rewald, *Caspar David Friedrich: Moonwatchers*, exh. cat. (New York: The Metropolitan Museum of Art, 2001).
9. James Roberts, *Introductory Lessons, with Familiar Examples in Landscape for the Use of Those who are Desirous of Gaining Some Knowledge of the Pleasing Art of Painting in Water Colours; To which are Added, Some Clear and Simple Rules, Exemplified by Suitable Sketches and More Finished Paintings . . . To which are added, Instructions for Executing Transparencies in a Style Both Novel and Easy* (London: J. and W. Smith, 1809); quoted in John Plunkett, "Light Work: Feminine Leisure and the Making of Transparencies," in *Crafting the Woman Professional in the Long Nineteenth Century: Artistry and Industry in Britain*, ed. Kyriaki Hadjiafxendi and Patricia Zakreski (Farnham, Surrey, U.K.: Ashgate, 2013), pp. 51–52.
10. Magic lanterns were rudimentary slide projectors that cast images by projecting light — usually from candles or kerosene lamps — through painted glass slides. Phantasmagorias were a form of popular theatrical entertainment that incorporated multiple magic lanterns to project spooky images of skeletons, ghosts, and demons. See Richard Balzer, *Optical Amusements: Magic Lanterns and Other Transforming Images; A Catalog of Popular Entertainments* (Watertown, Mass.: R. Balzer, 1987).
11. Specs., "Photography at Vienna: France and Italy," *British Journal of Photography*, August 29, 1873; quoted in Janet E. Buerger, "Carlo Naya, Venetian Photographer: The Archeology of Photography," *Image* 26, no. 1 (March 1983), p. 2.
12. Edward Steichen, *A Life in Photography* (Garden City, N.Y.: Doubleday, 1963), chap. 1, p. [2].
13. Richard Abel, "*A Trip to the Moon* as an American Phenomenon," in *Fantastic Voyages of the Cinematic Imagination: Georges Méliès's* Trip to the Moon, ed. Matthew Solomon (Albany: State University of New York Press, 2011), pp. 129–42.
14. Lisa Hartje Moura, "Souvenirs of Places Never Visited — The Moon" (master's thesis, Haute Ecole d'Arts Appliqués, Geneva, 2016), p. 57, http://masterthesis-maspaceandcommunication.com/travaux/01_LisaMourra/LisaMourra.pdf.
15. Georges Méliès, "Appendix: Reply to Questionary, 1930," in *Fantastic Voyages of the Cinematic Imagination*, p. 233.
16. Ron Miller, "Spaceflight and Popular Culture," chap. 26 in *Societal Impact of Spaceflight*, ed. Steven J. Dick and Roger D. Launius (Washington, D.C.: National Aeronautics and Space Administration, Office of External Relations, History Division, 2007), p. 506; https://history.nasa.gov/sp4801-chapter26.pdf.
17. National Aeronautics and Space Administration, *Apollo 11 Technical Air-to-Ground Voice Transcription*, Manned Spacecraft Center, Houston, July 1969, p. 588, https://www.jsc.nasa.gov/history/mission_trans/AS11_TEC.PDF.
18. David A. Kirby, "Science Consultants, Fictional Films, and Scientific Practice," *Social Studies of Science* 33, no. 2 (April 2003), p. 244.
19. Oberth and Ley were founding members, along with Werner von Braun, of the German Verein für Raumschiffahrt (Society for Space Travel), a group of engineers dedicated to developing liquid-fueled rockets.
20. Kirby, "Science Consultants, Fictional Films, and Scientific Practice," p. 250.
21. As Bonestell explained in 1977: "If I had painted the big forty-foot mural that I did for the Boston Museum of Science [with the mountains] beaten down that way, I don't think it would have looked very interesting, although it would have been correct. I tried to make it just as dramatic as I could"; quoted in Ron Miller and Frederick C. Durant III, *The Art of Chesley Bonestell* (London: Paper Tiger, 2001), pp. 26–28. The mural was removed from the Boston Museum in 1970 and entered the collection of the Smithsonian National Air and Space Museum in 1976.

22. "Fact Book," United Artists Corporation Records, Unprocessed Accession, 4; quoted in Bradley Schauer, *Escape Velocity: American Science Fiction Film, 1950–1982* (Middletown, Conn.: Wesleyan University Press, 2017), p. 37.

MOONSHOT

1. A video excerpt and transcript of the speech can be found on the John F. Kennedy Presidential Library and Museum website: https://www.jfklibrary.org/learn/about-jfk/historic-speeches/address-to-joint-session-of-congress-may-25-1961.
2. *Ogonyok* 45 (November 1, 1959), quoted in Iina Kohonen, *Picturing the Cosmos: A Visual History of Early Soviet Space Endeavor*, trans. Albion Butters and Tiina Hyytiäinen (Bristol, U.K.: Intellect, 2017), p. 29. The translation of the quotation has been slightly modified.
3. Yaroslav Golovanov, *Korolev: Fakty i mify* (Moscow: Nauka, 1994), p. 688; quoted in translation in Asif A. Siddiqi, "Cosmic Contradictions: Popular Enthusiasm and Secrecy in the Soviet Space Program," in *Into the Cosmos: Space Exploration and Soviet Culture,* ed. James T. Andrews and Asif A. Siddiqi (Pittsburgh: University of Pittsburgh Press, 2011), p. 54.
4. For an in-depth discussion of Soviet space propaganda, see Kohonen, *Picturing the Cosmos*.
5. Paul Haney (director of public affairs at the Manned Spacecraft Center from 1965 to 1969), "Spinning Space in the Cold War," *International Journal of Press/Politics* 3, no. 3 (June 1, 1998), pp. 126–31, quoted in David Meerman Scott and Richard Jurek, *Marketing the Moon: The Selling of the Apollo Lunar Program* (Cambridge, Mass.: MIT Press, 2014), pp. 20–21.
6. Paul Mandel, "With All the Cost and Risk, Why Go?" *Life*, April 27, 1962, p. 81.
7. *Aerojet-General Spacelines and Rocket Review*, 1962, pp. 4–5.
8. Edgar M. Cortright interviewed by Rich Dinkel, Hampton, Virginia, August 20, 1998, https://www.jsc.nasa.gov/history/oral_histories/CortrightEM/EMC_8-20-98.pdf, p. 25. In a hard landing, the spacecraft crashes into the terrain and is destroyed upon impact; in a soft landing, small rockets are fired before landing to slow the descent of the spacecraft so that, ideally, it survives the impact.
9. Ralph J. Turner, "Extraterrestrial Landscapes through the Eyes of a Sculptor," *Leonardo* 5, no. 1 (Winter 1972), p. 12.
10. The Lunar Orbiter photographs were superseded only by the even higher resolution images sent back by NASA's Lunar Reconnaissance Orbiter, first launched in 2009 and still circling the moon as of 2019.
11. "Space: A New Look at Copernicus," *Time*, December 9, 1966, p. 58.
12. Frank Borman, with Robert J. Serling, *Countdown: An Autobiography* (New York: Silver Arrow Books, William Morrow, 1988), p. 212.
13. Archibald MacLeish, "A Reflection: Riders on Earth Together, Brothers in Eternal Cold," *New York Times,* December 25, 1968, p. 1.
14. When *2001: A Space Odyssey* premiered at Houston's Windsor Cinerama Theater in April 1968, MGM distributed free passes to many NASA officials, including all the astronaut flight crews. See Scott and Jurek, *Marketing the Moon*, p. 14.
15. Ron Schick and Julia Van Haaften, *The View from Space: American Astronaut Photography 1962–1972* (New York: Clarkson N. Potter, 1988), p. 92.
16. For more on the cameras Glenn used in space, see Jennifer Levasseur, "Another Journey for John Glenn's Ansco Camera," Smithsonian National Air and Space Museum, May 27, 2011, https://airandspace.si.edu/stories/editorial/another-journey-john-glenn's-ansco-camera.
17. Schick and Van Haaften, *The View from Space*, p. 12.
18. Ibid., p. 32.
19. In preparation, many newspapers and magazines published instructions for amateurs who planned to photograph the moon landing from the television screen. See "Tips on Taking Photos from Television Sets," *New York Times*, July 20, 1969, p. 38.
20. All Apollo 11 photographs are available online at *Apollo 11 Image Library*, figure captions by Eric M. Jones and Ken Glover, https://www.hq.nasa.gov/alsj/a11/images11.html.
21. Brian Duff, letter to H. J. P. Arnold, April 2, 1987, quoted in Arnold, "Lunar Surface Photography: A Study of Apollo 11," in *History of Rocketry and Aeronautics: Proceedings of the Twentieth and Twenty-First History Symposia of the International Academy of Astronautics, Innsbruck, Austria, 1986; and Brighton, United Kingdom, 1987*, ed. Lloyd H. Cornett Jr., AAS History Series 15 (San Diego, Calif.: Published for the American Astronautical Society by Univelt, 1993), p. 278.
22. Duff, letter to Arnold, July 8, 1987, quoted in ibid., p. 281. According to Arnold (ibid., p. 276), Armstrong appears with his back to the camera in one shot of a panorama that Aldrin shot with the Hasselblad on the lunar surface (AS11-40-5886).
23. Duff, letter to Arnold, May 26, 1987, quoted in ibid., p. 278.
24. Richard W. Underwood, interview with Summer Chick Bergen, "NASA Johnson Space Center Oral History Project: Oral History Transcript," Houston, October 17, 2000, https://www.jsc.nasa.gov/history/oral_histories/UnderwoodRW/UnderwoodRW_10-17-00.pdf, p. 21.
25. Buzz Aldrin (@TheRealBuzz), Twitter, July 20, 2017, https://twitter.com/TheRealBuzz/status/888178962848993280.

ART AFTER APOLLO

1. Calvin Tomkins, *Off the Wall: Robert Rauschenberg and the Art World of Our Time* (Garden City, N.Y.: Doubleday, 1980), p. 288.
2. Anne F. Collins, "Art, Technology, and the American Space Program, 1962–1972," *Intertexts* 3, no. 2 (Fall 1999), https://www.thefreelibrary.com/Art%2C+technology%2C+and+the+American+space+program%2C+1962-1972.-a080849918; Tomkins, *Off the Wall*, pp. 286–89.
3. Quoted from James Dean, "The Artist and the Space Shuttle," unpublished essay, 1984, rev. 1990, in Collins, "Art, Technology, and the American Space Program."

4. Robert Rauschenberg and Billy Klüver, *E.A.T. News*, no. 2 (January 15, 1967), p. 1, quoted in Jaklyn Babington, *Stoned Moon: Robert Rauschenberg* (Canberra: National Gallery of Australia, 2010), p. 5.
5. Kenneth Tyler, Quantas Birthday Lecture, October 14, 1999, quoted in Babington, *Stoned Moon*, p. 7.
6. Babington, *Stoned Moon*, p. 7.
7. The painting was reproduced as a foldout in an article by Arthur C. Clarke, "Apollo and Beyond," *Look*, July 15, 1969, pp. 43–49.
8. Roland Barthes, *Camera Lucida: Reflections on Photography*, trans. Richard Howard, paperback ed. (New York: Hill and Wang, 1981), 40.
9. Forrest Myers, "History of the Moon Museum," unpublished essay, emailed by the artist to Mia Fineman, August 22, 2018.
10. See Neil M. Maher, *Apollo in the Age of Aquarius* (Cambridge, Mass.: Harvard University Press, 2017), pp. 1–10.
11. Gil Scott-Heron, "Whitey on the Moon," *Small Talk at 125th and Lenox*, Flying Dutchman — FDS-131, 1970.
12. Babington, *Stoned Moon*, pp. 8–9.
13. Giovanna Borasi and Mirko Zardini, eds., *Other Space Odysseys: Greg Lynn, Michael Maltzan, Alessandro Poli*, exh. cat. (Montreal: Canadian Centre for Architecture; Baden: Lars Müller, 2010), pp. 11–13, 64–91.
14. "La conquista della luna ha dimostrato l'immense possibilità che ha l'uomo ma ha dimostrato anche come il padrone si serva della scienza e della tecnica per aumentare il suo potere e il nostro sfruttamento." Manifesto of the R. Panzieri Workers Club, Marghera, July 21, 1969, *Quindici* 3, no. 19 (August 19, 1969), p. [41]; http://www.bibliotecaginobianco.it/flip/QND/03/1900b/#36.
15. Christina Hunter, "Mapping Space and Time," in *Nancy Graves Project & Special Guests*, ed. Brigitte Franzen and Annette Lagler (Ostfildern: Hatje Cantz, 2013), p. 113.
16. On Celmins's moon drawings, see Cécile Whiting, "'It's Only a Paper Moon': The Cyborg Eye of Vija Celmins," *American Art* 23, no. 1 (Spring 2009), pp. 36–55.
17. Dater recalled making this association only after seeing the negative; as both photographer and subject, she set up the shot and had just ten seconds to pose before the automatic shutter released.
18. Aleksandra Mir, email correspondence with Mia Fineman, September 12, 2018.
19. Namwali Serpell, "The Zambian 'Afronaut' Who Wanted to Join the Space Race," *New Yorker*, March 11, 2017, https://www.newyorker.com/culture/culture-desk/the-zambian-afronaut-who-wanted-to-join-the-space-race.

SELECTED BIBLIOGRAPHY

Andrews, James T., and Asif A. Siddiqi, eds. *Into the Cosmos: Space Exploration and Soviet Culture*. Pittsburgh: University of Pittsburgh Press, 2011.

Armstrong, Carol. *Scenes in a Library: Reading the Photograph in the Book, 1843–1875*. Cambridge, Mass.: MIT Press, 1998.

The Audiences of the Moon: Photographs from the Age of Lunar Exploration. Exh. cat. Austin: Laguna Gloria Art Museum, 1978.

Babington, Jaklyn. *Stoned Moon: Robert Rauschenberg*. Canberra: National Gallery of Australia, 2010.

Bajac, Quentin, Agnès de Gouvion Saint-Cyr, Denis Canguilhem, et al. *Dans le champ des étoiles: Les photographes et le ciel, 1850–2000*. Exh. cat. Paris: Editions de la Réunion des Musées Nationaux, 2000.

Barger, M. Susan, and William B. White. *The Daguerreotype: Nineteenth-Century Technology and Modern Science*. Washington, D.C.: Smithsonian Institution Press, 1991.

Benson, Michael. *Cosmigraphics: Picturing Space through Time*. New York: Abrams, 2014.

Blühm, Andreas. *Der Mond*. Exh. cat. Ostfildern: Hatje Cantz; Cologne: Wallraf-Richartz-Museum & Fondation Courboud, 2009.

———. *The Moon: "Houston, Tranquility Base Here. The Eagle Has Landed."* Exh. cat. Translated from German. Houston: Museum of Fine Arts, 2009.

Borman, Frank, with Robert J. Serling. *Countdown: An Autobiography*. New York: Silver Arrow Books, William Morrow, 1988.

Cashford, Jules. *The Moon: Myth and Image*. New York: Four Walls Eight Windows, 2002.

Cressy, David. "Early Modern Space Travel and the English Man in the Moon." *American Historical Review* 111, no. 4 (October 2006), pp. 961–82.

Ferris, Timothy. *Spaceshots: The Beauty of Nature Beyond Earth*. New York: Pantheon Books, 1984.

Gillespie, Sarah Kate. *The Early American Daguerreotype: Cross-Currents in Art and Technology*. Cambridge, Mass: MIT Press, 2016.

Hoffleit, Dorrit. *Some Firsts in Astronomical Photography*. Cambridge, Mass: Harvard College Observatory, 1950.

Jones, Eric M., and Ken Glover, eds. *Apollo Lunar Surface Journal*. https://www.hq.nasa.gov/alsj/.

Keller, Corey, ed. *Brought to Light: Photography and the Invisible, 1840–1900*. Exh. cat. San Francisco: San Francisco Museum of Modern Art, in association with Yale University Press, 2008.

Kemp, Martin. *Seen/Unseen: Art, Science, and Intuition from Leonardo to the Hubble Telescope*. Oxford: Oxford University Press, 2006.

Kohonen, Iina. *Picturing the Cosmos: A Visual History of Early Soviet Space Endeavor*. Translated by Albion Butters and Tiina Hyytiäinen. Bristol, U.K.: Intellect, 2017.

Lankford, John. "The Impact of Photography on Astronomy." In *The General History of Astronomy*, vol. 4, *Astrophysics and Twentieth-Century Astronomy to 1950*, pt. A, edited by Owen Gingerich, pp. 16–39. Cambridge: Cambridge University Press, 1984.

Levasseur, Jennifer. "Pictures by Proxy: Images of Exploration and the First Decade of Astronaut Photography at NASA." PhD diss., George Mason University, 2014.

Light, Michael. *Full Moon*. New York: Alfred A. Knopf, 1999.

Locke, Richard Adams. *The Celebrated "Moon Story": Its Origins and Incidents, with a Memoir by the Author and an Appendix . . . by William N. Griggs*. New York: Bunnell and Price, 1852.

———. *Moon Walk 1835: Was Neil Armstrong Really the First Man on the Moon?* Edited by C. W. Tazewell. 5th ed. Virginia Beach, Va.: W. S. Dawson Co., 1990.

Loewy, Maurice, and Pierre Puiseux. *Atlas photographique de la lune, publié par l'Observatoire de Paris*. 8 pts. Paris: Imprimerie Nationale, 1896–1904.

Maher, Neil M. *Apollo in the Age of Aquarius*. Cambridge, Mass.: Harvard University Press, 2017.

Miller, Ron, and Frederick C. Durant III. *The Art of Chesley Bonestell*. London: Paper Tiger, 2001.

Mills, Deborah J. "George Willis Ritchey and the Development of Celestial Photography." *American Scientist* 54, no. 1 (March 1966), pp. 64–93.

Montgomery, Scott L. *The Moon and the Western Imagination*. Tucson: University of Arizona Press, 1999.

Morgan, Hal, and Andreas Brown. *Prairie Fires and Paper Moons: The American

Photographic Postcard, 1900–1920. Boston: David R. Godine, 1981.

Moura, Lisa Hartje. "Souvenirs of Places Never Visited — The Moon." Master's thesis, Haute Ecole d'Arts Appliqués, Geneva, 2016. http://masterthesis-maspaceandcommunication.com/travaux/01_LisaMourra/LisaMourra.pdf.

Nicolson, Marjorie Hope. *Voyages to the Moon.* New York: Macmillan Co., 1948.

Norman, Daniel. "The Development of Astronomical Photography." *Osiris* 5 (1938), pp. 560–94.

Ordway, Frederick I., III, and Randy Liebermann, eds. *Blueprint for Space: Science Fiction to Science Fact.* Washington, D.C.: Smithsonian Institution Press, 1992.

Pang, Alex Soojung-Kim. "'Stars Should Henceforth Register Themselves': Astrophotography at the Early Lick Observatory." *British Journal for the History of Science* 30, no. 2 (June 1997), pp. 177–202.

Pérez González, Carmen, ed. *Selene's Two Faces: From 17th Century Drawings to Spacecraft Imaging.* Nuncius series 3. Leiden: Brill, 2018.

Phillips, Christopher. "'Magnificent Desolation': The Moon Photographed." In *Cosmos: From Romanticism to the Avant-Garde*, edited by Jean Clair, pp. 144–49. Exh. cat. Montreal: Montreal Museum of Fine Arts; Munich: Prestel, 1999.

Poole, Robert. *Earthrise: How Man First Saw the Earth.* New Haven: Yale University Press, 2008.

Rewald, Sabine. *Caspar David Friedrich: Moonwatchers.* Exh. cat. New York: The Metropolitan Museum of Art, 2001.

Robertson, Frances. "Science and Fiction: James Nasmyth's Photographic Images of the Moon." *Victorian Studies* 48, no. 4 (Summer 2006), pp. 596–623.

Rothermel, Holly. "Images of the Sun: Warren De la Rue, George Biddell Airy and Celestial Photography." *British Journal for the History of Science* 26, no. 2 (June 1993), pp. 137–69.

Rydal Jørgensen, Lærke, and Marie Laurberg, eds. *The Moon — From Inner Worlds to Outer Space.* Exh. cat. Humlebæk, Denmark: Louisiana Museum of Modern Art, 2018.

Schick, Ron, and Julia Van Haaften. *The View from Space: American Astronaut Photography 1962–1972.* New York: Clarkson N. Potter, 1988.

Scott, David Meerman, and Richard Jurek. *Marketing the Moon: The Selling of the Apollo Lunar Program.* Cambridge, Mass.: MIT Press, 2014.

Shostak, Anthony, ed. *Starstruck: The Fine Art of Astrophotography.* Exh. cat. Lewiston, Me.: Bates College Museum of Art, 2012.

Solomon, Matthew, ed. *Fantastic Voyages of the Cinematic Imagination: Georges Méliès's Trip to the Moon.* Albany: State University of New York Press, 2011.

Terpak, Frances. "Imaging the Moon." In Barbara Maria Stafford and Frances Terpak, *Devices of Wonder: From the World in a Box to Images on a Screen*, pp. 197–204. Los Angeles: Getty Research Institute, 2001.

Thomas, Ann, ed. *Beauty of Another Order: Photography in Science.* New Haven: Yale University Press, in association with National Gallery of Canada, Ottawa, 1997.

Thomas, Davis, ed. *Moon: Man's Greatest Adventure.* New York: Harry N. Abrams, 1970.

Trouvelot, E. L. *The Trouvelot Astronomical Drawings Manual.* New York: Charles Scribner's Sons, 1882.

Turkina, Olesya. *Soviet Space Dogs.* Translated by Inna Cannon and Lisa Wasserman. London: FUEL Design & Publishing, 2014.

Turner, Ralph J. "Extraterrestrial Landscapes through the Eyes of a Sculptor." *Leonardo* 5, no. 1 (Winter 1972), pp. 11–17.

Vida, István Kornél. "The 'Great Moon Hoax' of 1835." *Hungarian Journal of English and American Studies* 18, no. 1–2 (Spring–Fall 2012), pp. 431–41.

Whitaker, Ewen A. *Mapping and Naming the Moon: A History of Lunar Cartography and Nomenclature.* Cambridge: Cambridge University Press, 1999.

Wilder, Kelley. *Photography and Science.* London: Reaktion Books, 2009.

INDEX

Page numbers in *italics* refer to illustrations.

Ackermann, Rudolph, 74–75
Aerojet-General Corporation: *Mock-Up of Moon Suit in Action* (pl. 68), 107, *119*, 177; *Moon Mobile Carries Two Explorers, Life-Support Systems* (pl. 69), 107, *119*, 177
Aldrin, Edwin "Buzz": and Armstrong's absence from lunar images, 112, 184n22; Armstrong's photographs of, on lunar surface (pls. 83–87), 111–12, *131–34*, 178; and objects left on moon, 139; photographs by, documenting his lunar footprint (fig. 16a–c), 111, *112*, 138, 144
Almond, Darren: *Fifteen Minute Moon* (pl. 112), 145, *167*, 180
Alphonsus peak (pl. 70), 108, *120*
Anders, William: *Earthrise* (NASA Apollo 8; fig. 14, pl. 78), 109–10, *109*, *126*, 177–78
Apollo 1 mission (NASA), 14–15
Apollo 8 mission (NASA): *Earthrise* (Anders; fig. 14, pl. 78), 109–10, *109*, *126*, 177–78
Apollo 11 mission (NASA): Aldrin's photographs of his lunar footprint (fig. 16a–c), 111, *112*, 138, 144; *Apollo 11 Command and Service Modules Photographed from the Lunar Module in Orbit* (pl. 81), *129*, 178; *Apollo 11 Moon Landing on Television Screens* (pl. 82), 111, *130*, 178; *Apollo 11 Moon Shot, Cape Kennedy, Florida* (Winogrand, pl. 89), 137, *147*, 179; Armstrong's absence from lunar images, 112, 184n22; Armstrong's photographs of Aldrin on lunar surface (pls. 83–87), 111–12, *131–34*, 178; Armstrong's tribute to Jules Verne, 78; art inspired by, 138, 141–43, 144, 145; and Nasmyth's lunar landscapes, 25; photographic equipment taken on, 111; *President Richard M. Nixon Welcomes the Apollo 11 Astronauts aboard Recovery Ship USS Hornet* (pl. 88), *135*, 178; public opposition to, 140
Apollo 12 mission (NASA): *The Moon Museum* (pl. 93), 139–40, *150*, 179
Apollo 14 mission (NASA), 141
Apollo 15 mission (NASA): *Fallen Astronaut* (van Hoeydonck; pls. 94, 95), 140, *151*, 179
Apollo 16 mission (NASA): *Duke Family Photograph on Lunar Surface* (Duke, pl. 92), 139, *149*, 179
Apollo 17 mission (NASA): *Blue Marble* (Schmitt, pl. 79), 110, *127*, 178
Arago, François, 20–21
Archer, Frederick Scott, 22
Aristarchus crater (pls. 75, 76), *123*
Arlen, Harold, 78
Armstrong, Neil: *Buzz Aldrin on the Moon with Components of the Early Apollo Scientific Experiments Package* (pl. 83), 111, *131*, 178; *Buzz Aldrin on the Moon with the American Flag* (pl. 86), 111, *133*, 178; *Buzz Aldrin Walking on the Surface of the Moon* (pls. 85, 87), 111, 112, *132*, *134*, 144, 178; *Buzz Aldrin with Apollo 11 Lunar Module on the Moon* (pl. 84), 111, *131*, 178; lack of good lunar images of, 112, 184n22; and objects left on moon, 139; "one small step" line, 138; and television coverage of moon landing, 78, 111, 140

Balzac, Honoré de, 76, *76*
Barthes, Roland, 139
Baturin, Vasily M.: *Brothers of the Cosmos* (album, pls. 63–66), 105, 106, *114–17*, 177
Bayard, Emile-Antoine, and Alphonse-Marie-Adolphe de Neuville: illustrations in *From the Earth to the Moon followed by Around the Moon* by Jules Verne (pl. 56), 78, *96*, 176
Behn, Aphra, 72
Bell Laboratories, 139
Black, James Wallace. *See* Whipple, John Adams, and James Wallace Black
Black Panther Party, 143
Blake, William: *I want! I want!* from *For Children: The Gates of Paradise* (fig. 9), 74, *75*
Bliss Brothers Studio: *Gorge, A Trip to the Moon, Pan-American Exposition, Buffalo, New York* (pl. 52), 77–78, *90*, 176
Blunt, Charles F.: *The Moon's Phases*, in *Lecture on Astronomy* (pl. 9), *35*, 173
Bond, William Cranch, 22
Bonestell, Chesley: Boston Museum of Science mural, 79, 183n21; study for *A Lunar Landscape* (pl. 62), 79, *102–3*, 177
Borman, Frank, 109
Boston Museum of Science, 79, 183n21
Brett, John: *Gassendi's Crater on the Moon* (pl. 25), 25, *52*, 174; travels to Sicily, 182n32
British Association, 22, *23*

Calle, Paul, 138
Cameron, Julia Margaret: *Sir John Herschel* (fig. 2), *20*
Cape Kennedy (Cape Canaveral), Florida (pl. 89), 111, 137–38, *146*
cartes de visite (pls. 18, 22), 23, 25, *44–45*, *49*
Celmins, Vija: *Moon Surface (Luna 9) #1* (fig. 20), 141–42, *143*
Chamberlain, John (pl. 93), 139–40, *150*, 179
Charlesworth, Sarah: *Moon Watch* (pl. 111), 144–45, *166*, 180
Claudet, Antoine-François-Jean: *Multiple Exposures of the Moon* (pl. 13), 22, *39*, 173–74, 181n15
Clayton, Lenka: *Moon 25/12/2015 (& 1977)* (pl. 109), 144, *164*, 180
Cold War (space race), 15, 79, 105–6, 110, 144
Collins, Michael, 110
Columbia University Observatory (New York), 145
Comte, August, 19
Coney Island, New York, *77*, 78
Constructing the Model for Germany's "Moon Rocket" (pl. 60), 79, *100*, 177

Cooke, Hereward Lester, 138
Copernicus crater (pl. 72), 109, *122*
Cortis, Jojakim, and Adrian Sonderegger: *Making of AS11-40-5878 (by Edwin Aldrin, 1969)* (pl. 108), 144, *163*, 180
Craters of the Moon, Idaho (pl. 103), 142–43, *158*
Cyrano de Bergerac, 72

Daguerre, Louis-Jacques-Mandé, 20–21, 75
daguerreotypy, 20–22, 26
Dater, Judy: *Self-Portrait at Craters of the Moon* (pl. 103), 142–43, *158*, 180, 185n17
Dean, James (NASA), 138
De Biasi, Mario: photograph of protestors in front of a mock lunar module (fig. 18), 140, *140*
De La Rue, Warren: Flammarion's use of photographs by (pl. 22), 25, *49*, 174; *The Moon* (pl. 17), 22, *43*, 174; A. A. Turner's use of photographs by (pl. 18), 23, *44–45*, 174
De La Rue, Warren, and Robert Howlett: *The Moon* (pl. 19), 22–23, *46*, 174, 182n21
De La Rue, Warren, and William Henry Fox Talbot: *The Half Moon* (pl. 20), 23, *47*, 174, 182nn22–23
De Middel, Cristina: *Bambuit*, from the series The Afronauts (pl. 106), 143–44, *161*, 180
Dodd, Lamar, 138
dogs in space (pl. 63), 105, *114*
Doré, Gustave: *It Looked Round and Shining Like a Glittering Island*, in *Aventures du Baron de Munchhausen* by R. E. Raspe (pl. 42), *82*, 176
Draper, Henry: *Lunar Transparency* (pl. 16), *42*, 174
Draper, John William: *Moon* (pl. 10), 21, *36*, 173
Drinking with the Moon (pl. 55), 78, *94*, 176
Duff, Brian, 111–12
Duke, Charles: *Duke Family Photograph on Lunar Surface* (NASA Apollo 16, pl. 92), 139, *149*, 179
Dundy, Elmer "Skip," 77–78

earth, images of: *Blue Marble* (Schmitt, NASA Apollo 17; pl. 79), 110, *127*, 178; *Earthrise* (Anders, NASA Apollo 8; fig. 14, pl. 78), 109–10, *109*, *126*, 177–78; *Original Earthrise* (NASA Lunar Orbiter 1, fig. 13), *108*, 109
earthshine, 74
Eastman Kodak Company, 108–9, 110
eclipses (pls. 28–31), 18, 25–26, *55–58*, 182n32
Experiments in Art and Technology (E.A.T.), 138, 139

fairground attractions, 77–78
far side of moon (pls. 67, 73), 19, 106, *118*, *123*
Feoktistov, Konstantin (pl. 64), *115*
films (motion pictures): *Destination Moon* (Pichel, pl. 61), 79, *101*, 177; *A Trip to the Moon* (Méliès; pls. 48–51, 53), 71, 77, 78, *88*, *89*, *91*, 176; *2001: A Space Odyssey* (Kubrick), 110, 184n14; *Woman in the Moon* (Lang; pls. 58, 59), 79, *98*, *99*, 177
Fischli, Peter, 144
Flammarion, Camille, and Adolphe-Alexandre Martin: *The Moon*, after De La Rue, from *Astronomical Gallery, Photographs of Science and Art* (pl. 22), 25, *49*, 174
Fontana, Francesco: *Lunar Map*, in *New Observations of Heavenly and Earthly Objects* (pl. 2), *29*, 173
Fra Mauro region (pl. 101), 15, 141, *156*
French Academy of Sciences, 20, 27
French Astronomical Society, 25
Friedrich, Caspar David: *Two Men Contemplating the Moon* (pl. 44), 74, *84*, 176

Gagarin, Yuri (pl. 65), 105, *116*
Galileo Galilei: and telescope resolution, 106; *Two Drawings of Waxing Moon*, in *Starry Messenger* (pl. 1), 17–18, *28*, 173
Galluzzo, Leopoldo: *Other Discoveries Made on the Moon by Sigr. Herschell* (fig. 8, pl. 41), *73*, 74, *81*, 176
Gassendi, Pierre, 18
Gassendi's crater (pl. 25), 25, *52*
Gemini G. E. L. studio (Los Angeles), 138
Gemini program (NASA): *Ed White Extravehicular Activity* (McDivitt, pl. 80), 110–11, *128*, 178; overview of, 14
Glenn, John, 15, 110
Goddard, Robert, 78
Godwin, Francis: *The Man in the Moone* (fig. 7), 72, *72*
Golovanov, Yaroslav, 106
Gordon, Harry: *"Rocket" Dress* (pl. 96), 140, *152*, 179
Graves, Nancy: *IV Julius Caesar Quadrangle of the Moon* and *II Fra Mauro Region of the Moon*, from the series Lithographs Based on Geologic Maps of Lunar Orbiter and Apollo Landing Sites (pls. 100, 101), 141, *156*, 179–80
Great Exhibition (London, 1851), 22, 181n15
Great Forty-Foot Telescope (Herschel family, England; fig. 3), 19–20, *20*, 26
Great Moon Hoax (1835), 19–20, 74
Great Refractor (Harvard College Observatory, pl. 11), 22, 24, 26, *37*

Halley's Comet, 18, 78
Harbou, Thea von, 79
Hariot, Thomas, 181n1
Harpalus crater, 79
Harper, Sharon: *Moon Studies and Star Scratches, No. 6* (pl. 114), 145, *170*, 180
Harvard College Observatory (Cambridge, Mass.): Great Refractor (pl. 11), 22, 24, 26, *37*; and Humphrey's lunar daguerreotypes, 21; and Pickering's lunar atlas, 27; and Smith's artist book *Tidal*, 145; Trouvelot as staff member of, 25; Whipple's collaboration with, 22, 26
Hasselblad 70 mm camera (fig. 15a), 110–11, *111*, 112, 143, 184n22
Hazard, Allyn B., 107
Heinlein, Robert, 79
heliostats, 21, 181n13
Henry, Paul and Prosper: *Lunar Photograph, South Pole* (pl. 32), *59*, 175
Herschel, Caroline, 19
Herschel, Sir John: Cameron's portrait of (fig. 2), *20*; and Great Moon Hoax, 19–20, 74; on Nasmyth's lunar landscapes, 25; *View of the Telescope at Slough* (fig. 3), 19–20, *20*
Herschel, William, 19, 20, 26
Hevelius, Johannes: *Lunar Map*, in *Selenography* (pl. 6), 18, *32*, 173
Hoeydonck, Paul van: *Fallen Astronaut* sculpture (pls. 94, 95), 140, *151*, 179
Holden, Edward, 27
Howlett, Robert, 23, 182nn21–22. *See also* De La Rue, Warren, and Robert Howlett
Humboldt, Alexander von, 21
Humphrey, Samuel Dwight: *Multiple Exposures of the Moon* (pl. 12), 21, *38*, 173
Hunte, Otto: production sketch for the film *Woman in the Moon* (pl. 58), 79, *98*, 177
Hurd, Peter, 138

Illustrated London News, 22, *23*

Janssen, Jules, 21, 25, 27
Jet Propulsion Laboratory (Pasadena), 108

Kawada, Kikuji: *Artificial Moon Trail, Tokyo* (pl. 115), 145, *171*, 180
Kennedy, John F., 105, 106, 138
Kepler crater (pl. 74), *123*
Khrushchev, Nikita (pl. 66), *117*
Komarov, Vladimir (pl. 64), *115*

Korolev, Sergei, 105
Kubrick, Stanley, 110, 144, 184n14
Kuiper, Gerard, 107–8

Laika (canine cosmonaut, pl. 63), 105, *114*
Land Art, 142
Lang, Fritz: *Woman in the Moon* (film, pl. 59), 79, *99*, 177
Langenheim, William and Frederick: *Eclipse of the Sun* (pl. 30), 26, *57*, 174–75
Langlois, Henri, 77
Lawrence, H. A., and Charles Ray Woods: *Solar Eclipse from Caroline Island* (pl. 31), 26, *58*, 175
Le Morvan, Charles: *Systematic Photographic Map of the Moon* (pl. 39), 27, *68–69*, 175
Leviathan of Parsonstown (Rosse telescope, Ireland; fig. 6), 26, *26*
Ley, Willy, 79, 183n19
Lick Observatory (Mount Hamilton, Calif.): *Transparency of the Moon from Negatives Made at Lick Observatory* (pl. 36), 27, *63*, 175
Life (magazine), 106–7, 110, 111
literature of lunar travel, 71–74
Liverpool, 22, *23*
Locke, Richard Adams, 74
Loewy, Maurice, and Pierre Puiseux: *Photographic Atlas of the Moon* (pls. 37, 38), 27, *64–67*, 71, 109, 175
Look (magazine), 138, 185n7
Luna Park, Coney Island (fig. 11), *77*, 78
Lunar and Planetary Laboratory, Tucson (pl. 70), 107–8, *120*
Lunar Orbiter program (NASA): and Graves's lithographs, 141; vs. Lunar Reconnaissance Orbiter, 184n10; *Original Earthrise* (fig. 13), *108*, 109; overview of, 107, 108–9; photographs of lunar terrain (pls. 72–77), 109, *122–25*, 177

MacLeish, Archibald, 110
magic lanterns, 75, 183n10
Majocchi, Giovanni Alessandro, 25–26
"Man in the Moon" portraits (pl. 54), 78, *92–95*, 176
Man Ray: *The World, in Electricité* (pl. 43), 75, *83*, 176
mapping the moon: artistry vs. scientific veracity in, 18, 19, 24–25, 27; atlases, 18, 27, 71, 109, 141; daguerreotypes' use in, 20–22, 26; drawing vs. photography as favored medium in, 18, 22, 23, 25; Galileo as pioneer of, 17–18; in NASA's Lunar Orbiter program, 108–9, 141, 184n10; Soviet mapping of far side, 106
Mare Humorum (pl. 27), 25, *54*
Marey, Etienne-Jules, 21
Martin, Aldophe-Alexandre. *See* Flammarion, Camille, and Adolphe-Alexandre Martin
Mayall, John Jabez Edwin, 181n15
McCoy, John, 138
McDivitt, James: *Ed White Extravehicular Activity* (NASA Gemini 4, pl. 80), 110–11, *128*, 178
megalethoscopes, 75
Meigs, John: *To the Moon* (fig. 17), 138, *139*
Méliès, Georges: preparatory drawings for *A Trip to the Moon* (pls. 48–51), 77, *88*, *89*, 176; *A Trip to the Moon* (film, pl. 53), 71, 77, 78, *91*, 176
Mellan, Claude: *Full Moon*, *The Moon in Its First Quarter*, and *The Moon in Its Final Quarter* (pls. 3–5), 18, *30*, *31*, 173
Mercury program (NASA): camera carried by Schirra, 110; and Gordon's *"Rocket" Dress*, 140; overview of, 14; *Project Mercury Astronauts at Langley Air Force Base* (Morse, fig. 12), 106, *107*
Meudon Observatory (Paris), 25
Miller, Ron, 78
Mir, Aleksandra: *First Woman on the Moon* (pl. 104), 143, *159*, 180
moonlight, romanticization of, 74–76
Morghen, Filippo: *Pumpkins Used as Dwellings to Secure against Wild Beasts*, from *The Collection of the Most Notable Things Seen by John Wilkins* (pl. 40), 72–74, *80*, 176
Morieu, Eugène, 25
Morse, Ralph: *Project Mercury Astronauts at Langley Air Force Base* (fig. 12), 106, *107*
Mount Wilson Observatory (Los Angeles), 26, *60*
Muniz, Vik: *Memory Rendering of the Man on the Moon* (pl. 107), 144, *162*, 180
Muybridge, Eadweard, 22, 142
Myers, Forrest (pl. 93), 139–40, *150*, 179

Nadar: *Nadar with His Wife, Ernestine, in a Balloon* (pl. 57), 78, *97*, 177
NASA: Apollo program, 14–15, 106–7 (*see also specific Apollo missions*); Artists' Cooperation Program, 138; Gemini program (pl. 80), 14, 110–11, *128*, 178; Lunar Orbiter program (fig. 13, pls. 72–77), 107, 108–9, *108*, *122–25*, 141, 177, 184n10; Lunar Reconnaissance Orbiter, 144, 184n10; Mercury program (fig. 12), 14, 106, *107*, 110, 140; photographic equipment carried by astronauts, 110–11, *111*, 143; and premiere of *2001: A Space Odyssey*, 184n14; public relations and publicity of images, 106–7, 109–12, 137, 140; Ranger program, 107–8; Surveyor program (pl. 71), 107, 108, *121*, 177
Nasmyth, James: *An Ideal Sketch of "Pico"* and *Normal Lunar Crater*, in *The Moon: Considered as a Planet, a World, and a Satellite* (pls. 23, 24), 24–25, *50*, *51*, 174; plaster models of lunar surface (fig. 5), 24–25, *24*, 107–8
Naya, Carlo: *Night View of the Grand Canal, Venice* (pl. 46), 75–76, *86*, 176
Neuville, Alphonse-Marie-Adolphe de. *See* Bayard, Emile-Antoine, and Alphonse-Marie-Adolphe de Neuville
New York City: Columbia University Observatory, 145; Lyceum of Natural History, 21; Photogravure and Color Company, 27; Rutherfurd's lunar views from (pl. 21), 24, *48*
New York Sun (newspaper), 19, 74
Nikolayev, Andriyan (pl. 66), *117*
Nixon, Richard M. (pl. 88), *135*
Nkoloso, Edward Mukaka, 144
Novros, David (pl. 93), 139–40, *150*, 179

Oberth, Hermann, 78, 79, 183n19
Oldenburg, Claes (pl. 93), 139–40, *150*, 179
Orme, Edward, 74–75
O'Sullivan, Timothy, 142

Pach, Gustavus W.: *The Great Refractor* (pl. 11), 22, *37*, 173
Paik, Nam June, and Jud Yalkut: *Electronic Moon No. 2* (video, pl. 97), 141, *153*, 179
Pal, George, 79
Pan-American Exposition (Buffalo, N.Y., 1901; pl. 52), 77–78, *90*
Paris Observatory, 25, 27, 71
Peiresc, Nicolas-Claude Fabri de, 18
phantasmagorias, 75, 183n10
photographic technology: chronophotography, 21–22; daguerreotypy, 20–22, 26; dry-plate negatives, 26, 27; equipment carried by astronauts, 110–11, *111*, 143; Herschel's contributions to, 20; Lunar Orbiters' robotic imaging system, 108–9; nocturnal views simulated, 75–76; online photo sharing, 144; stereoviews of the moon, 22–23; Talbot's contributions to, 20, 23; wet-collodion process, 22, 23

Pichel, Irving: *Destination Moon* (film, pl. 61), 79, *101*, 177
Pickering, William H.: *Lunar Fancies*, in *The Moon: A Summary . . . with a Complete Photographic Atlas* (pl. 35), 27, *62*, 175
Pico (lunar mountain, pl. 23), 25, *50*
Poe, Edgar Allan, 74
Poli, Alessandro (Superstudio): *Highway with Sine Wave of Energy* and *New Rural Landscapes*, from the series Interplanetary Architecture (pls. 98, 99), 141, *154*, *155*, 179
Ponti, Carlo, 75–76
Popovich, Pavel (pl. 66), *117*
postcards (pls. 54, 55), 78, *92–95*, 176
Puiseux, Pierre. *See* Loewy, Maurice, and Pierre Puiseux

Ranger program (NASA), 107–8
Rauschenberg, Robert: at Apollo 11 launch, 137–38; drawing for *The Moon Museum* (pl. 93), 139–40, *150*, 179; *Sky Garden (Stoned Moon)* (pl. 90), 138, *147*, 179; *Untitled* (pl. 91), 138, *148*, 179
Ritchey, George W.: *Mirror for Mount Wilson Observatory Telescope* (pl. 33), 26, *60*, 175; *The Moon (10 Days Old)* (pl. 34), 26, *61*, 175
Rockwell, Norman, 138, 185n7
Rodin, Auguste, 76
Rosse, Mary, Countess of: photograph of Lord Rosse's telescope (fig. 6), 26, *26*
Rosse, William Parsons, 3rd Earl of, 26, *26*
Rowlandson, Thomas: *The Assignation* (pl. 45), 75, *85*, 176
Royal Astronomical Society (London), 22, 23, 182n23, 182n32
Russell, John: *A Drawing of a Part for the Map of the Moon* (pl. 7), 18–19, *33*, 173; *Lunar Planisphere, Flat Light* (pl. 8), 19, *34*, 173; *Selenographia* (lunar globe, fig. 1), 19, *19*, 27
Rutherfurd, Lewis Morris: Flammarion's use of photographs by, 25; *The Moon, New York* (pl. 21), 24, *48*, 174

Schirra, Walter, 110
Schmitt, Harrison: *Blue Marble* (NASA Apollo 17, pl. 79), 110, *127*, 178
Schroter's Valley (pl. 75), *123*
Scott, David: *Paul van Hoeydonck's "Fallen Astronaut" Sculpture on the Lunar Surface* (NASA Apollo 15, pl. 95), 140, *151*, 179
Scott-Heron, Gil, 140
Selene (moon goddess), 18, 72
selenography, 17–18, 27
Shames, Stephen: *The Moon Belongs to the People* (pl. 105), 143, *160*, 180
Smith, Kiki: *Tidal* (pl. 113), 145, *168–69*, 180
Sonderegger, Adrian. *See* Cortis, Jojakim, and Adrian Sonderegger
Soviet Union: canine cosmonauts (pl. 63), 105, *114*; female cosmonaut Tereshkova (pl. 64), 105, *115*; Gagarin and Vostok 1 spacecraft (pl. 65), 105, *116*; Khrushchev with cosmonauts (pl. 66), *117*; Luna program (pl. 67), 105–6, 108, *118*; photographic equipment carried by cosmonauts, 110; secrecy surrounding space program, 106, 107; space race between U.S. and, 15, 79, 105–6, 110, 144; Sputnik 1 (satellite), 13, 105; Zond program, 108
space suits (pl. 68), 79, 107, *119*
Sparks, Jared, 21
Steichen, Edward J.: *Balzac, Towards the Light, Midnight* (fig. 10), 76, *76*; *The Pond — Moonrise* (pl. 47), 76, *87*, 176
stereoviews (pl. 19), 22–23, 24, *46*, 182n21
Stuart, Michelle: *#7 Moon Tide* (pl. 102), 141–42, *157*, 180
Superstudio. *See* Poli, Alessandro (Superstudio)
Surveyor program (NASA): *Day 322, Survey U, Sectors 15 and 16* (pl. 71), 108, *121*, 177; overview of, 107, 108

Talbot, William Henry Fox, 20, 23
Talbot, William Henry Fox, and Warren De La Rue: *The Half Moon* (pl. 20), 23, *47*, 174, 182nn22–23
TASS: *The Far Side of the Moon* (USSR Luna 3, pl. 67), 106, *118*, 177
Taylor, Alfred Swaine, 181n15
telescopes: Charlesworth's diptych of moon and telescope (pl. 111), 144–45, *166*; clock drives for, 21, 22, 27; earliest instruments, 17, 106, 181n1; first, designed specifically for photography, 24; Great Forty-Foot Telescope (England, fig. 3), 19–20, *20*, 26; Great Refractor (Harvard College Observatory, pl. 11), 22, 24, 26, *37*; Leviathan of Parsonstown (Ireland, fig. 6), 26, *26*; mirror for Mount Wilson Observatory telescope (pl. 33), 26, *60*
television coverage of moon landings (pl. 82), 78, 108, 111, *130*, 140–41, 143, 184n19
Tereshkova, Valentina (pl. 64), 105, *115*
Thompson, Frederic, 77–78
Time (magazine), 109
Titov, Gherman, 110
transparencies, 74–75
Trouvelot, Etienne Léopold: *Interior of Crater*, in *Astronomical Notebook* (pl. 26), 25, *53*, 174; *Mare Humorum* and *Total Eclipse of the Sun*, from *The Trouvelot Astronomical Drawings Manual* (pls. 27, 28), 25, *54*, *55*, 174
Tsiolkovsky, Konstantin, 78
Turner, Austin Augustus: *Twelve Photographs of the Moon*, after De La Rue (pl. 18), 23, *44–45*, 174
Turner, Ralph: *High Relief of Alphonsus Peak* (pl. 70), 107–8, *120*, 177

UFA studio (Germany), 79
Umbrico, Penelope: *Everyone's Moon 2015-11-04 14:22:59* (video, pl. 110), 144, *165*, 180
Underwood, Richard, 110, 112
U.S. Geological Survey: Geologic Atlas of the Moon, Julius Caesar Quadrangle (fig. 19), 141, *142*; and NASA Surveyor video images, 108
USSR. *See* Soviet Union

Venice (pl. 46), 75–76, *86*, 138
Verne, Jules, 26, 78, *96*
Von Braun, Werner, 183n19
Vouet, Simon, 18

Warhol, Andy (pl. 93), 139–40, *150*, 179
Weiss, David, 144
Whipple, John Adams: on capturing the moon, 17; *Partial Eclipse of the Sun* (pl. 29), 26, *56*, 174; *View of the Moon* (pl. 14), 22, *40*, 174
Whipple, John Adams, and James Wallace Black: *The Moon* (pl. 15), 22, *41*, 174
White, Edward (pl. 80), 110–11, *128*
Wilkins, John, 72–74
Williams, J. and A.: photograph of the moon exhibited at Liverpool (fig. 4), 22, *23*
Winogrand, Garry: *Apollo 11 Moon Shot, Cape Kennedy, Florida* (pl. 89), 137, *146*, 179
women in space, 79, 105, 142–43
Woods, Charles Ray. *See* Lawrence, H. A., and Charles Ray Woods
Wright brothers, 13, 78

Yalkut, Jud. *See* Paik, Nam June, and Jud Yalkut
Yegerov, Boris (pl. 64), *115*

Zambian space program, 143–44

PHOTOGRAPH CREDITS

© Darren Almond, 1998. Image © The Metropolitan Museum of Art: pl. 112
© Vija Celmins. Digital Image © The Museum of Modern Art/Licensed by SCALA/Art Resource, NY: fig. 20
© Estate of Sarah Charlesworth, courtesy the Estate and Paula Cooper Gallery: pl. 111
Courtesy Collection Cinémathèque Française: pl. 58
© Lenka Clayton. Image © The Metropolitan Museum of Art: pl. 109
Courtesy Coney Island History Project: fig. 11
© Jojakim Cortis and Adrien Sonderegger: pl. 108
© Judy Dater, 1981. Image © The Metropolitan Museum of Art: pl. 103
© Mario De Biasi: fig. 18
© Cristina De Middel: pl. 106
Courtesy Electronic Arts Intermix (EAI), New York: pl. 97
© Friedrich-Wilhelm-Murnau-Stiftung, Wiesbaden; courtesy Kino Lorber, Inc.: pl. 59
© Nancy Graves Foundation/Licensed by VAGA, New York. Photo courtesy Imaging Department © President and Fellows of Harvard College: pls. 100, 101
© Sharon Harper, courtesy Rick Wester Fine Art: pl. 114
Courtesy Harvard University Archives: pl. 11
© Paul van Hoeydonck: pl. 94
Courtesy Houghton Library, Harvard University: fig. 7; pls. 26, 56
© Illustrated London News Ltd./Mary Evans: fig. 4
© Institute of Historical Survey Foundation, Mesilla Park, NM: fig. 17
© Kikuji Kawada, courtesy L. Parker Stephenson Photographs. Image © The Metropolitan Museum of Art: pl. 115
Courtesy Hans P. Kraus, Jr., Inc., New York: pls. 19, 20
© 2011 Lobster Films/Foundation Groupama Gan por le Cinéma/Foundation Technicolor pour le Patrimoine du Cinéma: pl. 53
Courtesy Lunar and Planetary Institute: fig. 19
© Man Ray 2015 Trust/Artists Rights Society (ARS), NY/ADAGP, Paris. Image © The Metropolitan Museum of Art, photo by Hyla Skopitz: pl. 43
Georges Méliès/Collection Cinémathèque Française: pls. 48–51
Image © The Metropolitan Museum of Art: fig. 2 and pls. 3–5, 18, 30, 40, 44, 46, 55, 57, 72–75
Image © The Metropolitan Museum of Art, photo by Anna-Marie Kellen: pl. 96
Image © The Metropolitan Museum of Art, photo by Mark Morosse: pls. 2, 42, 71
Image © The Metropolitan Museum of Art, photo by Hyla Skopitz: pls. 23, 24
Image © The Metropolitan Museum of Art, photo by Juan Trujillo: pls. 70, 76, 77
© Aleksandra Mir: pl. 104
Ralph Morse/The LIFE Picture Collection: fig. 12
© Vik Muniz. Image © The Metropolitan Museum of Art: pl. 107
© Museum of the History of Science, University of Oxford: fig. 3
© Forrest Myers: pl. 93
NASA: fig. 16a–c
NASA/LOIRP: fig. 13
Courtesy National Air and Space Museum, photo by Eric F. Long: fig. 15a, b
© National Maritime Museum, Greenwich, London: fig. 1
Courtesy Photography Collection, the New York Public Library: fig. 14 and pls. 22, 27, 28, 36, 52, 78–80, 92
© Robert Rauschenberg Foundation: pl. 91
© Robert Rauschenberg Foundation and Gemini G.E.L. Photo by Ben Blackwell: pl. 90
Photo © Lew Reid: pl. 16
© The Royal Society: fig. 6
Courtesy Science Museum/Science and Society Picture Library: fig. 5
© Stephen Shames, 1970. Image © The Metropolitan Museum of Art, photo by Mark Morosse: pl. 105
© Kiki Smith, courtesy Pace Gallery. Image © The Metropolitan Museum of Art, photo by Katherine Dahab: pl. 113
Courtesy Smithsonian Libraries, Washington, D.C.: fig. 8 and pls. 1, 9, 41
© 2019 The Estate of Edward Steichen/Artists Rights Society (ARS), New York. Image © The Metropolitan Museum of Art: fig. 10 and pl. 47
© Michelle Stuart. Image © The Metropolitan Museum of Art, photo by Mark Morosse: pl. 102
© Archivio Superstudio: pls. 98, 99
© Penelope Umbrico, courtesy Bruce Silverstein Gallery, New York: pl. 110
Courtesy the Wade Williams Collection, licensed through Corinth Films, Inc.: pl. 61
© The Estate of Garry Winogrand, courtesy Fraenkel Gallery, San Francisco. Image © The Metropolitan Museum of Art: pl. 89
Courtesy John G. Wolbach Library, Harvard College Observatory: pls. 12, 14, 29